Lecture Notes in Mechanical Engineering

Series Editors

Francisco Cavas-Martínez, Departamento de Estructuras, Universidad Politécnica de Cartagena, Cartagena, Murcia, Spain

Fakher Chaari, National School of Engineers, University of Sfax, Sfax, Tunisia

Francesco Gherardini, Dipartimento di Ingegneria, Università di Modena e Reggio Emilia, Modena, Italy

Mohamed Haddar, National School of Engineers of Sfax (ENIS), Sfax, Tunisia

Vitalii Ivanov, Department of Manufacturing Engineering Machine and Tools, Sumy State University, Sumy, Ukraine

Young W. Kwon, Department of Manufacturing Engineering and Aerospace Engineering, Graduate School of Engineering and Applied Science, Monterey, CA, USA

Justyna Trojanowska, Poznan University of Technology, Poznan, Poland

Francesca di Mare, Institute of Energy Technology, Ruhr-Universität Bochum, Bochum, Nordrhein-Westfalen, Germany

Lecture Notes in Mechanical Engineering (LNME) publishes the latest developments in Mechanical Engineering—quickly, informally and with high quality. Original research reported in proceedings and post-proceedings represents the core of LNME. Volumes published in LNME embrace all aspects, subfields and new challenges of mechanical engineering. Topics in the series include:

- Engineering Design
- Machinery and Machine Elements
- Mechanical Structures and Stress Analysis
- Automotive Engineering
- Engine Technology
- Aerospace Technology and Astronautics
- Nanotechnology and Microengineering
- Control, Robotics, Mechatronics
- MEMS
- Theoretical and Applied Mechanics
- Dynamical Systems, Control
- Fluid Mechanics
- Engineering Thermodynamics, Heat and Mass Transfer
- Manufacturing
- Precision Engineering, Instrumentation, Measurement
- Materials Engineering
- Tribology and Surface Technology

To submit a proposal or request further information, please contact the Springer Editor of your location:

China: Ms. Ella Zhang at ella.zhang@springer.com
India: Priya Vyas at priya.vyas@springer.com
Rest of Asia, Australia, New Zealand: Swati Meherishi at swati.meherishi@springer.com
All other countries: Dr. Leontina Di Cecco at Leontina.dicecco@springer.com

To submit a proposal for a monograph, please check our Springer Tracts in Mechanical Engineering at http://www.springer.com/series/11693 or contact Leontina.dicecco@springer.com

Indexed by SCOPUS. All books published in the series are submitted for consideration in Web of Science.

More information about this series at http://www.springer.com/series/11236

Peter F. Pelz · Peter Groche
Editors

Uncertainty in Mechanical Engineering

Proceedings of the 4th International Conference on Uncertainty in Mechanical Engineering (ICUME 2021), June 7–8, 2021

 Springer

 TECHNISCHE UNIVERSITÄT DARMSTADT

SFB 805
Control of Uncertainty in Load-Carrying Structures in Mechanical Engineering

Editors
Peter F. Pelz
Chair of Fluid Systems
Technische Universität Darmstadt
Darmstadt, Germany

Peter Groche
Institute for Production Engineering
and Forming Machines
Technische Universität Darmstadt
Darmstadt, Germany

ISSN 2195-4356   ISSN 2195-4364 (electronic)
Lecture Notes in Mechanical Engineering
ISBN 978-3-030-77255-0   ISBN 978-3-030-77256-7 (eBook)
https://doi.org/10.1007/978-3-030-77256-7

This Springer imprint is published by the registered company Springer Nature Switzerland AG
The registered company address is: Gewerbestrasse 11, 6330 Cham, Switzerland

Preface

Uncertainty is ubiquitous. Even though crisis like Covid-19 discloses the uncertainty within the product development and usage phase of a high variety of industries, methods to master this uncertainty are still not widely used.

An interdisciplinary and international group of researchers and industry members met at the 4th International Conference of Uncertainty in Mechanical Engineering (ICUME) in June 2021 to present and discuss their research to master uncertainty with its many facets and to enable a transfer of the obtained results.

Even though a conference lives from its interactions, the ICUME 2021 was held virtually, caused by the Covid-19 crisis restrictions.

The conference was organized by researchers from the Collaborative Research Center (CRC) 805 at Technische Universität Darmstadt (TU Darmstadt), which conducted interdisciplinary research on the topic of uncertainty in mechanical engineering. The long history of CRC805 with 12 years, starting in March 2009 and ending in March 2021, showed the importance of the pioneering approaches to master uncertainty.

The conference series on uncertainty in mechanical engineering was initiated in 2011 and has evolved since then. It focusses on the design and usage of mechanical engineering systems but also attracts researcher from different domains, like mathematics, law, linguistics, and history. Therefore, the editorial team partitioned the conference proceedings in five parts to reflect the interdisciplinarity. These parts are:

- mastering uncertainty by digitalization,
- resilience,
- uncertainty in production,
- uncertainty quantification, and
- optimization under uncertainty.

The part "mastering uncertainty by digitalization" summarizes contributions that specifically use digital approaches to master uncertainty. The interplay between CAD, ontologies, and linear programming as well as the treatment of semantic uncertainty and model uncertainty is presented.

The part "resilience" presents contributions that explicitly consider the resilience of engineering systems with a focus on general methodological developments to derive resilient technical systems, as well as focused approaches to design more resilient water supply systems. Here, the design of water supply systems for high-rise buildings and water supply networks in cities is presented.

The chapter "uncertainty in production" presents contributions that focus on uncertainty in productions systems, like deep rolling or tapping. Furthermore, legal uncertainties are also considered.

The chapter "uncertainty quantification" presents multiple approaches to quantify and master uncertainty for multiple engineering systems, like for instance wind turbines or transmissions, and the last chapter "optimization under uncertainty" presents approaches to optimize and quantify uncertainty for truss-like structures.

We thank all authors and presenters on behalf of the conference organizers and the local scientific committee. We also thank all reviewers for their valuable feedback and the German Research Foundation (Deutsche Forschungsgemeinschaft (DFG)) for their funding.

The editors hope to meet the interest of a broad readership with the selection of the following contributions and like to motivate for further investigations.

Peter F. Pelz
Peter Groche

Committees

Local Scientific Committee

R. Anderl	TU Darmstadt, Germany
P. Groche	TU Darmstadt, Germany
H. Kloberdanz	TU Darmstadt, Germany
M. Kohler	TU Darmstadt, Germany
T. Melz	TU Darmstadt, Germany
P. Pelz	TU Darmstadt, Germany
M. Pfetsch	TU Darmstadt, Germany
M. Schäffner	TU Darmstadt, Germany
C. Schänzle	TU Darmstadt, Germany
S. Ulbrich	TU Darmstadt, Germany
M. Weigold	TU Darmstadt, Germany
J. Wendt	TU Darmstadt, Germany

International Scientific Committee

L. Altherr	University of Applied Science Münster, Germany
S. Atamturktur	Penn State University, USA
S. Duncan	University of Oxford, UK
R. Engelhardt	Continental AG, Germany
A. Fügenschuh	BTU Cottbus, Germany
M. Görtan	Hacettepe University, Turkey
P. Kolár	Czech Technical University, Czech Republic
M. Kuchlbauer	Friedrich Alexander Universität Erlangen-Nürnberg, Germany
F. Liers	Friedrich Alexander Universität Erlangen-Nürnberg, Germany
U. Lorenz	Universität Siegen, Germany
D. Moens	KU Leuven, Belgium

R. Platz	Penn State University, USA
B. Scharte	ETH Zürich, Switzerland
S. Thöns	Lund University, Sweden
D. Vandepitte	KU Leuven, Belgium
J. Yanagimoto	University of Tokyo, Japan
M. Zäh	Technical University of Munich, Germany

Local Organizing Committee

N. Brötz	TU Darmstadt, Germany
P. Groche	TU Darmstadt, Germany
P. Leise	TU Darmstadt, Germany
P. Pelz	TU Darmstadt, Germany
M. Rexer	TU Darmstadt, Germany
A. Schmitt	TU Darmstadt, Germany
D. Wagner	TU Darmstadt, Germany

Contents

Mastering Uncertainty by Digitalization

Ontology-Based Calculation of Complexity Metrics for Components in CAD Systems

Moritz Weber$^{(\boxtimes)}$ ⑩ and Reiner Anderl ⑩

Technische Universität Darmstadt, Otto-Berndt-Straße 2, 64287 Darmstadt, Germany
m.weber@dik.tu-darmstadt.de

Abstract. The high complexity of assemblies and components in Computer-Aided Design (CAD) leads to a high effort in the maintenance of the models and increases the time required for adjustments. Metrics indicating the complexity of a CAD Model can help to reduce it by showing the results of changes. This paper describes a concept to calculate metrics aiming to describe the extent of complexity of components in CAD systems based on an ontology-based representation in a first step. The representation is initially generated from CAD models using an automated process. This includes both a boundary representation and the history of the feature-based design. Thus, the design strategy also contributes to measuring the complexity of the component so that the same shape can lead to different complexity metrics. Semantic rules are applied to find patterns of the design and to identify and evaluate various strategies. Different metrics are proposed to indicate the particular influence factors of complexity and a single measure for the overall complexity. Furthermore, the influencing factors can also be used to allow the designer to see how to reduce the complexity of the component or assembly.

Keywords: Complexity · CAD · Ontology · OWL2

1 Introduction

The complexity in mechanical design increases, and consequently also the effort required to maintain and change components in systems for computer-aided design (CAD). A complexity metric can help to make the complexity in mechanical design more manageable. It enables designers, project managers, and controllers to estimate the cost and time needed for design and change tasks better. However, no standardized or universally accepted measure for the assessment of the complexity of CAD models exists [1]. This paper proposes a method to calculate such metrics. The aim is to calculate a suite of different metrics to provide and provide it to a designer. He can use this information to identify the opportunities to minimise the complexity of the design. As an addition, a single metric is thereby accessible by a fusion of the metrics of the suite.

For the conceptual design, a definition of complexity often found in the literature (e.g. [2–5]) and firstly stated by Corning [6] is used: Three key factors determine the complexity of a system:

© The Author(s) 2021
P. F. Pelz and P. Groche (Eds.): ICUME 2021, LNME, pp. 3–11, 2021.
https://doi.org/10.1007/978-3-030-77256-7_1

(1) Individuals: A complex system comprises numerous individual parts (or items, assets, components).
(2) Relations: There are many relations (or interaction, dependencies) between the various parts.
(3) Complicatedness: The parts create combined effects that are to predict and often novel or surprising (e.g. nonlinear or chaotic).

These three statements imply that the complexity increases with the number of parts and relations and decreases with the predictability of the compounds or their effects. Considering the three parts of the definition, the suitability of graph-oriented databases is assumed. Ontologies appear especially suitable for the representation and calculation of complexity, since all three parts are representable in a proven knowledge base model. In this paper, it is aimed to evaluate the complexity not only of the final shape or model but also the design strategy, which is applied to obtain it. The complexity of the production of components is not considered because this needs further knowledge about the available machines and other circumstances of production.

In literature, it is distinguished between the shape or design complexity and CAD complexity [2, 5]. The first is based on the complexity of the appearance and the visible features of the result, whereas the latter is based on the actual CAD embodiment of it. For shape complexity, Rossignac distinguishes between five different types [7]: Algebraic complexity metrics the degree of polynomials required to represent the form exactly. Topological complexity metrics the existence of non-multiple singularities, holes, or self-cuts, or the number of handles and elements. Morphological complexity measures smoothness and feature size. Combinatorial complexity measures the number of vertices in polygonal meshes. Representational complexity metrics indicate the size of a compressed model.

2 Related Work

There are different works that investigate an assessment of the complexity of products and product models. Große Austing [8] measures the complexity of general product models. Besides CAD models, this includes other models and documents like source code and requirement documents. For the calculation, graph-based representations are used, which need to be generated manually. The weighting factors of the nodes are obtained by regression. The work aims to build an estimation model for the time and effort needed to create the particular product model.

Chase and Murty [2, 5] differentiate between design complexity and CAD complexity. For design complexity, they adopted a method introduced by Stiny and Gips [9], which uses the length of the generative specification. For CAD complexity, they use a method, which counts the number of usages of specific design techniques and objects as well as the file size. The CAD complexity is indicated by a list of these values and not a single value.

Johnson, Valverde et al. [10] use different approaches to measure the complexity of CAD models of components objectively and compare them with subjective ratings by test persons. For the objective evaluation, their methods are using topologic and geometric properties. They utilise the number of faces and the ratio of the surface area of a model and a sphere with the same volume. Furthermore, they use the number of used features and the complexity of specific features. The best results showed a method which uses the ratio between the volume of the component and its bounding box.

Matthieson et al. [11–13] propose a complexity metric for assemblies. They present a new convention for modelling the physical architecture of assemblies as graphs. For the calculation of a complexity measure, they use graph-theoretic metrics. In their model, they use the part count, the average path length, and the path length density to estimate the assembly time, including a standard deviation as an uncertainty measure.

Besides the assessment of the complexity of CAD models, there are works proposing methods to evaluate the complexity of ontologies which can also be applied in the context of this paper.

Zhang et al. [14, 15] propose a method which mainly uses the quantity, ratio, and correlativity of concepts and relations as well as their hierarchy. They calculate a set of different measures to assess the complexity of a given ontology.

Zhang et al. [16] propose another set of metrics inspired by software complexity metrics. They base all their metrics on the graphical representation of the ontology and measure the complexity on class and ontology level.

3 Concept

The method for the calculation of the complexity metrics comprises three steps, which are described in higher detail in the following and depicted in Fig. 1. Section 3.1 describes the concept for the automated conversion from CAD models of components into an ontology-based representation. Therefore, an ontology is used to describe all parts of the entities of components. Sections 3.2–3.4 present different complexity metrics categorised in the three key factors of complexity of a system. Section 3.5 demonstrates these metrics on two components and two design strategies. The Chapter concludes with an outlook to methods to calculate a single measure as a rough indication of the overall complexity of the components or assemblies in Sect. 3.6.

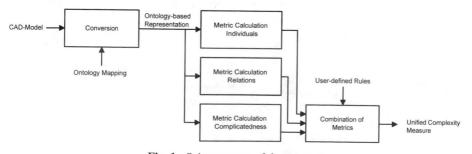

Fig. 1. Subprocesses of the concept

3.1 Ontologies

The concept uses ontologies to structure the discrete entities of the component independently from the CAD-program used. For the adaption of the internal structure of the various CAD-Programs, mappings must be developed. The proposed Ontology forms the Terminological Box (TBox) of the information model. The converted CAD models form the Assertional Box (ABox). All metrics are therefore calculated using only the ABox. The TBox is used to convert the parts from the format used by the various CAD-Programs to a uniform structure achieve comparability.

Figure 2 shows the hierarchy of the concepts of the ontology. The ovals represent the different concepts, and the arrows represent inheritances. Triangles indicate concepts hidden in the figure. The Hierarchy is divided in three major Parts: It uses the Boundary Representation (BRep) as well as the feature-based representation. Reference Attributes form the third part of the ontology. This way, the ontology-based information model represents all topologic and geometric entities of the CAD model as well as the design strategy and history. Therefore, this information can be used to evaluate the complexity of the CAD model of a component. The design of the component ontology uses the ontology proposed by Tessier and Wang [17] as one part of the base. The entities which describe the BRep model are taken from the ontology introduced by Perzylo, Somani et al. [18] and the OntoSTEP ontology introduced by NIST [19]. These ontologies were combined and modified to be more suitable for the aim of complexity analysis.

Features, Sketch Features, and Reference Attributes are formalised to represent the entities used to create the model and referenced to the respective BRep entities. Semantic rules help to identify patterns in the design and strategies. Since the use of an ontology-based information model, it is easier to find patterns and determine the compliance to design rules independently from the program used. These can be used to modify the single complexity metric proposed in Sect. 3.6.

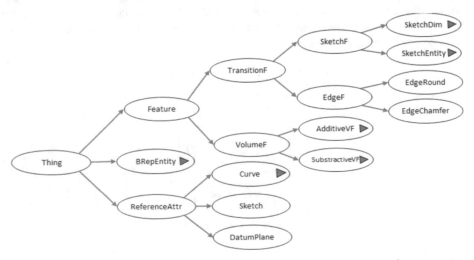

Fig. 2. Part of the concept hierarchy of the proposed component ontology

3.2 Metrics for Individuals

There are two main metrics for the number of individuals of CAD models, which can be divided further, the first being the Number of Instances (NoI) and the second being the Number of Properties (NoP). Both form the nodes and leaves in the graph-based information model, so they are a significant part in the size of the information model.

Number of Instances. The NoI is defined as the quantity of instances of all classes described in Sect. 3.1. It is dividable in the Number of Features (NoI_F), the Number of BRep Entities (NoI_B) and the Number of Reference Attributes (NoI_R) so that:

$$NoI = NoI_F + NoI_B + NoI_R \tag{1}$$

The numbers are defined as the number of instances of their respective classes and subclasses in the ABox. Furthermore, the number of distinct features (NodF) influences the complexity as well because the range of feature to be known by users or designers increases.

Number of Properties. The NoP is the number of specifications defined for features and reference attributes during design of the CAD models. These can be numeric values (NoP_V) as well as character strings (NoP_S). The numeric values can also use variable parameters for parametric design. So, the Number of Parameters (NoPm) and the NoP_V which are specified using parameters ($NoP_{V,Pm}$) are also crucial for the complexity of the model. The Ratio of numerical values not using parameters is defined as:

$$R_{Pm,V} = 1 - \frac{NoP_{V,Pm}}{NoP_V} \tag{2}$$

$R_{Pm,V}$ is the only measure proposed, where bigger values indicate a smaller complexity.

3.3 Metrics for Relations

Equivalently to Sect. 3.2, this part of the complexity can be indicated by the Number of Relations (NoR) between different instances in the information model. Pairs of inverse relations are counted as one relation. A special type of relation is the parent-child relation between a feature or reference attribute and the features or reference attributes used for its creation. The number of these relations is called Number of Parent-Child-Relations (NoR_C). As an Addition, the longest path from the root node to a child node is given by LP. It describes the maximum number of predecessors a node in the ontology-based representation has. Analogous to NodF, the number of different relation types is referred to as NodR.

3.4 Metrics for the Complicatedness

The complicatedness is the most crucial influence factor for the complexity of a system. If there are only simple relations between the different individuals, the entire system is easily predictable and applied to 3D CAD models easily changeable and understandable.

The complicatedness increases with the number of subsystems one subsystem influences and how complicated these influences are. Therefore, three metrics are calculated.

The complicatedness of the structure can be described by the mean number of parent-child-relations of the features to other features (MoR$_{C,F}$) and is calculated by:

$$MoR_{C,F} = \frac{NoR_C}{NoI_F} \tag{3}$$

where NoR$_F$ is the number of relations, with features in it.

Because all relations in an ontology are directed, it is feasible to calculate all instances influenced by one instance by following all relations from an instance. The mean number of instances influenced by an instance in the ontology is given by Moni.

Of interest is also the Mean number of numeric Properties per Feature (MoP$_{V,F}$) because it indicates the ratio of features created with the help of mirroring and patterns which decrease the complexity. It is defined as:

$$MoP_{V,F} = \frac{NoP_V}{NoI_F} \tag{4}$$

3.5 Examples

For exemplification and clarification of the proposed metrics, two components shown in Fig. 3 are used. The first is a cuboid with three different edge lengths and three edge fillets, each with the same radius, for which a parameter is used.

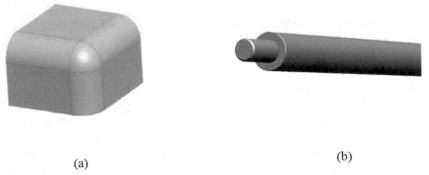

(a)   (b)

Fig. 3. Two example components: (a) Cuboid with three rounded edges (b) Rod with threaded ends

The second component is one of the members of the upper truss of the CRC805 demonstrator which is an abstracted airplane landing gear. (For a detailed description of the see [20]). It is designed as a long cylinder with a smaller coaxial cylinder on both ends. This cylinder is threaded on the outside. Two Chamfers are on the edges of the cylinders. This validation inspects two distinctive design strategies. In the first all feature besides the large cylinder are mirrored to get a symmetrical rod, in the second not. Instead, parameters are used to define all values of both cylinders.

Table 1 shows selected metrics for both components. It is visible, that the greater number of features of the threaded rod lead to a higher complexity in the areas of individuals and relations. The design strategy using parameters instead of mirroring decreases the complexity in the subarea of Individuals lightly, since no mirroring plane is needed but increases the number of property values and–of course–parameters. In the subarea it changes all metrics with $MoR_{C,F}$ is lower since all features are only direct children of only one other feature. Then again it increases $MoP_{V,F}$ because of the features not only being copies of other features and therefor have numerical values. A final assessment of overall complicatedness depends on preferences and company guidelines.

Table 1. Selected metrics for the example components and design strategies

	Individuals		Relations			Complicatedness		
Component (strategy)	NoI_F	NoI_{Ra}	NoR_C	NoP_V	$NoPm$	$R_{Pm,V}$	$MoR_{C,F}$	$MoP_{V,F}$
Rounded Cuboid	4	1	4	6	1	0.67	1.0	1.5
Threaded Rod (Mirror)	10	2	15	8	0	1.00	1.5	0.8
Threaded Rod (Pm)	9	1	9	14	6	0.57	1.0	1.55

3.6 Fusion of the Metrics

To give a rough overview over the complexity combinations of the metrics proposed in Sects. 3.2–3.4 a single measure is calculated. This fusion is influenced by the purpose of the measure and its target group. At this point, it is possible to use corporate design guidelines. For example, discrepancies from rules for the number of elements in a sketch or the general size of designs can be considered. The overall complexity metric depends strongly on the viewpoint and the company guidelines, as complexity also comes with using distinctive design strategies in one company or even one component. A consistent design strategy in one company helps designers to understand and change components. The weighting of the proposed metrics enables the rating of the compliance to design rules. The single measure can therefore be used as an assessment of the design without deeper knowledge. It can be used as a first indication of the time needed to understand the design idea and for the subsequent changes of it. This is particularly advantageous in agile development, where the approximate time for a task must be known as early and as precise as possible priorly.

4 Conclusions

There is no broadly accepted measure to indicate the complexity of CAD models [1]. However, the assessment of the complexity helps to control the complexity of models and therefore to minimise the effort and time needed to maintain and change models if required. This paper proposes a concept for a method to describe and calculate metrics for the complexity of assemblies and components in CAD systems. It therefore utilises an ontology-based information model as an intermediate.

The first step is to convert the internal model structure of the CAD System to the ontology-based information model. Two different general ontologies are the basis for the conversion of components and assemblies. The information model is then enriched with information obtained by application of semantic rules and is tested for validity and integrity by reasoning.

Based on this ontology-based representation of the component, a set of metrics regarding the three subareas of complexity are calculated. This set of metrics can be used to reduce the complexity of the model by indicating the influence factors. Thereby, it eventually helps to reduce the time needed to understand and change the design. Based on these numbers, a single measure is calculated as a rough overview of the complexity of the model.

The results of the concept help the designer and are also helpful in controlling and other departments. With the single measure as an indication for the complexity of a CAD model, it is possible to estimate better the difficulty and time needed to change the component or assembly.

Acknowledgement. The authors like to thank the Deutsche Forschungsgemeinschaft (DFG, German Research Foundation) for funding this project within the Sonderforschungsbereich (SFB, Collaborative Research Center) 805 "Control of Uncertainties in Load-Carrying Structures in Mechanical Engineering" – project number: 57157498.

References

1. Amadori, K., Tarkian, M., Ölvander, J., Krus, P.: Flexible and robust CAD models for design automation. Adv. Eng. Inform. **26**, 180–195 (2012)
2. Chase, S., Murty, P.: Evaluating the complexity of CAD models as a measure for student assessment. In: Clayton, M., Velasco, V., Guillermo, P. (eds.) Eternity, Infinity and Virtuality in Architecture: Proceedings of the 22nd Annual Conference of the Association for Computer-Aided Design in Architecture, pp. 173–182. Catholic University, Washington D.C. (2000)
3. Sinha, K., et al.: Structural Complexity and Its Implications for Design of Cyber-Physical Systems. Boston (2014)
4. Xia, B., Chan, A.P.C.: Measuring complexity for building projects: a Delphi study. Eng. Constr. Archit. Manage. **19**(1), 7–24 (2012)
5. Murty, P., Chase, S., Nappa, J.: Evaluating the Complexity of CAD Models in Education and Practice (1999)
6. Corning, P.A.: Complexity is just a word! Technological Forecasting and Social, pp. 197–200 (1998)
7. Rossignac, J.: Shape complexity. Vis. Comput. **21**, 985–996 (2005)
8. große Austing, S.: Komplexitätsmessung von Produktmodellen. Oldenburg (2012)
9. Stiny, G., Gips, J.: Algorithmic aesthetics. Computer Models for Criticism and Design in the Arts. Berkley (1978)
10. Johnson, M.D., Valverde, L.M., Thomison, W.D.: An investigation and evaluation of computer-aided design model complexity metrics. Comput.-Aided Des. Appl. **15**, 61–75 (2018)

11. Mathieson, J.L., Wallace, B.A., Summers, J.D.: Assembly time modelling through connective complexity metrics. Int. J. Comput. Integr. Manuf. **26**, 955–967 (2013)

12. Miller, M.G., Mathieson, J.L., Summers, J.D., Mocko, G.M.: Representation: structural complexity of assemblies to create neural network based assembly time estimation models. In: ASME 2012 International Design Engineering Technical Conferences and Computers and Information in Engineering Conference, pp. 99–109. American Society of Mechanical Engineers Digital Collection

13. Namouz, E.Z., Summers, J.D.: Complexity connectivity metrics – predicting assembly times with low fidelity assembly cad models. In: Abramovici, M., Stark, R. (eds.) Smart Product Engineering, pp. 777–786. Springer, Heidelberg (2013)

14. Zhang, D., Ye, C., Yang, Z.: An evaluation method for ontology complexity analysis in ontology evolution. In: Staab, S., Svátek, V. (eds.) Managing Knowledge in a World of Networks, pp. 214–221. Springer, Heidelberg (2006)

15. Zhe, Y., Zhang, D., Chuan, Y.E.: Evaluation metrics for ontology complexity and evolution analysis. In: 2006 IEEE International Conference on e-Business Engineering (ICEBE'06), pp. 162–170 (2006)

16. Zhang, H., Li, Y.-F., Tan, H.B.K.: Measuring design complexity of semantic web ontologies. J. Syst. Softw. **83**, 803–814 (2010)

17. Tessier, S., Wang, Y.: Ontology-based feature mapping and verification between CAD systems. Adv. Eng. Inform. **27**, 76–92 (2013)

18. Perzylo, A., Somani, N., Rickert, M., Knoll, A.: An ontology for CAD data and geometric constraints as a link between product models and semantic robot task descriptions. In: 2015 IEEE/RSJ International Conference on Intelligent Robots and Systems (IROS), pp. 4197–4203 (2015)

19. Krima, S., Barbau, R., Fiorentini, X., Sudarsan, R., Sriram, R.D.: Ontostep: OWL-DL ontology for step. National Institute of Standards and Technology, NISTIR 7561 (2009)

20. Feldmann, R., Gehb, C.M., Schäffner, M., Matei, A., Lenz, J., Kersting, S., Weber, M.: A detailed assessment of model form uncertainty in a load-carrying truss structure. In: Mao, Z. (ed.) Model Validation and Uncertainty Quantification, vol. 3, pp. 303–314. Springer International Publishing, Cham (2020)

Towards CAD-Based Mathematical Optimization for Additive Manufacturing – Designing Forming Tools for Tool-Bound Bending

Christian Reintjes[1(✉)], Jonas Reuter[2], Michael Hartisch[1], Ulf Lorenz[1], and Bernd Engel[2]

[1] Chair of Technology Management, University of Siegen, Unteres Schloss 3, 57072 Siegen, Germany
[2] Chair of Forming Technology, Institute of Production Engineering, University of Siegen, Breite Straße 11, 57076 Siegen, Germany
{christian.reintjes,jonas.reuter,michael.hartisch,ulf.lorenz, bernd.engel}@uni-siegen.de

Abstract. The trend towards flexible, agile, and resource-efficient production systems requires a continuous development of processes as well as of tools in the area of forming technology. To create load-adjusted and weight-optimized tool structures, we present an overview of a new algorithm-driven design optimization workflow based on mixed-integer linear programming. Loads and boundary conditions for the mathematical optimization are taken from numerical simulations. They are transformed into time-independent point loads generating physical uncertainty in the parameters of the optimization model. CAD-based mathematical optimization is used for topology optimization and geometry generation of the truss-like structure. Finite element simulations are performed to validate the structural strength and to optimize the shape of lattice nodes to reduce mechanical stress peaks. Our algorithm-driven design optimization workflow takes full advantage of the geometrical freedom of additive manufacturing by considering geometry-based manufacturing constraints. Depending on the additive manufacturing process, we use lower and upper bounds on the diameter of the members of a truss and the associated yield strengths. An additively manufactured flexible blank holder demonstrates the algorithm-driven topology design optimization.

Keywords: Adjustable forming tool surfaces
Mixed integer linear programming · Additive manufacturing
Tool-bound bending · Lightweight forming tools

1   Introduction

Increasing mass customization and product complexity combined with shorter product life cycles require agile, flexible, and smart production systems in manufacturing technology [13]. In addition, future studies on the topic of manufacturing

© The Author(s) 2021
P. F. Pelz and P. Groche (Eds.): ICUME 2021, LNME, pp. 12–22, 2021.
https://doi.org/10.1007/978-3-030-77256-7_2

technology are required to be subordinated to the maxim of resource efficiency. In forming technology, the forming tools play a key role as the link between semi-finished products and machines and directly impact the flexibility of a forming process [2]. To be more precise, kinematic forming processes such as three-roll-push bending [3] or incremental swivel bending [11] have inherent flexibility due to their shape-giving tool movement. In contrast, tool-bound processes like stamping are limited regarding an achievable variety of geometries.

State-of-the-art forming tools are typically solid and oversized steel parts generating an unnecessarily high level of energy consumption for the tool production along the entire value chain and in the operation of the tools. This research gap can be addressed by combining lightweight construction with topology optimization to obtain an efficient design tool for forming tool development. On account of the fact that Additive Manufacturing (AM) methods enable the fabrication of complex-shaped and topology-optimized tools [2]—in comparison to conventional manufacturing methods—the combination of lightweight construction, topology optimization, and AM is of significant interest.

Xu et al. [12] show that a blank holder's weight can be decreased by 28.1% using topology optimization methods with a negligible impact on structural performance. Burkart et al. [1] point out that their achieved weight reduction of a blank holder by over 20% using topology optimization can reduce dynamic press loads by 40% resulting in an extended process window with shorter cycle times. Besides the established and in industrial finite element software implemented continuum topology optimization methods based on Solid Isotropic Material with Penalization Method (SIMP) [10], also algorithm-driven optimization based on mathematical programming [5] can be used for early-stage design optimization of truss-like lattice structures. Reintjes and Lorenz [7] show a large-scale truss topology optimization of additively manufactured lattice structures based on the high performance of commercial (mixed-integer) linear programming software like CPLEX.

Considering lightweight construction and topology optimization, this paper presents a new algorithm-driven optimization workflow for additively manufactured forming tools, mainly consisting of mathematical programming, numerical topology optimization, and verification via numerical simulation. We distinguish strictly between the rigid-body equilibrium of forces calculated via a mixed-integer linear program and a verification of the results via a linear-elastic and a non-linear-elastic numerical analysis. Based on the algorithm-driven optimization workflow we optimize a demonstrator tool of a segmented blank holder. Finally, we give an outlook on how an optimized lattice structure can be used as a mechanism for in-process modification of local surface geometry and local structural stiffness.

2   Mathematical Optimization and the Application to a Segmented Blank Holder

Within the Centre of Smart Production Design Siegen (SMAPS), we investigate sensoric and actuatoric forming tools with the aim of self-adjustable surfaces.

Targeting the adaption of contact pressure distribution, dynamic compensation of part springback, and change of geometry for part variant diversity, different scales of surface adjustment are needed, as illustrated in Fig. 1 left [4]. As a vision, future forming tools will have the self-adjusting capability to control material flow and react to changing process conditions. To this end, a deep understanding of the forming process itself, sensor and actuator integration, as well as a force transmitting tool structure that is able to change surface geometry and stiffness locally, is necessary. A simple demonstrator for such a flexible tool is shown in Fig. 1 right. The segmented blank holder consists of a housing with thread holes at the bottom (3), a cover (1), and a segmented inlay structure for force transmission (2). The surface adjustment can be realized by the infeed of one screw per segment. Tests were carried out with different arrangements and infeeds of the screws. The basic proof of concept was done by measurement of the surface deformation using Gom ARAMIS, which showed different surface profiles dependent on the screw setup [4]. We examine how such a force transmitting inlay

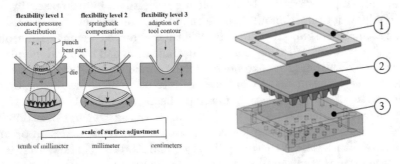

Fig. 1. Flexibility levels of forming tools (left) and demonstrator of a segmented blank holder (right)

can be generated using truss-like lattice structures generated by algorithm-driven design optimization. First, a linear static finite element simulation using Altair Optistruct was performed to obtain the load case for mathematical optimization. We assume that the insert is loaded by a screw force of $F_{screw} = 4.5$ kN and a contact pressure between workpiece and inlay, resulting in the process force $F_{process} = 13.5$ kN, see Fig. 2. The reaction load is the contact pressure p_{cover} between the cover and the inlay. After a transformation of the stress given in the Finite Element Analysis (FEA) into linear constraints (point loads), we get a formulation suitable for a Mixed-Integer Linear Program (MILP) inclusive of physical uncertainty in the parameters of the optimization model.

3   CAD-Based Mathematical Optimization

The design process of complex truss-like lattice structures in Computer-Aided Design (CAD) is inefficient and limits the number of parts (members) to be

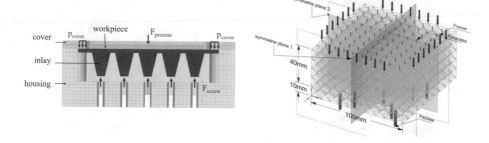

Fig. 2. Load case for the FEA (left) and point loads for the MILP (right)

automatically built in a CAD model [8,9]. For the reasons stated, transforming large-scale mathematical optimization results into a CAD model is not a straightforward task. Our first research concerning this problem found that using an Autodesk Inventor Professional Add-In, we were able to generate 6084 round structural elements (volume bodies) in 20 h and 24 min [6,7]. To further improve computational efficiency, in order to be able to define a part as a structural element rather than only as a volume body and allow an easy geometry preparation for numerical analysis, we developed a direct CAD creation (Ansys SpaceClaim 2020 R2) in addition to a history-based CAD creation (Autodesk Inventor Professional). Our Ansys SpaceClaim Add-In *construcTOR*, see Fig. 3, allows CAD engineers to generate algorithm-driven design iteration studies within the Ansys Workbench. The Add-In involves a Graphical User-Interface (GUI) and bidirectional linkage to CPLEX 12.6.1, see Fig. 3, such that no profound knowledge about mathematical optimization is needed. To avoid local stress peaks at the intersection of members during numerical analysis, we post-process the intersection of members. For this purpose, a solid sphere (near-side body only) merging into the members with a diameter at least equal to the diameter of the member with the largest cross-section is added. The large number of parts given in Table 1 details that we were able to lift the limitation dictated by history-based modeling, see [7]. Besides, a significant reduction in computational time and memory usage depending on the type of implementation, geometrical complexity of the member's cross-section and instance size exist. We compared the execution time and memory usage divided into the generation of the members and the faceting of Ansys SpaceClaim 2020 R2. In both cases, we used the beam class of the SpaceClaim API V19 and our own implementation as volume bodies.

Mixed-Integer Linear Program for Truss Optimization

In order to formally represent the ground structure (see Fig. 2) an undirected graph $G = (V, E)$ is used with vertices (frictionless joints) and connecting edges (straight and prismatic members). Additionally, a set of bearings $B \subset V$ must be specified. Note that the vertices are fixed in space, as angles between two possible members and distances between joints matter in our modeling approach. We additionally require that the resulting structure is symmetrical with respect to two symmetry planes, see Fig. 2. We use the function $R : E \to E$, mapping edge e to

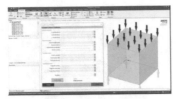

(a) Create MILP within
ANSYS SpaceClaim

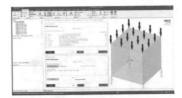

(b) Bidirectional linkage to
CPLEX

(c) Generate truss-like
structure from best solution

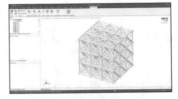

(d) Post-Processing of the
intersections

Fig. 3. CAD-integrated mathematical optimization of lattice structures using the con-strucTOR GUI

Table 1. Benchmark of different implementation types[a]

Implementation	Number of parts	Beam section	Time [s]			Memory usage [MB]	
			Generation	Faceting	Overall	Generation	Faceting
Beam class	15000	Circle	223	6879	7102	1290	1292
		Square	228	6817	7045	1414	1412
Volume body	400000	Circle	2354	9204	11558	7404	17966
		Square	2810	10884	13694	15714	25610

[a]The calculations were performed on a workstation with an Intel Xeon E5-2637 v4 (3,5 GHz), 64 GB RAM and an NVIDIA GeForce RTX 2080 (8 GB RAM).

its representative $R(e)$ in order to enforce that the members at edges e and $R(e)$ share the same cross-sectional area with respect to the given symmetry. Due to manufacturing restrictions a member must have a minimum cross-sectional area. Therefore, we use a binary variable x_e to indicate the existence of a member at edge $e \in E$ with a specified minimum cross-sectional area and a continuous variable a_e to specify its additional (optional) cross-sectional area. The continuous variable n_e represents the normal force in a member at edge e and r_b specifies

Table 2. Variables

Symbol	Definition
$\mathbf{x} \in \{0,1\}^E$	x_e: indicator, whether a member is present at edge e
$\mathbf{a} \in \mathbb{Q}_+^E$	a_e: additional (optional) cross-sectional area of a member e
$\mathbf{r} \in \mathbb{Q}^{B \times 3}$	r_b^d: bearing reaction force at b in spatial direction $d \in \{x,y,z\}$
$\mathbf{n} \in \mathbb{Q}^E$	n_e: normal force in member present at edge e

Table 3. Sets and Parameters

Symbol	Definition
V	Set of vertices
$E \subseteq V \times V$	Set of edges
$I : V \to 2^E$	$I(v) = \{e \in E \mid v \in e\}$: Set of edges incident to vertex v
$B \subseteq V$	Set of bearings
$L_e \in \mathbb{Q}_+$	Length of edge e
$A_{\min} \geq 0$	Minimum cross-sectional area of a member
$A_{\max} \geq 0$	Maximum cross-sectional area of a member
σ_y	Yield strength of the cured material
$S \geq 1$	Factor of safety
$\mathbf{F} \in \mathbb{Q}^{V \times 3}$	F_v^d: external force at vertex v in spatial direction $d \in \{x, y, z\}$
$\mathbf{V}(v, v') \in \mathbb{Q}^3$	Vector from $v \in V$ to $v' \in V$ (corresponding to lever arm)
$R : E \to E$	$R(e)$: edge representing edge e due to symmetry

the bearing reaction force of bearing b. The variables and parameters used in our model are given in Tables 2 and 3, respectively. We use bold letters when referring to vectors. With respect to the considered application, the external forces $\mathbf{F}$ are taken from numerical simulations of the blank holder ($F_{process}$ and p_{cover}) and the bearing reaction forces r are corresponding to F_{screw}.

$$\min \sum_{e \in E} L_e \left(A_{\min} \cdot x_{R(e)} + a_{R(e)} \right) \tag{1}$$

$$\text{s.t.} \ S|n_e| \leq \sigma_y \left(A_{\min} \cdot x_{R(e)} + a_{R(e)} \right) \qquad \forall e \in E \tag{2}$$

$$\sum_{e \in I(b)} n_e^d + F_b^d + r_b^d = 0 \qquad \forall b \in B, d \in \{x, y, z\} \tag{3}$$

$$\sum_{e \in I(v)} n_e^d + F_v^d = 0 \qquad \forall v \in V \setminus B, d \in \{x, y, z\} \tag{4}$$

$$a_{R(e)} \leq (A_{\max} - A_{\min}) x_{R(e)} \qquad \forall e \in E \tag{5}$$

$$\sum_{v \in V} \mathbf{V}(b, v) \times \mathbf{F}_v + \sum_{b' \in B} \mathbf{V}(b, b') \times \mathbf{r}_{b'} = \mathbf{0} \quad \forall b \in B \tag{6}$$

$$\sum_{v \in V} \mathbf{F}_v + \sum_{b \in B} \mathbf{r}_b = \mathbf{0} \tag{7}$$

$$\mathbf{x} \in \{0, 1\}^E, \ \mathbf{a} \in \mathbb{Q}_+^E, \ \mathbf{r} \in \mathbb{Q}^{B \times 3}, \ \mathbf{n} \in \mathbb{Q}^E \tag{8}$$

The Objective Function (1) aims at minimizing the volume of the resulting stable and symmetric complex space truss considering the external static load case. Constraint (2) ensures that the local longitudinal stress in a member must not exceed the member's yield strength taking into account a factor of safety. Constraints (3) and (4) ensure the static equilibrium at each vertex of the structure.

The decomposition of n_e into its components n_e^d with respect to each direction in space $d \in \{x, y, z\}$ is attained by standard vector decomposition, exploiting the invariant spatial and angular relationships due to the invariant ground structure. Variables indicating an additional cross-sectional area are bound to be zero by Constraint (5) if no member is present. Constraints (6) and (7) define the equilibrium of moments by resolution of the external forces and ensure, in combination with Constraints (3) and (4), that the resulting structure is always a static system of purely axially loaded members. In particular, the cross product $\mathbf{V}(b, v) \times \mathbf{F}_v$ is the moment caused by the external force $\mathbf{F}_v$ on bearing b with lever arm $\mathbf{V}(b, v)$. Analogously, $\mathbf{V}(b, b') \times \mathbf{r}_{b'}$ is the moment about bearing b caused by the bearing reaction force at b'. For the case of the segmented blank holder, see Fig. 2, solutions for $A_{min} = \{0.79, 3.14, 7.07\}$ mm^2 are shown in Fig. 4. Table 4 displays the computational results[1]. For our experiments we consider a basic vertex distance of 10 mm and the material aluminum with yield strength $\sigma_y = 0.19$ GPa.

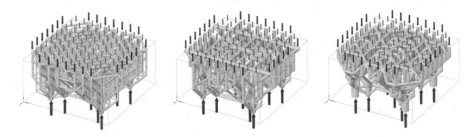

Fig. 4. $A_{min} =$ (left) 0.79 mm^2, (middle) 3.14 mm^2, (right) 7.07 mm^2

Table 4. Computational results

A_{min} [mm^2]	A_{max} [mm^2]	Best found [mm^3]	Bound [mm^3]	Gap [%]	Runtime [s]	First found time [s]	First found value [mm^3]
0.79	78.54	22815	22714	0.44	969828	7193	23756
3.14	78.54	33622	23822	29.15	1032300	2029	49233
7.07	78.54	56377	27192	51.77	362779	3214	86809

4   Finite Element Analysis and Shape Optimization

To validate the mathematical optimization results, linear static FEAs are performed using Altair OptiStruct. The load case is analogous to the load case

---

[1] The calculations were executed on a workstation with an Intel Xeon E5-2637 v3 (3,5 GHz) and 128 GB RAM using CPLEX Version 12.6.1 restricted to a single thread.

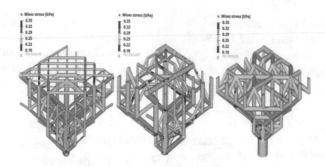

Fig. 5. Comparison of the FEA of three lattice structures generated by MILP

shown in Fig. 4. The geometries showed in Fig. 4 are discretized with solid elements of type CTETRA with a nominal element edge length of 0.5 mm. Note that through this volumetric mesh each node of the lattice structure can transmit rotary moments, which is contrary to the assumptions of the MILP model. Another difference between both models is the material behavior: While the MILP model cannot consider constitutive material equations without costly linearization, a linear-elastic material (MATL1) is implemented in the FEA model with an elastic modulus of aluminum of $E = 70$ GPa. The results of the FEAs are shown in Fig. 5, whereby for simplicity reasons, we take advantage of the double symmetry and visualize just a quarter of the model. We see that the stresses in all three models are, in general, below the yield strength of $\sigma_y = 0.19$ GPa. From this we conclude that the design suggestion by mathematical optimization is a solution with good mechanical performance and geometrical properties for this load case. Nevertheless, it turns out that some higher stressed positions exist. To overcome this problem, we suggest adding an FEA based free-shape optimization to the algorithm-driven design process. To this end, high stressed areas are identified whose shape OptiStruct is allowed to change, as exemplary shown in Fig. 6 for one lattice node. In the initial state (Fig. 6 left) there are maximum von Mises stresses of about 0.5 GPa. The objective of the optimization is to move the grid points of the finite element mesh, which are defined in the design region, in normal direction of this surface until the upper bound stress

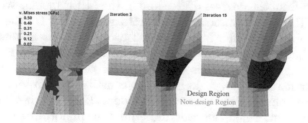

Fig. 6. Shape optimization: (left) v. Mises stress in the initial state, (middle) geometry of the node after 5 iterations, (right) geometry after 15 iterations

constraint of 0.19 GPa is satisfied. After 3 iterations (Fig. 6 middle) the surface is slightly shaped and after 15 iterations (Fig. 6 right) we see the final geometry of the lattice node, where the upper bound stress constraint of 0.19 GPa is satisfied. Consequently, the mechanical strength of the structure is given after this optimization.

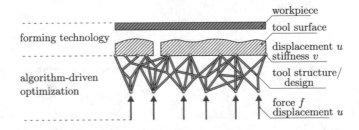

Fig. 7. Forming tool with in-process adjustable active fool surfaces

5   Conclusion and Outlook

We investigated an algorithm-driven optimization workflow for designing additively manufactured lightweight forming tools using the example of a flexible blank holder. To this end, an interactive CAD-tool was used for pre- and postprocessing the solution of a MILP optimization for truss-like lattice structures. As a minimum cross-sectional area is essential due to design restrictions in AM and symmetry can be exploited to effectively optimize structural systems, we introduced a MILP model considering continuous cross-sectional areas of the lattice members and two planes of symmetry. Finally, finite element based simulations and shape optimizations were performed to validate and improve the design suggestions supported by the preceding CAD-based mathematical optimization. Our research has highlighted that CAD-based mathematical optimization is an efficient and reliable tool for preliminary designing truss-like lattice structures for forming tools. Using finite element shape optimizations, highly stressed areas can be geometrically modified, resulting in an overall usable design. However, there is still a need for discussion that the degrees of freedom of a lattice node in the FEA differ from the degrees of freedom in the MILP model. We claim that a node in the MILP model cannot transmit rotary moments. On the contrary, due to the postprocessing of the MILP optimization solutions to merged volumes and the consequently volumetric meshing, a lattice node in the FEA can transmit rotary moments. This fact is one reason for the stress peaks in the FEA. Another reason for the stress peaks is that no constitutive material equations and no geometry are implemented in the MILP model. Therefore, it cannot take local stresses into account, which, however, underlines the importance of our workflow. Further work needs to be done to establish a component library, including joints for our Ansys SpaceClaim Add-In *construcTOR*. As shown in Fig. 7, we are currently investigating forming tools with in-process adjustable active tool surfaces to control material flow. Based on analysis of the interaction between local tool surface properties and the forming result, we will define process-time dependent, necessary displacement, and stiffness at the links between force transmitting lattice structure and tool surface. A new method based on our workflow will be investigated to fulfill these requirements. We will build mechanical mechanisms for adjustable surfaces and structural stiffness through technical joints instead of a solid volume at a lattice node or variable-length lattice members.

References

1. Burkart, M., Liewald, M., Wied, J., Todzy, T., Hartmann, M., Müller, M.: Optimization of a part holder design considering dynamic loads during return stroke of tool and ram. Procedia Manuf. **47**, 861–866 (2020)
2. Cao, J., Brinksmeier, E., Fu, M., Gao, R.X., Liang, B., Merklein, M., Schmidt, M., Yanagimoto, J.: Manufacturing of advanced smart tooling for metal forming. CIRP Ann. **68**(2), 605–628 (2019)
3. Groth, S., Engel, B., Langhammer, K.: Algorithm for the quantitative description of freeform bend tubes produced by the three-roll-push-bending process. Prod. Eng. Res. Devel. **12**(3–4), 517–524 (2018)
4. Kuhnhen, C., Knoche, J., Reuter, J., Al-Maeeni, S.S.H., Engel, B.: Hybrid tool design for a bending machine. In: 14th Conference on Intelligent Computation in Manufacturing Engineering CIRP ICME (2020)
5. Reintjes, C., Hartisch, M., Lorenz, U.: Design and optimization for additive manufacturing of cellular structures using linear optimization. In: Operations Research Proceedings 2018: Selected Papers of the Annual International Conference of the German Operations Research Society (GOR), pp. 371–377 (2019)
6. Reintjes, C., Lorenz, U.: Mixed integer optimization for truss topology design problems as a design tool for am components. In: International Conference on Simulation for Additive Manufacturing, vol. 2, pp. 193–204 (2019)
7. Reintjes, C., Lorenz, U.: Bridging mixed integer linear programming for truss topology optimization and additive manufacturing. Optim. Eng. Spec. Issue Tech. Oper. Res. (2020)
8. Rosen, D.W.: Design for additive manufacturing: a method to explore unexplored regions of the design space. In: Annual Solid Freeform Fabrication Symposium, vol. 18, pp. 402–415 (2007)
9. Rosen, D.W.: Computer-aided design for additive manufacturing of cellular structures. Comput. Aid. Des. Appl. **4**(5), 585–594 (2013)
10. Saadlaoui, Y., Milan, J.-L., Rossi, J.-M., Chabrand, P.: Topology optimization and additive manufacturing: comparison of conception methods using industrial codes. J. Manuf. Syst. **43**, 178–186 (2017)
11. Sörensen, P.F., Mašek, B., Wagner, M.F.-X., Rubešová, K., Khalaj, O., Engel, B.: Flexible manufacturing chain with integrated incremental bending and Q-P heat treatment for on-demand production of AHSS safety parts. J. Mater. Process. Technolo. **275** (2020). Article ID. 116312
12. Xu, D., Chen, J., Tang, Y., Cao, J.: Topology optimization of die weight reduction for high-strength sheet metal stamping. Int. J. Mech. Sci. **59**(1), 73–82 (2012)
13. Yang, D.-Y., Bambach, M., Cao, J., Duflou, J., Groche, P., Kuboki, T., Sterzing, A., Tekkaya, A.E., Lee, C.: Flexibility in metal forming. CIRP Ann. **67**(2), 743–765 (2018)

Development of an Annotation Schema for the Identification of Semantic Uncertainty in DIN Standards

Jörn Stegmeier[1][✉], Jakob Hartig[2], Michaela Leštáková[2], Kevin Logan[2], Sabine Bartsch[1], Andrea Rapp[1], and Peter F. Pelz[2]

[1] Institute of Linguistics and Literary Studies, Technische Universität Darmstadt, Dolivostraße 15, 64293 Darmstadt, Germany
joern.stegmeier@tu-darmstadt.de
[2] Chair of Fluid Systems, Technische Universität Darmstadt, Otto-Berndt-Straße 2, 64287 Darmstadt, Germany

Abstract. This paper presents the results of a pilot study carried out in cooperation between Linguistics and Mechanical Engineering, funded by the collaborative research centre (CRC) 805 "Beherrschung von Unsicherheit in lasttragenden Systemen des Maschinenbaus". Our goal is to help improve norm compliant product development and engineering design by focusing on ambiguous language use in norm texts (= "semantic uncertainty"). Depending on the country and product under development, industry standards may be legally binding. Thus, standards play a vital role in reducing uncertainty for manufacturers and engineers by providing requirements for product development and engineering design. However, uncertainty is introduced by the standards themselves in various forms, the most notable of which are the use of underspecified concepts, modal verbs like *should*, and references to texts which contain semantically uncertain parts. If conformity to standards is to be ensured, the person using the standards must interpret them and document the interpretation. In order to support users in these tasks, we
1. developed an annotation schema which allows the identification and classification of semantically uncertain segments of standards,
2. used the schema to create a taxonomy of semantic uncertainty in standards,
3. developed a proof-of-concept information system.

The results of this project can be used as a starting point for automated annotation. The information system alerts users to semantically uncertain segments of standards, provides background information, and allows them to document their decisions how to handle the semantically uncertain parts.

Keywords: Information system · Taxonomy · Semantic uncertainty

© The Author(s) 2021
P. F. Pelz and P. Groche (Eds.): ICUME 2021, LNME, pp. 23–34, 2021.
https://doi.org/10.1007/978-3-030-77256-7_3

1   Introduction

Standards and Their Role in Product Development. Technical standards helped with rationalisation and quality management of the production of goods in the 20th century by organising and standardising the shape, size and design of products and processes in a meaningful way [25]. Today a plethora of international, national and regional organisations develop and publish technical standards to unify rules for the exchange of information, ensuring compatibility and reducing the variety of products, services, interfaces and terms [22]. Technical standards therefore play a role in many processes in the manufacturing industry as well as in product development processes.

The application of standards is voluntary, but can be mandatory by law or contract [22]. In all cases non-compliance with standards, at least in the European Union, is associated with high risks for manufacturers since in the case of product liability the burden of proof is on the manufacturer. When compliant with norms, the burden of proof is reversed [30]. To ensure compliance, standards have to be written clearly and concisely [5]. This is in stark contrast to the findings in [9]. Among users of technical standards there is a considerable lack of knowledge of how technical standards must be interpreted.

We attribute this difference to the need of technical standards to be applicable for a wide range of contexts, situations and new technical developments.

Uncertainty in Standards. While the main purpose of standards is to unambiguously regulate products and product development, they can not be entirely strict. One the one hand, there are aspects which defy complete strictness, such as design or different solutions to a problem which yield the same result. On the other hand, standards need to allow for innovation, which is only possible with a certain degree of flexibility and thus rules out complete strictness. However, standard compliance is only achievable if any and all uncertain parts are resolved and the solution is not only documented but also communicated to all persons involved.

Uncertainty in technical standards is foremost a lack of information and, hence, a lack of knowledge which makes resolving it primarily a matter of researching and understanding further information. Resolving uncertain parts adds to the to-do list and should be addressed in an early stage of the project to ensure compliance. Identifying and classifying uncertain parts in standards should be regarded as a form of division of labor. It is less time consuming to have a dedicated team analyze and annotate all standards relevant for a project than having each engineer go through them on their own.

Example. The phrase 'allgemein anerkannte Regeln der Technik' [*generally acknowledged rules of technology*] is a good example for uncertainty that arises through ambiguous language use. It hinges on various assumptions:

1. There are rules of technology,
2. there is a kind of review process for these rules the result of which has merit for everybody,

3. there is a possibility to know which rules of technology are considered to be generally acknowledged.

The phrase leaves the reader in a state of uncertainty, since it does not provide enough information to know which specific way of behaviour is part of the generally acknowledged rules and which is not. Only if there were a closed list of accepted rules of technology would this phrase not be uncertain. Since such a list would stand in the way of innovation, it cannot be provided even if it could be compiled. From this perspective, this phrase is also a good example for the need of uncertainty in technical standards. The authors of technical standards are completely aware of this phrase's ambiguity as is evident from DIN 45020 [8] where 'acknowledged rule of technology' is defined as 'technical provision acknowledged by a majority of representative experts as reflecting the state of the art' [8, entry 1.5] and 'state of the art' is defined as 'developed stage of technical capability at a given time as regards products, processes and services, based on the relevant consolidated findings of science, technology and experience' [8, entry 1.4]. Both definitions do not provide specific enough information to decide without further steps how to handle a given task.

Scope and Aims. The project was designed as a pilot study which means that proof-of-concept took precedence over depth. The project's main aim was to develop an annotation schema for uncertainty in the language of DIN standards, a taxonomy of uncertainty based upon it, and an information system which provides access to the categorized instances of uncertainty. Annotating has a long-standing tradition in the humanities and can be regarded both as a part of knowledge acquisition and as a scholarly primitive [17,29]. Basically any form of data enrichment, from writing notes in the margin of a manuscript to computationally classifying sentences or words, can be regarded as annotation. Developing an annotation schema is an iterative process in which classes and subclasses are created based upon concrete instances in the documents (see Sect. 3 for some details on the process). It makes sense to use the same environment for both annotating and the development of the annotation schema. We used the application Inception for both tasks [14]. The backend for the information system is a MySQL database where we stored information about the documents as well as the annotated instances of ambiguous language use. We chose the series DIN 1988, consisting of the parts DIN 1988-100, DIN 1988-200, DIN 1988-300, DIN 1988-500, DIN 1988-600 since these standards play a role in the work of the CRC 805, see e.g. [16].

2   Meaning, Knowledge, and Uncertainty

Words and Meaning. There are numerous theories and approaches concerning meaning in language which are subsumed (for an overview, see [2,3,21,23]). One of the most seminal models of the relationship between words and meaning is the 'semiotic triangle' [21, p. 11] (see Fig. 1).

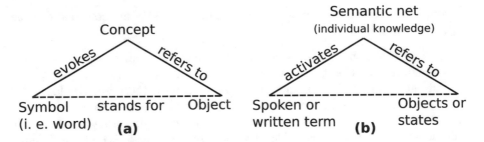

Fig. 1. Relationship between words and meaning. The *semiotic triangle* in (a) refers to language as a whole while the adaptation in (b) aims at an individual language user.

There is no direct connection between words and objects in the world. Words do not mean anything by themselves, rather, they trigger or activate parts of the knowledge store in our mind. The word *tree* does not contain a tree, it evokes the concept of a tree in the mind of the language user which is an abstraction of and a reference to the trees or a specific tree in the world. The semiotic triangle, which is also the basis for the general principles regarding concepts and terms in DIN 2330 [7], aims to illustrate the relationships between words and meaning in language in general, i.e. language as a system. However, language and language use (communication) are interdependent [2, p. 360]. On an individual level, words and their meaning are handled by the 'mental lexicon', which 'can be regarded as an individual network containing different kinds of personalized information on known words' [28, p. 6]. This also means that 'a word does not simplistically relate to a concept [...], but to a network of interrelated and overlapping distinct "senses", related to background world-knowledge' [19, p. 12] or, in other words, a semantic net.

For the purposes of this project, we understand uncertainty as a condition a) in which it is impossible to comply with the standards and b) which necessitates further steps of knowledge acquisition (see Fig. 2 below). We further consider this kind of uncertainty to be a result of ambiguous language use in technical standards.

Uncertainty enters language in various forms, the most notable of which are polysemy and underspecification. Polysemy occurs when a term activates multiple nodes of the network in the mental lexicon at once, for example the term 'mouse'. For a modern user of English, there are at least two concepts or senses activated upon hearing or reading this term. 1. *rodent*. 2. *peripheral computer device*. Usually, polysemy is resolved by taking into account the neighbouring terms (co-text) or the communicative setting (context) [13, cf. p. 7 f.].

Language, Knowledge, and Knowledge Acquisition. Even though language as whole can be regarded as a system shared and shaped by its users, the realms where individual language users are active are subsystems of language as a whole. These subsystems are formed and determined by (combinations of) sociodemographic factors like age, region, education, and, most notably for our purposes, occupation, specialization, and experience (these phenomena are studied in detail in sociolinguistics [18], and LSP, languages for special purposes, [15]). Hence, the knowledge and 'senses' available in an individual's mental lexicon are in part determined by the same factors. Specific fields of knowledge like linguistics or engineering create and constantly reshape their own specialized subsystem of language as a whole in order to accurately denote objects and how they relate to each other (mathematics and formal logic can be regarded as a part of these specialized subsystems or as subsystems in their own right). The constant reshaping brings about a shift in meaning for some words and phrases since the concepts they refer to undergo change. For a member of a specific field to keep track of theses shifts in meaning, constant knowledge acquisition is in order.

For our purposes, we draw on [1,24] and regard knowledge acquisition to be a cognitive process which involves the following steps: Sources need to be found and (after evaluation) used to gather data presumed to be pertinent to the project in question. The data needs to be pre-processed (both computationally and cognitively) to transform it into information which in turn can be cognitively understood, which results in knowledge. The newly acquired knowledge needs to be applied, which entrenches it into the mind and adds to the explicit and implicit knowledge. All of these steps draw on previous knowledge which is why we regard knowledge acquisition to be an ongoing iterative process (see Fig. 2).

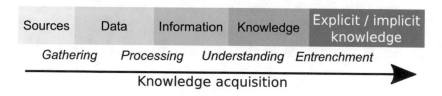

Fig. 2. Knowledge acquisition.

3   Taxonomy of Uncertainty

The taxonomy is the result of iteratively identifying and annotating (= assigning a class of uncertainty) instances of ambiguous language use in the technical standards. Identifying uncertain parts hinged upon the definition of uncertainty given above in Sect. 1, namely the answer to the question whether there was information missing in a sentence or the co-text of the sentence. Within each iteration, we inspected the emerging classes of uncertainty to ensure that they accurately reflected all instances of ambiguous language use and that they were sufficiently distinct from each other to avoid overlap. Both, the final annotations schema and the final annotations were validated by one last round of annotating, carried out by three engineers. Even though we focused on uncertainty arising from language use, we knew from previous experience with technical standards that there is at least one class of uncertainty which arises from conflicting knowledge rather than from lack of information conveyed by the text of a technical standard. Consider the following example: An engineer who is familiar with a specific technical standard operates on the knowledge already present in his mind but is not aware that there is a newer version of the technical standard available in which something has changed. Let's assume that the changes themselves are unambiguous but in conflict with the previous version of the standard. This constellation leads to uncertainty which is independent from language use. Therefore we distinguish evident uncertainty from hidden uncertainty as first sub-classes of uncertainty and regard evident uncertainty to be any form of uncertainty that arises from language use.

Our analysis of the standards yielded the following classes of uncertainty (Fig. 3):

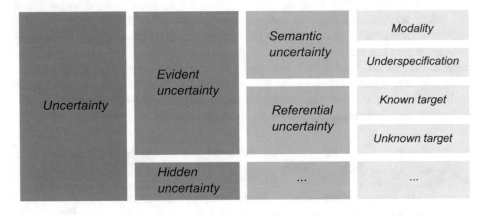

Fig. 3. Taxonomy of uncertainty.

Uncertainty that is grounded in terms and phrases is either modal or under-specified in nature. Modal uncertainty arises (intentionally) from any use of 'should' or 'can' leaving the decision which steps to take up to the standard user. Underspecification comprises any other case of ambiguous language use, ranging from phrases like 'the generally acknowledged rules of technology' to single words like 'bedürfen' in the following example: 'Dies gilt insbesondere für Apparate, die einer regelmäßigen Inspektion und Wartung bedürfen.' ['In particular, this applies to devices that are in need of scheduled inspection and maintenance.'] [6, p. 38]. To resolve the uncertainty, the maintenance needs for each device have to be checked. The instances of ambiguous language found in the technical standards comprise a *vocabulary of uncertainty* which will be the basis for the enhancements described below in Sect. 5. For a more detailed account of the taxonomy see [27].

4   Information System

Based on the taxonomy of uncertainty, we developed a proof-of-concept information system, which is targeted at engineers who work in a project where technical standards play a crucial role and annotating the documents is part of the project work. It is designed to provide the following features:

- a description of the taxonomy used to categorize the uncertain parts
- an overview over all standards that are relevant for the project
- a list of all uncertain parts of the annotated standards with the possibility to take notes
- inbuilt additional information on specific underspecified concepts
- possibility to add project specific information like for example instances of hidden uncertainty

Description of the Taxonomy of Uncertainty. The information system provides a detailed description of the taxonomy which offers the possibility to add project specific information. This is especially targeted at users who would like to re-define (parts of) the taxonomy or use project specific examples for the description to improve the project's internal communication and understanding.

Overview Over Standards Used. The overview is rendered as a network graph generated by the relationships between technical standards and a) their references to other technical standards which are listed as 'normative references' in each document, and b) other documents pertinent to the uncertain parts of the technical standards in question.

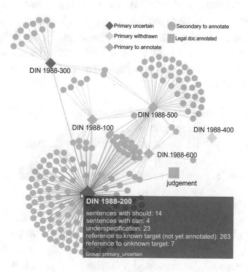

Fig. 4. Standards referenced by primary standards (edited screenshot of information system).

It not only shows which documents are linked to each other but also gives information about the group a document belongs to and about the annotation results (see Fig. 4). The groups are freely configurable to match the needs of a specific project. For our study we chose the following categories:

- *primary to annotate*: a technical standard directly pertinent to a given project
- *primary uncertain*: a technical standard directly pertinent to a given project which has been already annotated and contains uncertain parts
- *primary withdrawn*: a technical standard that is no longer valid but part of the series directly pertinent to a given project
- *secondary to annotate*: a not yet annotated technical standard which is linked to a primary document
- *legal doc annotated*: legal documents that contain information which helps to resolve some of the uncertain parts in the technical standards (here: a judgment)

As is evident from the categories, the information system is not only targeted at managing technical standards (= sources of uncertainty) but also any other documents which contain useful information. As an example for this, we chose a judgment which deals with a case where a newly installed drinking water system needed to be cleaned repeatedly and with enormous effort because the thread cutting agent used for cutting the pipes did not adhere to regulations [26]. We included this judgment for its descriptions of the steps taken to clean the pipes because they can be understood as an instance of following the 'generally acknowledged rules of technology'.

List of Classified Instances of Ambiguous Language Use. The core functionality of the information system is to display all uncertain parts in a structured way and provide a possibility to take notes on how to deal with specific instances of uncertainty in the technical standards in question. The default view shows all instances of all classes of uncertainty for all annotated technical standards. The tables on the top of the page provide links to more specific queries. Currently, these can be used to display

1. all instances of all classes of uncertainty found in a specific technical standard (first column of left table in Fig. 5)
2. all instances of a specific class of uncertainty found in a specific technical standard (second column of left table in Fig. 5),
3. and all instances of a specific class of uncertainty (first column of right table in Fig. 5).

The screenshot in Fig. 5 shows an excerpt of all uncertain items annotated as 'underspecified'. To limit this to underspecified items found in DIN 1988-200 the user just needs to click on *underspecification*.

Any specifications can be accessed via the link provided by the information system. The specifications provide a summary as well as an excerpt of the original document, and a link to the original document.

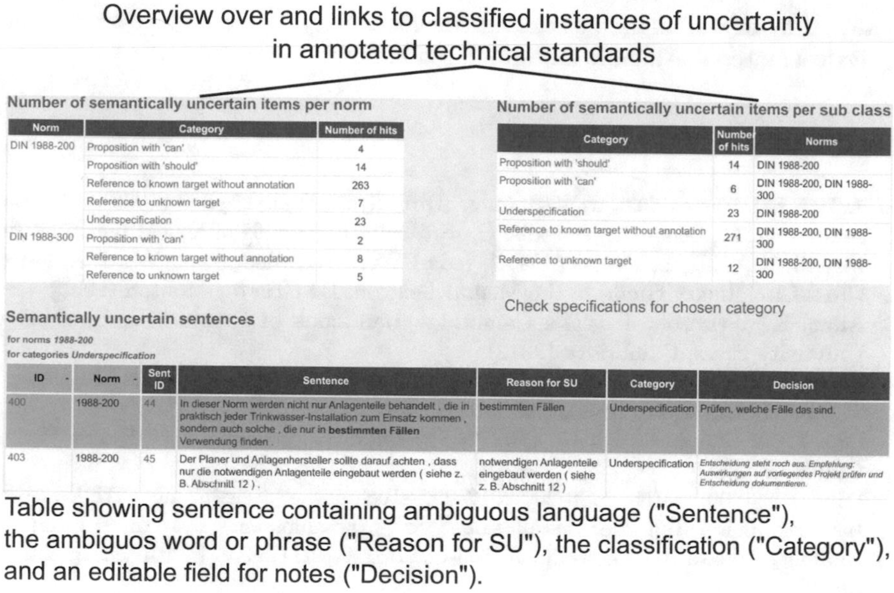

Overview over and links to classified instances of uncertainty in annotated technical standards

Number of semantically uncertain items per norm

Norm	Category	Number of hits
DIN 1988-200	Proposition with 'can'	4
	Proposition with 'should'	14
	Reference to known target without annotation	263
	Reference to unknown target	7
	Underspecification	23
DIN 1988-300	Proposition with 'can'	2
	Reference to known target without annotation	8
	Reference to unknown target	5

Number of semantically uncertain items per sub class

Category	Number of hits	Norms
Proposition with 'should'	14	DIN 1988-200
Proposition with 'can'	6	DIN 1988-200, DIN 1988-300
Underspecification	23	DIN 1988-200
Reference to known target without annotation	271	DIN 1988-200, DIN 1988-300
Reference to unknown target	12	DIN 1988-200, DIN 1988-300

Check specifications for chosen category

Semantically uncertain sentences

for norms *1988-200*
for categories *Underspecification*

ID	Norm	Sent ID	Sentence	Reason for SU	Category	Decision
400	1988-200	44	In dieser Norm werden nicht nur Anlagenteile behandelt, die in praktisch jeder Trinkwasser-Installation zum Einsatz kommen, sondern auch solche, die nur in **bestimmten Fällen** Verwendung finden.	bestimmten Fällen	Underspecification	Prüfen, welche Fälle das sind.
403	1988-200	45	Der Planer und Anlagenhersteller sollte darauf achten, dass nur die notwendigen Anlagenteile eingebaut werden (siehe z. B. Abschnitt 12).	notwendigen Anlagenteile eingebaut werden (siehe z. B. Abschnitt 12)	Underspecification	Entscheidung steht noch aus. Empfehlung: Auswirkungen auf vorliegendes Projekt prüfen und Entscheidung dokumentieren.

Table showing sentence containing ambiguous language ("Sentence"), the ambiguos word or phrase ("Reason for SU"), the classification ("Category"), and an editable field for notes ("Decision").

Fig. 5. Display of uncertain items in the information system.

5   Conclusion and Outlook

In the future, we will enhance the project in two ways. On the one hand, we will further develop the taxonomy of uncertainty and on the other hand, we will focus on automation, especially on automated annotation. To develop the taxonomy in a suitable manner, we will create a gold standard of correctly annotated instances of uncertainty, which means that we will annotate a larger number of carefully chosen technical standards. Both, determining the number of annotated instances and determining which technical standards to annotate requires time and consideration. The number of annotated instances needs to be high enough to yield significant results for rule-based automated annotation. The technical standards to annotate need to be representative for a given field of mechanical engineering and balanced with regard to aspects like document type, for example national vs. international codes. This brief outline of how we will proceed follows the best practices for corpus linguistic projects (for a more detailed account, cf. the section on methodological considerations in [4]). The gold standard of annotations will in turn allow us to make use of recent developments in computational linguistics with regard to automated classification and annotation, especially trainable classification systems like the ones provided by Inception [14]. Additionally, resources made available by lexicographical projects will be used to automatically retrieve synonyms for the instances of uncertainty (possible resources include for example [10–12, 20]. After evaluation with regard to their context dependent meanings, these synonyms will be used to extend the vocabulary of uncertainty and, hence, the lexical material available for automated annotation.

References

1. Ackoff, R.L.: From data to wisdom. J. Appl. Syst. Anal. **16**, 3–9 (1989)
2. Acquaviva, P., Lenci, A., Paradis, C., Raffaelli, I.: Models of lexical meaning. In: Pirrelli, V., Plag, I., Dressler, W.U. (eds.) Word Knowledge and Word Usage: A Cross-Disciplinary Guide to the Mental Lexicon. De Gruyter Mouton (2020)
3. Allan, K., Jaszczolt, K.: The Cambridge Handbook of Pragmatics. Cambridge University Press, Cambridge (2012)
4. Biber, D., Reppen, R.: The Cambridge Handbook of English Corpus Linguistics. Cambridge University Press, Cambridge (2015)
5. DIN 820-2 Standardization - Part 2: Presentation of documents. Beuth Verlag GmbH (2018)
6. DIN 1988-200: DIN 1988-200:2012-05, Technische Regeln für Trinkwasser-Installationen – Teil 200: Installation Typ A (geschlossenes System) – Planung, Bauteile, Apparate, Werkstoffe; Technische Regel des DVGW. Beuth Verlag GmbH (2012)
7. DIN 2330: DIN 2330:2013-07. Begriffe und Benennungen: Allgemeine Grundsätze. Concepts and Terms: General principles. Concepts et termes: Principes généraux. Beuth Verlag GmbH (2013)
8. DIN 45020: DIN 45020:2007-03. Normung und damit zusammenhängende Tätigkeiten – Allgemeine Begriffe (ISO/IEC Guide 2:2004); Dreisprachige Fassung EN 45020:2006 Standardization and related activities – General vocabulary (ISO/IEC Guide 2:2004); Trilingual version EN 45020:2006 Normalisation et activités connexes – Vocabulaire général (ISO/IEC Guide 2:2004); Version trilingue EN 45020:2006. Beuth Verlag GmbH (2007)
9. Drechsler, S.: Kompetenzbedarfe von Maschinenbauingenieuren in Bezug auf Richtlinien, Normen und Standards zur Ausübung ihrer beruflichen Tätigkeit. Ph.D. thesis, Karlsruher Institut für Technologie (2016)
10. Fankhauser, P., Kupietz, M.: DeReKoVecs. IDS, Institut für deutsche Sprache Mannheim (2017)
11. Hamp, B., Feldweg, H.: GermaNet - a Lexical-Semantic Net for German. In Proceedings of ACL workshop Automatic Information Extraction and Building of Lexical Semantic Resources for NLP Applications, pp. 9–15 (1997)
12. Heid, U., Schierholz, S., Schweickard, W., Wiegand, H.E., Gouws, R.H., Wolski, W.: Das Digitale Wörterbuch der Deutschen Sprache (DWDS). DE GRUYTER, Berlin (2010)
13. Henrich, V.: Word Sense Disambiguation with GermaNet. Dissertation, Universität Tübingen (2015)
14. Klie, J.-C., Bugert, M., Boullosa, B., Castilho, R.E., de Gurevych, I.: The inception platform: machine-assisted and knowledge-oriented interactive annotation. In: Proceedings of the 27th International Conference on Computational Linguistics: System Demonstrations. Association for Computational Linguistics, pp. 5–9 (2018)
15. Humbley, J., Budin, G., Laurén, C.: Languages for Special Purposes: An International Handbook. De Gruyter Mouton, Berlin (2018)

16. Philipp, L., Lena, C.A., Pelz, P.F.: Energy-efficient design of a water supply system for skyscrapers by mixed-integer nonlinear programming. In: Operations Research Proceedings 2017: Selected Papers of the Annual International Conference of the German Operations Research Society (GOR), Freie Universiät Berlin, Germany, 6–8 Sept 2017, S. 475–481, Springer (2017). ISBN 978-3-319-89920-6

17. Meister, J.C.: From TACT to CATMA or a mindful approach to text annotation and analysis. In: Rockwell, G., Sinclair, S. (eds.) Festschrift for John Bradley (Forthcoming)

18. Mesthrie, R.: The Cambridge Handbook of Sociolinguistics. Cambridge University Press, Cambridge (2011)

19. Murphy, G.L.: The Big Book of Concepts, 1. MIT Press paperback ed. MIT Press, Cambridge [u.a.] (2004)

20. Naber, D.: OpenThesaurus: ein offenes deutsches Wortnetz. In: Gesellschaft für Linguistische Datenverarbeitung (Germany), Fisseni, B. (ed.) Sprachtechnologie, mobile Kommunikation und linguistische Ressourcen: Beiträge zur GLDV-Tagung 2005 in Bonn. Lang, New York, Frankfurt am Main (2005)

21. Ogden, C.K., Richards, I.A.: The Meaning of Meaning. Harcourt, Brace & World, New York (1923)

22. Feldhusen, J., Grote, K.H. (eds.): Pahl/Beitz Konstruktionslehre: Methoden und Anwendungerfolgreicher Produktentwicklung. Springer Vieweg, Berlin (2013)

23. Riemer, N.: The Routledge Handbook of Semantics. Routledge, London (2016)

24. Rowley, J.: The wisdom hierarchy: representations of the DIKW hierarchy. J. Inf. Sci. **2**, 163–180 (2007). https://doi.org/10.1177/0165551506070706

25. Spur, G., Krause, F.: Das virtuelle Produkt: Management der CAD-Technik. Hanser, München (1997)

26. Kullmann, S., Dressler, L.M.: VI ZR 229/93 (1994)

27. Stegmeier, J. Hartig, J., Bartsch, S., Leštáková, M., Logan, K., Rapp, A., Pelz, P.F.: Linguistic analysis of technical standards to identify uncertain language use. In: Groche, P., Pelz, P.F., et al. (eds.) Mastering Uncertainty in Mechanical Engineering, Springer (Forthcoming)

28. Trautwein, J.: The Mental Lexicon in Acquisition: Assessment, Size and Structure. Universität Potsdam, Potsdam (2019)

29. Unsworth, J.: Scholarly Primitives: what methods do humanities researchers have in common, and how might our tools reflect this? (2000)

30. Wendt, J., Oberländer, M.: Product Compliance: Neue Anforderungen an sichere Produkte. In: Zeitschrift für Energie- und Technikrecht (ZTR), vol. 2016, no. 02, S. 62–70. Pedell (2017). ISSN: 2224-6819

Mastering Model Uncertainty by Transfer from Virtual to Real System

Nicolas Brötz$^{(\boxtimes)}$, Manuel Rexer, and Peter F. Pelz

Technische Universität Darmstadt, Chair of Fluid Systems,
64287 Darmstadt, Germany
{nicolas.broetz,manuel.rexer,peter.pelz}@fst.tu-darmstadt.de
https://www.fst.tu-darmstadt.de

Abstract. Two chassis components were developed at the Technische Universität Darmstadt that are used to isolate the body and to reduce wheel load fluctuation.

The frequency responses of the components were identified with a stochastic foot point excitation in a hardware-in-the-loop (HiL) simulation environment at the hydropulser. The modelling of the transmission behaviour influence of the testing machine on the frequency response was approximately represented with a time delay of 10 ms in the frequency range up to 25 Hz. This is considered by a Padé approximation. It can be seen that the dynamics of the testing machine have an influence on the wheel load fluctuation and the body acceleration, especially in the natural frequency of the unsprung mass. Therefor, the HiL stability is analysed by mapping the poles of the system in the complex plane, influenced by the time delay and virtual damping.

This paper presents the transfer from virtual to real quarter car to quantify the model uncertainty of the component, since the time delay impact does not occur in the real quarter car test rig. The base point excitation directly is provided by the testing machine and not like in the case of the HiL test rig, the compression of the spring damper calculated in the real-time simulation.

Keywords: Test rig · Stability · Model uncertainty · Time delay · Active Air Spring · Fluid Dynamic Vibration Absorber

1 Introduction

Developing new products or technologies is always at a high level of risk. To minimise the latter, it is essential evaluating the function and quality, cf. Pelz et al. [1], of the innovation as early as possible in the development process and to examine the interaction with the overall system. Early evaluation also corresponds to agile product development. One method to implement agile product development in the design process is HiL [2]. In HiL experiments, the newly developed component is integrated into a virtual system, thus enabling an accelerated or shortened development time.

P. F. Pelz and P. Groche (Eds.): ICUME 2021, LNME, pp. 35–44, 2021.
https://doi.org/10.1007/978-3-030-77256-7_4

HiL was first used for aerospace applications [3]. Since 1980 design engineering makes use of HiL for the development of vehicle components [4]. We use HiL to validate the two chassis components developed at the chair of fluid systems at the Technische Universität Darmstadt, (i) the Active Air Spring (AAS) [1] and (ii) the Fluid Dynamic Vibration Absorber (FDVA) [1]. The tests presented in this paper are performed with the FDVA. It intends to reduce wheel load fluctuation by transmitting the vibration energy of the wheel to the structural extension. The hydraulically translated oil mass, pumped by a piston from one chamber to another via ducts, represents the inductance. The inductance is connected to a compliance, thus a coil spring [5]. Figure 1 shows the FDVA on the right hand side. We will not focus on the component but on the two test rigs used to validate it in a dynamic system.

HiL is not without uncertainty, especially uncertainty due to boundary conditions and time delay appear. In this paper, we highlight the time delay. Batterbee et al. came to the conclusion that HiL test rig dynamics is to degrade performance results of a damper at higher frequencies [6]. Research is done for variability in time delays by Guillo-Sensano et al. [7]. Also, the interface location for HiL with time delay is an object of research by Terkovics et al. [8]. All the research is done because the time delay leads to an instability of the system. Therefore, the need of time delay reduction or compensation is high. Osaki et al. use a simple compensation strategy by adding a virtual damping to the system [9].

In this paper we have a look at the model uncertainty of the FDVA in a dynamic quarter car system. We look at the impact of our HiL stability with time delay of 10 ms [10]. To analyze the stability we evaluate the poles of the system with varying virtual damping and time delay. Knowing the effects of time delay, we discuss the problem of time delay compensation and describe the possibility of quantifying the FDVA's model uncertainty by finally introducing a quarter car test rig that does not have the time delay.

2   Hardware-in-the-Loop Test Rig with Time Delay

A common method to validate a hardware component in a complex system is to use HiL. The benefit for that is a reduced effort in manufacturing the system the component is used in. We build such a test rig, see Fig. 1, to validate the simulation models of suspension components and the components themselves.

The HiL test rig consists of connected real-time simulation and hardware [10]. In the real-time simulation a quarter car is simulated, that is reduced to a wheel mass m_w, a body mass m_b and a linear tire model with stiffness k_t and damping b_t. The quarter car model is excited via the input variable, the road excitation z_r, mapping the drive over a federal highway at 100 km/h. The outputs of the quarter car model are the body and wheel displacement z_b, z_w. Both values are subtracted and fed into the controller of the uni axial test rig. This moves the hydraulic cylinder to stimulate the FDVA in a deflection controlled manner. The signal transmission is impacted with a time delay that we discuss in the following. At the top of the FDVA we measure the force F and feed it back to the quarter car model. Besides this force F a virtual damper with the damping constant $\tilde{b}_b$ acts between the wheel and body mass. Table 1 shows the parameters of the quarter car.

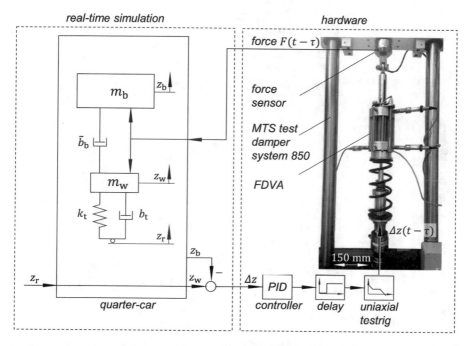

Fig. 1. Hardware-in-the-Loop test rig with quarter car simulation and MTS test damper system with integrated Fluid Dynamic Vibration Absorber

The validated model of the FDVA [5,11] is now tested in a dynamic system environment. The measurement results are evaluated by the amplifications of the frequency response, which is common for suspension components. The virtual damping $\tilde{b}_b$ was implemented to test the AAS that needed an additional damper. The FDVA should be tested without any additional virtual damping $\tilde{b}_b$ because the FDVA on its own can reduce the amplification of the wheel movement in the wheel eigenfrequency [5]. But reducing the virtual body damping from 1140 Ns/m to 700 Ns/m gets the system unstable. Having a look at the frequency response in Fig. 2, the measurement and simulation of the FDVA in the HiL test rig differ. You can find the amplification between wheel load and road excitation on the left and on the right you see the amplification between body acceleration and road excitation. The amplifications of wheel eigenfrequency at 13 Hz differ by more than 400%. Thus, we have to eliminate the source leading to this model uncertainty at the wheel eigenfrequency. Therefore we have a look at the stability of the system.

Table 1. Quarter car parameters

Parameter	Variable	Value
Body mass	m_b	290 kg
Wheel mass	m_w	40 kg
Body stiffness	k_b	31 500 N/m
Body damping	b_b	1 140 Ns/m
Tyre stiffness	k_t	200 000 N/m
Tyre damping	b_t	100 Ns/m
Time delay	τ	10 ms

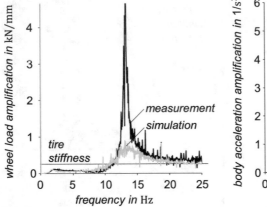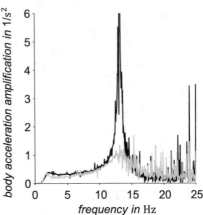

Fig. 2. HiL quarter car results with FDVA driving over a federal highway at 100 km/h. The FDVA with two opened ducts has an eigenfrequency at 8 Hz

2.1   HiL Stability

The HiL test rig controller is a black box. The PID parameters can be changed, but there is no possibility to detect the different time delays inside of it or even reduce them. Based on this we have to accept that there is an overall time delay $\tau = 10$ ms. To analyse the impact of the time delay τ and the virtual damping $\tilde{b}_b$ we investigate the equations of motion for the quarter car model

$$m_b \ddot{z}_b(t) = F(t - \tau) + \tilde{b}_b \left[\dot{z}_w(t) - \dot{z}_b(t) \right], \tag{1}$$

$$\begin{aligned} m_w \ddot{z}_w(t) = k_t \left[z_r(t) - z_w(t) \right] + b_t \left[\dot{z}_r(t) - \dot{z}_w(t) \right] \\ - F(t - \tau) - \tilde{b}_b \left[\dot{z}_w(t) - \dot{z}_b(t) \right], \end{aligned} \tag{2}$$

$$F(t - \tau) = k_h \left[z_w(t - \tau) - z_b(t - \tau) \right] + b_b \left[\dot{z}_w(t - \tau) - \dot{z}_b(t - \tau) \right]. \tag{3}$$

We transform Eqs. (1) and (2) in first order differential equations and use a linearized force, see Eq. (3), to evaluate general impacts. In MATLAB we analyse the poles of the resulting state space model, see Fig. 3. The first and second pole of the HiL system are shown. By increasing the virtual damping constant $\tilde{b}_b$ the poles move left and the system gets more stable, but we want to reduce the virtual damping to investigate the real damping. By adding more time delay all poles move to the right and the real term of one pole gets positive and gets the system unstable.

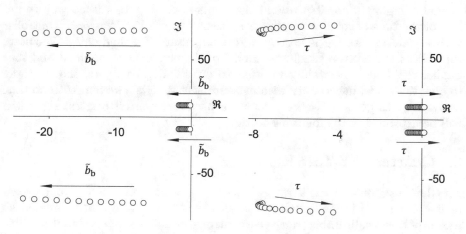

Fig. 3. HiL system poles depending on the simulated damping constant (left) and on the time delay (right)

Thus, the HiL system gets unstable at a time delay $\tau = 10\,\text{ms}$. The HiL system instability is shown in Fig. 4. The Bode and Nyquist plot show that next to the wheel eigenfrequency at $13\,\text{Hz}$ where the phase reaches $-180\,\text{deg}$, the magnitude is above $0\,\text{dB}$. For better visualisation the Nyquist Plot shows the Nyquist locus circles clockwise around the point $[-1,0]$. Therefor we have a negative damping that leads to the amplification at the wheel eigenfrequency shown in Fig. 2.

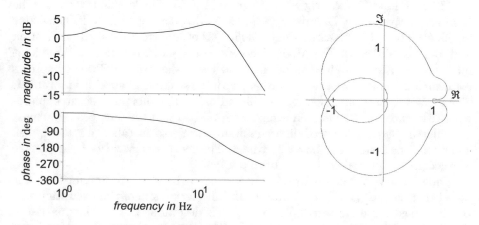

Fig. 4. Bode diagram (left) and Nyquist plot (right) for the unstable HiL quarter car with input z_r and output z_w at a time delay $\tau = 10\,\text{ms}$

2.2   Time Delay Compensation Problems for Model Validation

Time delay must be compensated to stabilize a HiL system. If we are testing a damper, there should be no virtual damping investigating real damping, but in our case, the system gets unstable. The simple compensation strategy by adding virtual damping is possible to stabilise the system, but this is no way to get lose of the time delay. For non-linear systems like the FDVA a state controller with a state monitoring is necessary to compensate the time delay. The state controller itself needs a simulation model of the nonlinear component and thus implies the model uncertainty we want to investigate. Therefore, the only way to quantify model uncertainty with measurement data is a system with no time delay. For this purpose, we have set up a hardware quarter car test rig, witch will be explained in the following section.

3   Quarter Car Test Rig

The developed quarter car test rig consists of a servo-hydraulic foot point excitation and a load frame on which the masses of the test rig can be guided by means of height-adjustable transverse control arms. This setup makes it possible to integrate a variety of axle kinematics into the test rig. The servo-hydraulic foot point excitation system from Form+Test includes a power supply, a valve block and a cylinder that can be moved with displacement or force control. The power unit has an output of 22 kW and provides a pressure of 280 bar. The installed cylinder has a maximum force of 25 kN at a maximum velocity of 0.7 m/s. The cylinder stroke is limited to 250 mm, which can be measured via the integrated stroke sensor, cf. Table 2. Furthermore, a force sensor is mounted on the piston rod to determine the wheel load. All measured variables of the system are transmitted to the measurement data recording. An interface of the controller is available for specifying the cylinder path.

Figure 5 shows the test rig with integrated FDVA. The kinematics used is the Modular Active Spring Damper System (MAFDS) developed as a demonstrator in the Collaborative Research Centre 805. Pelz et al. give a detailed description and its possibilities [1]. The MAFDS consists of a bar structure and three joint modules that absorb all lateral forces of the suspension system. Two coil springs connected in parallel are used as a wheel with the stiffness k_t specified in Table 1. The masses of the system can be flexibly adjusted. For this purpose, steel plates with 10 kg each are mounted on a support frame, so that the required body mass is achieved. Steel weights of different masses are also installed for the wheel. Therefore, the system can be easily tuned. The centres of gravity of each degree of freedom are located centrally above the cylinder.

Table 2 shows the sensors available for measuring the state quantities. The wheel force as well as all accelerations and displacements of the excitation and the two masses are captured. The velocity is determined by a combination of the derivative of the displacement and integration of the acceleration. Laser distance sensors with digital interfaces are used for precise measurement and low-noise signal transmission. The acceleration sensors are three similar piezoelectric sensors. The data acquisition is carried out by means of a MicroLabBox from dSpace, where it is also possible to specify any excitation z_r of the system. Since this is a real-time simulation environment, complex state estimators such as Kalman filters or controllers for active and semi-active systems can be realised.

Table 2. Sensors of the quarter car test rig

Sensor	Label	Variable	Range	Linearity
Body acceleration	IMI Sensors 626B02	$\ddot{z}_b$	$\pm 98\,\mathrm{m/s^2}$	$0.98\,\mathrm{m/s^2}$
Wheel acceleration	IMI Sensors 626B02	$\ddot{z}_w$	$\pm 98\,\mathrm{m/s^2}$	$0.98\,\mathrm{m/s^2}$
Road acceleration	IMI Sensors 626B02	$\ddot{z}_r$	$\pm 98\,\mathrm{m/s^2}$	$0.98\,\mathrm{m/s^2}$
Wheel force	GBR Serie-dr	F_w	$\pm 10\,\mathrm{kN}$	$10\,\mathrm{N}$
Road displacement	MTS RH	z_r	$0\ldots275\,\mathrm{mm}$	$0.04\,\mathrm{mm}$
Tire deflection	RIFTEK RF605-65/250	Δz_w	$65\ldots315\,\mathrm{mm}$	$0.25\,\mathrm{mm}$
Suspension deflection	RIFTEK RF605-65/250	Δz_b	$5\ldots315\,\mathrm{mm}$	$0.25\,\mathrm{mm}$

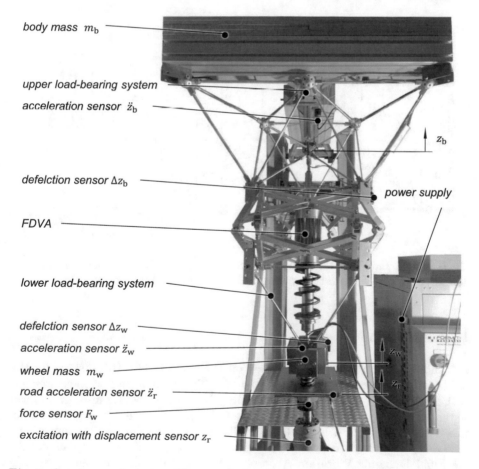

Fig. 5. Quarter car test rig with servo hydraulic foot point excitation and installed sensors

3.1   FDVA Results in the Quarter Car Test Rig

When we now use this quarter car test rig to validate the FDVA model in the dynamic quarter car we can perform tests without an additional body damper. The actual qualitative comparison between measurement and simulation for the quarter car test rig is possible, see Fig. 6. There is a good agreement between the two lines in the body acceleration amplification. The wheel load amplification shows a difference at higher frequencies, resulting from the mass of the connection of the force sensor and the wheel spring. This mass has an inertia that leads to data uncertainty. With the use of the tire deflection sensor and measurement of the tire spring stiffness the wheel load $F_{w,c} = \Delta z_w k_t$ can be calculated. The wheel load amplification of this soft sensor calculated wheel load shows a good agreement to the simulation, see Fig. 6. The model uncertainty is reduced to a minimum, because there is no simulated model for the hardware quarter car. Thus the basis to validate our FDVA model in a dynamic system is given with this test rig.

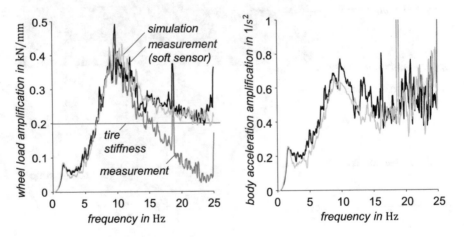

Fig. 6. FDVA results in the quarter car test rig driving over a federal highway at 100 km/h. The FDVA with two opened ducts has an eigenfrequency at 8 Hz

4   Conclusion

We studied the effect of the time delay on the stability of a HiL test rig for the validation of suspension component models. The simulation and measurement of the HiL test rig differ. Model uncertainty in the HiL test rig appears in form of the time delay. Therefore we analysed the poles of the state space model of a quarter car with a linear force feedback. With this understanding of the time delay impact we conclude that especially for non linear components like the Fluid Dynamic Vibration Absorber, there is only the possibility to build a real quarter car test rig to quantify model uncertainty. The quarter car test rig in form of hardware is described in detail. We are able to quantify a good agreement between simulation and measurement with this test rig. A detailed quantification of the FDVA model uncertainty by using the real quarter car test rig is part of further research.

Acknowledgement. Funded by the Deutsche Forschungsgemeinschaft (DFG, German Research Foundation) – Project number 57157498 – SFB805.

References

1. Pelz, P.F., Groche, P., Pfetsch, M.E., Schaeffner, M.: Mastering Uncertainty in Mechanical Engineering. Springer Nature (2021)
2. Hedrich, P., Brötz, N., Pelz, P.F.: Resilient product development - a new approach for controlling uncertainty. Appl. Mech. Mater. **885** (2018)
3. Isermann, R., Jochen, S., Stefan, S.: Hardware-in-the-loop simulation for the design and testing of engine-control systems. Control Eng. Pract. **7**(5), 643–653 (1999)
4. Schuette, H., Waeltermann, P.: Hardware-in-the-loop testing of vehicle dynamics controllers-a technical survey. SAE Trans. 593–609 (2005)
5. Brötz, N., Hedrich, P., Pelz, P.F.: Integrated Fluid Dynamic Vibration Absorber for Mobile Applications. Universitätsbibliothek der RWTH Aachen (2018)
6. Batterbee, D.C., Sims, N.D.: Hardware-in-the-loop simulation of magnetorheological dampers for vehicle suspension systems. Proc. Inst. Mech. Eng. Part I: J. Syst. Control Eng. **221**(2), 265–278 (2007)
7. Guillo-Sansano, E., Syed, M., Roscoe, A.J., Burt, G., Coffele, F.: Characterization of time delay in power hardware in the loop setups. IEEE Trans. Ind. Electron. **13**, 4237 (2020)
8. Terkovics, N., Neild, S.A., Lowenberg, M., Szalai, R., Krauskopf, B.: Substructurability: the effect of interface location on a real-time dynamic substructuring test. Proc. R. Soc. A: Math. Phys. Eng. Sci. **472**, 2192 (2016)
9. Osaki, K., Atsushi, K., Masaru, U.: Delay time compensation for a hybrid simulator. Adv. Robot. **24**(8–9), 1081–1098 (2010)
10. Lenz, E., Hedrich, P., Pelz, P.F.: Aktive Luftfederung - Modellierung, Regelung und Hardware-in-the-Loop-Experimente. Forschung im Ingenieurwesen **82**(3), 171–185 (2018)
11. Brötz, N., Pelz, P.F.: Bayesian Uncertainty Quantification in the Development of a New Vibration Absorber Technology. Model Validation and Uncertainty Quantification, vol. 3, pp. 19-26. Springer, Cham (2020)

44   N. Brötz et al.

Resilience

Potentials and Challenges of Resilience as a Paradigm for Designing Technical Systems

Philipp Leise[1](✉), Pia Niessen[2], Fiona Schulte[3], Ingo Dietrich[1], Eckhard Kirchner[3], and Peter F. Pelz[1]

[1] Department of Mechanical Engineering, Chair of Fluid Systems, Technische Universität Darmstadt, Darmstadt, Germany
{philipp.leise,ingo.dietrich,peter.pelz}@fst.tu-darmstadt.de

[2] Department of Mechanical Engineering, Ergonomics and Human Factors, Technische Universität Darmstadt, Darmstadt, Germany
pia.niessen@iad.tu-darmstadt.de

[3] Department of Mechanical Engineering, Product Development and Machine Elements, Technische Universität Darmstadt, Darmstadt, Germany
{schulte,kirchner}@pmd.tu-darmstadt.de

Abstract. The resilience paradigm constitutes that systems can overcome arbitrary system failures and recover quickly. This paradigm has already been applied successfully in multiple disciplines outside the engineering domain. For the development and design of engineering systems the realization of this resilience concept is more challenging and often leads to confusion, because technical systems are characterized by a lower intrinsic complexity compared to, e.g., socio-technical systems. The transfer of the resilience paradigm to technical systems though also offers high potential for the engineering domain. We present results from four-year research on transferring the resilience paradigm to the engineering domain based on mechanical engineering systems and summarize relevant design approaches to quantify the potentials of this paradigm. Furthermore, we present important challenges we faced while transferring this paradigm and present the lessons learned from this interdisciplinary research.

Keywords: Resilience · Technical system · Engineering · Uncertainty · Design methodology

1 Introduction

An increasing trend to higher product varieties leads to more and more complex production systems [1]. Furthermore, factors of global competition, sustainable product design and digitalization intensify the competition and time to market pressure on technological developments and at the same time increase the complexity of processes and products. This is caused by changing consumer behavior

© The Author(s) 2021
P. F. Pelz and P. Groche (Eds.): ICUME 2021, LNME, pp. 47–58, 2021.
https://doi.org/10.1007/978-3-030-77256-7_5

and technological changes. The supply and demand situation today often has to be answered much faster and more versatile. These changes lead to a need for adaptation for products, systems, and companies with their processes. However, the increasing complexity is not only evident for the market with its participants, but also takes on other dimensions such as infrastructures, networks, and supply. A greater need for coordination must also be mastered. Managing the increased complexity thus poses challenges for the design of systems at the various levels.

One possibility to master the increasing uncertainty is the paradigm of *resilience*. In this paradigm the considered technical system, production system, or supply-chain system is able to master, learn from, and adapt to disruptions within their lifetime, which were not considered explicitly within the design process.

This approach requires a change in an engineer's mindset, as engineers are trained to design systems and products in a deterministic process, where the definition of requirements happens at the beginning and covers only specific disruptions. This deterministic view leads to a reductive design approach, which means reducing or omitting the existing uncertainty that arises during the usage phase or within production. Traditionally, if the uncertainty in the production or usage period cannot be neglected, the system only has to respond to changing conditions in a robust way. "A robust system proves to be insensitive or only insignificantly sensitive to deviations in system properties or varying usage" [2, Glossary]. The mentioned deviation is often compensated by impinging a safety factor, and thus supersizing, which allows the system to withstand the changing properties without any impact on the system's functionality. However, a more sophisticated approach has been developed, too, referred to as Robust Design [3], [2, Section 3.3], [2, Section 3.5].

On the contrary, the resilience paradigm augments this traditional point of view, cf. [2, Section 3.5], by accepting the fact that most systems face unforeseeable disruptions within their lifetime.

The principle progression of a system's functional performance over time for a resilient behavior is shown in Fig. 1. The performance decreases after the onset of the (severe) disruption, but is kept above the required minimum performance f_{min}. After a period of time, which often depends on abating of the disruption the system's functional performance recovers at least to a certain extent.

To derive a comprehensive understanding of the abilities the resilience paradigm can provide for a *technical system*, we provide a brief definition. We define a technical system compliant with standard definitions of mechanical and mechatronic systems as shown by [6, Chapter 1]. Here, the term *system* describes the "totality of all elements considered" [2, Glossary]. It is delimited from the environment by its system boundary and usually consists of multiple subsystems. "Setting a system boundary defines the (...) product" [2, Glossary], which is developed. A technical system fulfills one or more predefined functions.

In the following we only consider mechanical and mechatronic systems, which usually consist of a mechanical structure and a predefined number of actuators and sensors, which are required to fulfill the predefined function. Here, we refer to single components like pumps or pistons as well as more complex,

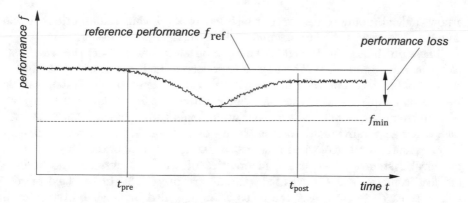

Fig. 1. Exemplary progression of the functional performance f of a system showing resilient behavior. A minimum performance is defined by f_{min}. At the time t_{pre} the (severe) disruption starts, while at the time t_{post} the new performance level is reached again. The system is able to master the disruption based on an adaptation and conceivably a learning procedure. Furthermore, the system does not fall below the minimum performance f_{min}. This example is adapted from the classic *resilience triangle* approach shown by [4] and [5].

e.g. load-carrying systems [2, Section 3.6]. Examples would be transmission systems, industry-scale fluid distribution systems, chemical plants or brake systems in vehicles. These shown technical systems distinguish themselves from socioeconomic systems by being rather complicated than complex systems [7]. This reduced complexity leads to challenges in the adaption of the resilience paradigm, since the reduced complexity yields less flexibility to adapt to disruptions.

In the following, we present results obtained in an interdisciplinary group from the engineering, mathematics and psychology domain. The group developed methodologies and reference systems to apply the paradigm of resilience in the mechanical engineering domain. Subsequently, we outline a concept of resilience in load-carrying systems and derive key functionalities each resilient system might fulfill according to our current point of view. Furthermore, we point out the challenges and potentials for a wide adoption of resilience in the engineering domain.

2 Overview of Resilience Concepts

Resilience is a paradigm widely used in different disciplines cf. [8]. It is derived from the Latin word *resilire*, which can be translated with "bounce back", [9, p. 184]. This translation of the origin only describes a very small part of resilience concepts and misleads the understanding as general systems should not only return to the state before the occurrence of disruptions, but learn from the endured experiences.

An extended view that can be seen as a major step within resilience research is given by the significant contribution of Holling in 1973 [10]. Holling enforced a new understanding of resilience, which led to significant contributions in the domain of ecology, socio-ecology and socio-technical system design.

These previously mentioned systems can be summarized under the term *complex adaptive systems* [11]. These systems consist of multiple agents that can act on disruptions based on their intrinsic motivation. Therefore, the system can be seen as an adaptive system. Its behavior is often non-linear, affected by agents with different goals and abilities, and often leads to unexpected outcomes. As each agent acts individually the complexity of systems, like socio-technical, ecological [10,12], or sociological systems, is far more pronounced than in mere technical systems. Nevertheless, in practice of technical systems the borders between complex and complicated systems are fuzzy [13]. In the field of reliability research for instance, the focus is on so-called high-reliability systems, such as nuclear power plants. These systems are extensively known but still classified as rather complex because unpredictable interdependencies can occur. Here, researchers try to design resilience as a safety paradigm. These systems are also understood as socio-technical, i.e. both the technical components and the human being is understood as an acting and reacting part of the system. Other systems, as for instance a star-shaped robot developed by Bongard and Lipson [14] can be described as a complicated system, which means extensive influences have an impact on the system, but it is theoretically ascertainable and predictable.

All systems have in common that resilience must be measured with the help of specific metrics to distinguish a more resilient system from a reference system. Therefore, the research in engineering has mostly focused so far on the definition of meaningful resilience metrics. This leads to a high number of metrics, which were proposed in the literature, as shown for instance by [15,16]. Most of these metrics related to technical systems were developed and used for network-like structures that can be represented by a mathematical graph. Examples are for instance water or electricity supply systems. In the graph representation, network properties like k-shortest paths [17] are considered as metrics to measure the resilience in case of rare events like component failures. In this approach systems are mostly considered as quasi-static, and they should fulfill a predefined minimum functionality even in the event of arbitrary system failures. To derive a resilient design of the underlying graph representation, they are improved algorithmically or in multiple iterations. For instance, by using a simulation-based approach [18].

Furthermore, Thoma et al. [7] criticize that much of the work in the area of technical systems has so far been too much conceptual. They see engineering research as having an obligation to go even further into the design of systems at all levels and to generate more concrete designs and solutions.

3   Our Approach—Definition, Resilience Functions and Metrics

In 2017 a group of roughly ten mechanical engineers and mathematicians, supported from 2019 on by one psychologist, started to work within the *Collaborative*

Research Center (CRC) 805 on resilience of technical systems. After looking into other scientific domains and their approaches, it became clear to us that there was a discrepancy between complex socio-technical systems the resilience community worked on and the rather complicated systems typical (mechanical) engineers face in their daily work. Thus, we derived a definition of resilience specifically for technical systems, [5], [2, Section 6.3]:

> *A resilient technical system guarantees a predetermined minimum of functional performance even in the event of disturbances and failures of system components, and a subsequent possibility of recovering.*

Resilience, from our point of view, is considered as complementary to robustness approaches, which are conventionally used for designing load-carrying systems in mechanical engineering.

Especially, for complicated systems, like technical systems, the resilience must already be considered within the design phase. Additionally, a resilient design of technical subsystems in combination with a resilience-considering design strategy can result in a composition of more resilient systems. Based on the system boundary even complex systems can then be considered within our approach.

Furthermore, relying on the work of Hollnagel [19, 20], we define *resilience functions* that a technical system needs to have: *monitoring, responding, learning, anticipating.*

In addition, we have derived a set of resilience metrics specifically for technical systems [2, 5], which allow quantifying resilience. We used those metrics to quantify the resilience of a by-wire car brake system [21], a water supply system [22], a dynamic vibration absorber [2, Section 6.3.6], a pumping system [23], a joint break [5], and a truss topology design [5].

4   Design of Resilient Technical Systems

After knowing what a resilient technical system seems to be, and how it can be evaluated, the question "How to design a resilient system" remains. In this section, we present practical implications and examples of more resilient technical system designs.

4.1   Practical Implications

Resilient technical systems cannot be seen detached from the conventional approaches for system design like the Robust Design approach. Some functions conventionally designed systems provide, and the models they are described with, also contribute to the description and development of resilient technical systems.

Besides this, common definitions in the resilience community like for instance "stress" and "shock", cf. [24], can be transferred to the mechanical engineering domain, where it is known as *disturbances* and component *failures*.

The application of the resilience paradigm results in an integration of the product design and the product usage phase [2, Section 7.2.3].

Furthermore, resilient technical systems can handle disturbances and/or failures by applying at least the first two of the already introduced four resilience functions *monitoring, responding, learning* and *anticipating,* cf. [2, Section 6.3.2]. For instance a system measures its current state and changes accordingly, if it detects a deviation from the "normal" state. This is also known from *fault detection and diagnosis,* cf. [25].

If the system fails completely, usually a human intervention is intended, which enables the system to achieve the final desired state. For either a change of the system itself, seen as its response to the monitored data, or the intervention of a human operator require the system's ability to (self-)adapt [26].

More resilient technical system designs also integrate a *learning* procedure to enable the system to learn from the endured disturbances and/or failures and the success of measures and strategies to handle the disruption. Learning can be understood as a reduction of model and data uncertainty through permanent model identification and adaptation during the life of a product. A further property also found in the resilience community is the possibility to *anticipate.* Anticipation is a predictive process (and system) change with the aim of reducing uncertainty. Thus, further more sophisticated controller strategies, like known from *adaptive control* [27], are suitable for resilient technical systems.

For systematic design of systems the general product development process according to VDI 2221 [28] can be applied to both mere robust design and more resilient design. Especially, both design methods necessitate the definition of requirements at the beginning and the design is supposed to be suitable for disruptions due to uncertainty, whereby resilience allows mastering uncertainty to a further extend than robustness.

Resilience design however requires an extension of the conventional design methods and models, as a central aspect of resilient behavior is the purposeful adaptivity of the system and a superior structure that specifies the resilience strategy for potential disruptions [26]. The models and methods for robust design are not necessarily able to describe a system's adaptivity. Thus, we developed additional models and extensions of known models.

The *resilience application model* is applicable for analyzing and comparing systems according to their resilience level and properties, but also for the synthesis of resilient properties in systems cf. [29]. It comprises the resilience characteristics, behavior, the considered disruption, and potential correlating signals for the description of the system and influencing factors.

A central model in conventional systematic design processes is the functional structure model, cf. [30, p. 242 ff.]. The model describes systems in a determined and inflexible way in its original form. We extended the model with representations for disrupted sub-functions, redundancy, adaptivity within the system, and a superordinate resilience function structure to make it applicable for the development of resilient systems, cf. [21].

4.2   Example Systems

So far, we presented a methodological approach to resilience of technical systems. In the following we will present three selected examples from research within the CRC 805, to present a path towards the resilient design of technical systems.

By-Wire Car Brake System. In by-wire car brake systems resilient approaches are realized already. This system includes a car's braking mechanism from the brake pedal's signal to the deceleration of the wheels and also comprises assistant systems like the anti-lock braking system. The brake system can be disturbed by a decrease of the board net voltage, which serves as the energy source for several subsystems of the car including the brake system. This scenario can, e.g., occur when the battery temperature is low and another subsystem, that requires high currents, like the engine starter, is running. The resilient functionality addresses this disruption by shutting down less important subsystems, like the assistant systems, in case of a decrease in the voltage level to keep up a minimum functionality to maintain the opportunity of braking, cf. [21]. As braking is highly safety relevant for cars *only braking* can be defined as the minimum functionality of the brake system. To be able to respond to a voltage decrease, monitoring of the voltage itself is required. For a more sophisticated resilience functionality further influencing parameters of the board net voltage like the battery temperature need to be detected. The monitored data could then be interpreted by the computer system, enable an anticipation of the upcoming voltage decrease and allow to initiate the response before a possible disruption occurs [21]. For monitoring of all parameters of interest multiple sensors are required. Another subsystem of cars that supports the resilience approach, e.g. for the brake system, is the automated start-stop. Making the monitored data of the automated start-stop available for, e.g., the brake system could enable more sophisticated resilient properties with little additional effort for implementing the monitoring.

Water Supply System. An optimization-based approach to design a resilient water supply system for high-rise buildings is given in [22]. To supply all levels in a high-rise building with fresh water, usually pumping systems are required. In the given example, the authors developed an algorithmic approach to consider the failure of up to three arbitrary pump failures and still derive energy- and investment-efficient system designs of decentralized water supply systems that can fulfill a predefined minimum functionality, as shown in Fig. 1. They used a Mixed-Integer Nonlinear Program and derived system designs that are more energy- and cost-efficient than classically designed systems with a comparable given resilience property. Furthermore, the given approach computes a control strategy in case pump failures occur.

Pumping System. A more resilient pumping system was derived in [23] and [2, Section 6.3.8]. It uses the previously mentioned four functions of resilient systems as a starting point. For each function one or more algorithmic approaches were developed. A subset has also been practically evaluated at the developed pumping system test rig to assure the transfer and applicability to real systems.

A specific focus was set on a system design that is on the one hand complicated and at the same time able to improve its functional performance if previously unseen disturbance patterns occur. The underlying algorithms are based on model identification, time series analysis and forecasting methods, which are commonly used within machine learning. These approaches can enable a more resilient system behavior, since they allow to increase the flexibility and to learn from endured experiences.

5   Challenges and Potentials

Next to the shown understanding of resilient systems and first design approaches, we also present challenges and potentials of this new paradigm.

5.1   Challenges

The realization of resilience in mechanical engineering poses a bunch of challenges due to the intrinsic properties of technical systems, their development, and usage.

Scope. Engineers tend to have a deterministic view. To understand a given problem set, engineers first define the system boundaries. Disruptions lying within the defined boundaries are considered while developing a solution, others are neglected. The concept of arbitrary disruptions is hard to grasp for engineers. If arbitrary disruptions are taken into account, two things can happen: i) the development is slowed down because of too many "but if's", ii) the system design becomes "over-engineered", thus being cost inefficient.

Adaptivity. The engineering approach to deal with complex systems is to break them down into subsystems, making each of the subsystems less complex. The flexibility and adaptivity of these subsystems is low. Without these properties however, the recovery of the functional performance (Fig. 1) after a disruption is hard to achieve. This applies especially for purely mechanical systems. If something breaks, it usually does not regain it's initial performance level.

Methodology. Engineering science has produced a high amount of methods and methodologies for product development and system design. Resilience being a paradigm, tends to be waived because there is a high uncertainty on how to achieve resilience within technical systems. So far systems have been analyzed and synthesis approaches have been deduced on an abstract level. The system analysis showed that resilience approaches already exist in current systems, especially mechatronic systems, like the mentioned by-wire car brake system. This provides example-based guidelines for the realization of resilience [21,29]. Yet, the systematic approaches need to be completed to a comprising resilience design methodology and evaluated by application to actual developments. Furthermore, the resilience design methodology requires further empirical testing.

Robustness. The distinction between robustness and resilience remains a challenge for engineers, especially discussing specific systems. Robust Design is well known in the engineering domain, and includes many aspects of the resilience paradigm, cf. [31–33], [32] and [33].

Stakeholders. The typical context, in which technical systems are developed ,is a customer relationship. The customer defines requirements, the supplier defines a specification of what he is able to deliver. Ideally, after negotiating, both stakeholders know, what they can expect and what they have to deliver. After delivery, the specifications are either met or they are not fulfilled. The introduction of arbitrary disruptions into these requirements-specification domain is challenging, because it implies uncertainty for both stakeholders. Furthermore, the state-of-the-art for production processes is to define performance measurements, which are fixed. This goes back to Henry Ford and the so called "Austauschbau" [2, Chapter 2]. Theses fixed performance measurements lead to a conflict with self-adaptive systems. During further research it is important to meet those challenges to successfully establish the resilience paradigm in the engineering domain, cf. [2, Chapter 3], [2, Section 5.1.1] and [2, Chapter 7].

5.2   Potentials

The resilience paradigm offers potentials to master uncertainty for technical system designs in a rapidly changing environment. Hence, the interest in this field is evolving. For instance cities enforce the resilience of their infrastructure [34]. In 2020 the Covid-19 pandemic disclosed the vulnerability of global production and supply chains. These developments will affect technical systems as well. To increase attention on the topic within engineering domain use cases, the following possible potentials are emerging:

Flexibility. With a focus on resilience, more flexibility [2, Section 3.5] can be created for processes and products. This results from the fact that systems are no longer designed deterministically, but that changes can always be made.

New Mechanism of Actions. Through new systems, mechanisms can be explored and tested that were not previously considered in the usual way.

Learning from Errors. By integrating learning as a property of the technical system, it is possible to better analyze errors and malfunctions and learn from them. This can lead to a successive improvement of the systems. Thereby, especially highly safety relevant systems can be addressed because resilience enables a reduction of the risk of failure and thus an increase in the safety level.

6   Conclusion

The resilience paradigm differs from existing approaches to master uncertainty in the engineering domain. Typically, engineers try to identify uncertainty and design a system as robust as necessary. Today, one cannot say whether a resilient design might result in an even increased performance at similar effort. Nevertheless, addressing the paradigm of resilience is an important task for engineers. They are in a position to develop technical systems for the future—for a future, in which there is a high demand for resilient systems due to crises such as climate change or Covid-19. However, it is important to understand the deeper implications of the resilience paradigm. This includes that there is not one but a variety of possibilities to make a system resilient and that resilient systems do not have to absorb every potential disruption—it is even more important to strengthen the system to master likely ones. Especially in specific domains, such as critical infrastructure, a resilient technical system design can be beneficial. In other technical domains, a resilient design will not be required. Therefore, the context of the technical system is important and must always be considered. Furthermore, resilience should be understood as a process and not only as an output. While having resilience as an objective in mind during a product development, it can lead to solutions that have not been considered in advance. In order to approach the concept of resilience, it is therefore indispensable to have an interdisciplinary exchange.

References

1. Efthymiou, K., Pagoropoulos, A., Papakostas, N., Mourtzis, D., Chryssolouris, G.: Manufacturing systems complexity review: challenges and outlook. Proc. CIRP **3**, 644–649 (2012)
2. Pelz, P.F., Groche, P., Pfetsch, M., Schäffner, M.: Uncertainty in Mechanical Engineering. Springer, to appear (2021)
3. Phadke, M.S.: Quality Engineering Using Robust Design, Prentice Hall PTR, Englewood Cliffs (1995)
4. Tierney, K., Bruneau, M.: Conceptualizing and measuring resilience: a key to disaster loss reduction. TR News **250** (2007)
5. Altherr, L.C., et al.: Resilience in mechanical engineering-a concept for controlling uncertainty during design, production and usage phase of load-carrying structures. Appl. Mech. Mater. **885**, 187–198 (2018)
6. Isermann, R.: Mechatronic Systems: Fundamentals. Springer, New York (2007)
7. Thoma, K., Scharte, B., Hiller, D., Leismann, T.: Resilience engineering as part of security research: definitions, concepts and science approaches. Eur. J. Secur. Res. **1**(1), 3–19 (2016)
8. Scharte ,B., Thoma, K.: Resilienz–Ingenieurwissenschaftliche Perspektive. In: Multidisziplinäre Perspektiven der Resilienzforschung, pp. 123–150. Springer (2016)
9. Voss, M., Dittmer, C.: Resilienz aus katastrophensoziologischer Perspektive. In: Multidisziplinäre Perspektiven der Resilienzforschung, pp. 179–197. Springer (2016)
10. Holling, C.S.: Resilience and stability of ecological systems. Ann. Rev. Ecol. Syst. **4**(1), 1–23 (1973)
11. Holland, J.H.: Complex adaptive systems. Daedalus **121**(1), 17–30 (1992)
12. Folke, C.: Resilience: the emergence of a perspective for social-ecological systems analyses. Glob. Environ. Change **16**(3), 253–267 (2006)
13. Hiermaier, S., Scharte, B.: Ausfallsichere Systeme, pp. 295–310. Springer, Heidelberg (2018)
14. Bongard, J., Zykov, V., Lipson, H.: Resilient machines through continuous self-modeling. Science **314**(5802), 1118–1121 (2006)

15. Hosseini, S., Barker, K., Ramirez-Marquez, J.E.: A review of definitions and measures of system resilience. Reliab. Eng. Syst. Saf. **145**, 47–61 (2016)
16. Shin, S., et al.: A systematic review of quantitative resilience measures for water infrastructure systems. Water **10**(2), 164 (2018)
17. Lorenz, I.-S., Pelz, P.F.: Optimal resilience enhancement of water distribution systems. Water **12**(9), 2602 (2020)
18. Klise, K.A. , Murray, R., Haxton, T.: An overview of the water network tool for resilience (WNTR). Technical Reports, Sandia National Lab. (SNL-NM), Albuquerque, NM, USA (2018)
19. Hollnagel, E.: Rag - The Resilience Analysis Grid. Resilience Engineering in Practice. A Guidebook. Ashgate, Farnham (2011)
20. Hollnagel, E.: FRAM: The Functional Resonance Analysis Method: Modelling Complex Socio-Technical Systems. CRC Press, Boca Raton (2017)
21. Schulte, F., Engelhardt, R., Kirchner, E., Kloberdanz, H.: Beitrag zur Entwicklungsmethodik für resiliente Systeme des Maschinenbaus. In: Krause, D., Paetzold, K., Wartzack, S. (eds.) DFX 2019: Proceedings of the 30th Symposium Design for X, 18-19 September 2019. The Design Society (2019)
22. Altherr, L.C., Leise, P., Pfetsch, M.E., Schmitt, A.: Resilient layout, design and operation of energy-efficient water distribution networks for high-rise buildings using MINLP. Optim. Eng. **20**(2), 605–645 (2019)
23. Leise, P., Breuer, T., Altherr, L.C., Pelz, P F.: Development, validation and assessment of a resilient pumping system. In: Comes, T., Hölscher, C. (eds.) Proceedings of the Joint International Resilience Conference, JIRC 2020, pp. 97–100. University of Twente, Enschede (2020)
24. Miller, F., et al.: Resilience and vulnerability: complementary or conflicting concepts? Ecol. Soc. **15**(3) (2010)
25. Isermann, R.: Fault-Diagnosis Systems: An Introduction from Fault Detection to Fault Tolerance. Springer, Heidelberg (2005)
26. Schlemmer, P.D., Kloberdanz, H., Gehb, C.M., Kirchner, E.: Adaptivity as a property to achieve resilience of load-carrying systems. Appl. Mech. Mater. **885**, 77–87 (2018)
27. Åström and Björn Wittenmark, K.J.: Adaptive Control. Courier Corporation (2013)
28. Verein Deutscher Ingenieure.: VDI 2221 Methodik zum Entwicklen und Konstruieren technischer Systeme und Produkte. VDI Verlag (1993)
29. Schulte, F., Kirchner, E., Kloberdanz, H.: Analysis and synthesis of resilient load-carrying systems. In: Proceedings of the Design Society: International Conference on Engineering Design, vol. 1, no. 1, pp. 1403–1412 (2019)
30. Jörg, F., Grote, K.-H. (eds.): Pahl/Beitz Konstruktionslehre: Methoden und Anwendung erfolgreicher Produktentwicklung, 8th edn. Springer, Heidelberg (2013)
31. Engelhardt, R.A.: Uncertainty Mode and Effect Analysis: heuristische Methodik zur Analyse und Beurteilung von Unsicherheiten in technischen Systemen des Maschinenbaus. Dissertation, Technische Universität Darmstadt, Darmstadt (2013)
32. Freund, T.: Konstruktionshinweise zur Beherrschung von Unsicherheit in technischen Systemen. Dissertation, Technische Universität Darmstadt, Darmstadt (2018)
33. Mathias, J.: Auf dem Weg zu robusten Lösungen: Modelle und Methoden zur Beherrschung von Unsicherheit in den Frühen Phasen des Produktentwicklung. Dissertation, Technische Universität Darmstadt, Darmstadt (2015)
34. Resilient Cities Network. https://resilientcitiesnetwork.org. Accessed 30 Dec 2020

Modelling of Resilient Coping Strategies within the Framework of the Resilience Design Methodology for Load-Carrying Systems in Mechanical Engineering

Fiona Schulte[✉], Hermann Kloberdanz, and Eckhard Kirchner

Darmstadt University of Technology, Darmstadt, Germany
schulte@pmd.tu-darmstadt.de

Abstract. During the development of load-carrying systems uncertainty caused by nescience can be handled applying resilience design. With this systematic approach, in addition to robust design, resilient system properties can be achieved. The resilience design methodology comprises new and extended models and methods. The central aspect of resilient properties is an adaptivity of the system. The procedure for resilience design starts with choosing a 'general coping strategy' appropriate for the design task. Based on this, a more detailed 'system coping strategy' is developed. This concrete strategy is based on the resilience functions responding, monitoring, anticipating and learning. The coping strategies always contain the function 'responding' because it represents the actual adaption of the system. The central, most abstract synthesis model for developing robust and resilient systems is the functional structure model. In this model the system functions and their interconnection by signals, material and energy flows are depicted. However, the realisation of resilience properties requires additional signals and flows. Hitherto, the functional structure for robust systems is static, whereas adaptivity requires flexible control of functions and flows. Therefore, an extension of the functional structure model is proposed to be able to depict the resilient system coping strategy and adaptivity. Within the resilient system the coping strategy is depicted by adaption functions based on the four resilience functions. Via an introduced interface and an enabler-structure the adaption functions are connected to the robust functional structure. The application of the proposed extension is illustrated by the example of a by-wire car brake system.

Keywords: Resilience design · Functional structure · Adaptivity

1 Introduction

The concept of resilience is known from living organisms and socioeconomic systems. It is based on the efficient use of resources. Living organisms are compelled to get by with limited resources and thus cannot be resistant to any kind of disruption, like injuries, because this would require excessive access to resources. Hence, robustness only evolved for common events, while for seldom extreme situations natural organisms

© The Author(s) 2021
P. F. Pelz and P. Groche (Eds.): ICUME 2021, LNME, pp. 59–69, 2021.
https://doi.org/10.1007/978-3-030-77256-7_6

evolutionary developed the ability to continue living with a reduced capability, adapt to lasting condition changes and recover from disruptions. Therefore, efficient strategies to cope with disruptions have evolved. Human beings, e.g., suffer from severe symptoms during influenza illnesses but do survive and fully recover [1].

Resilience offers potential for technical systems, too. Accepting, that a technical system cannot be robust, i.e. able to withstand, towards all disruptions, resilient behaviour in load-carrying systems would guarantee essential system functions, while non-substantial functions may fail. Furthermore, the system would be able to recover, when disruptions decline, as defined by the Collaborative Research Centre 805 for resilience in load-carrying systems [2]. In contrast to natural organisms for technical systems coping strategies to achieve resilient behaviour have to be planned during the development process. The general strategies are similar to natural coping strategies [1].

A high potential is estimated for highly safety relevant systems like, e.g., car brake systems [3] and upon occurrence of unknown or neglected uncertainty. Events addressing uncertainty caused by nescience, that are expected to cause severe disruptions, which cannot be covered by the system's robustness, are presumed. The objective of realising resilience in load-carrying systems is to deal with those severe disruptions and prevent risks due to complete system failures. In this context resilience also offers economic potential. Disruptions, that are neglected during the robust design process, because of too high economical effort regarding resources and considering the low exposure to the disruption, can be addressed using the resilience concept [1, 3].

2   Fundamentals

Based on resilience theories from other fields of research, basics for resilience engineering have been devised. However, to be able to realise resilient technical systems, related fields of research have to be considered, too. In case of load-carrying systems the vulnerability analysis can serve as a support for resilience engineering by pointing out the system's weaknesses and focusing on the crucial influences regarding resilient properties. Thus, vulnerability is understood as a partially complementing approach to resilience, here. Similarly, resilience design is understood as an extension of robust design. A robust system is even comprehended as a prerequisite for resilience design. Thus, also robust design and its methodologies in general as well as the inherent robust properties of a technical system have to be taken into account. In a resilience design methodology for developing comprehensive resilience concepts for load-carrying systems, robustness, vulnerability and resilient properties as well as their close interrelation have to be embedded.

2.1   Resilience

Resilience in technical systems describes the system's ability to "[…] guarantee […] a predetermined minimum of functional performance even in the event of disturbances or failure of system components, and a subsequent possibility of recovering at least the set point function [2]." As mentioned above resilience design is understood as an extension of the robust design methodology. While a robust system is designed to withstand distur-bances in a predefined range of the influencing parameters without significant reduction in its functionality, resilience design aims at mastering disruptions, like extreme vari-ations of the influencing parameters. To address disruptions beyond those handled by robust design, a resilient system adapts to the disturbed conditions and accepts a reduc-tion of the functionality, as long as the essential minimum functional performance is still provided. Consequently, robust design is applied for known uncertainty in a system, while resilience allows to handle disruptions beyond the common and well known range of the influencing parameters. Resilience Design therefore comprises adjusted methods of robust design and particular new models, procedures and methods.

For analysing a system's resilience properties a resilience application model has been developed [4]. It comprises the resilient behaviour of a technical system, which is depicted by the progression of the system's functional performance over time, showing the system's reaction to a disruption. To describe the system completely, it is also required to look at the static resilience characteristics and related metrics that describe the func-tional performance depending on influencing parameters according to [2]. For resilience engineering the interdependency of the properties with the disruption progression and potential correlating signals are considered, additionally [4].

Furthermore, functional resilience characteristics, also referred to as resilience func-tions, allow to describe a system's resilient properties. The functional resilience char-acteristics [2, 5] are based on the four abilities of resilience Hollnagel [6] postulated: Responding, monitoring, anticipating and learning. To realise the characteristic learn-ing in technical systems a human operator or an artificial intelligence system would be required. Since neither can be assumed in most technical systems, only responding, monitoring and anticipating are taken into account in the following [4]. The degree a system is able to apply the functional resilience characteristics to a disruption, pro-vides information about the system's resilience level [6]. Simple systems only feature responding, which describes the system's reaction to disruptions, whereas more sophisti-cated systems are characterised by monitoring the disruption progression and correlating signals and anticipating of an upcoming disruption by interpreting the monitored data [4, 5].

2.2 Vulnerability

According to Turner et al. [7] vulnerability can be defined as "[…] the degree to which a system […] is likely to experience harm due to exposure to a hazard […]" [7]. To use the knowledge about the system's vulnerability, first it needs to be identified. One approach to do so, is a vulnerability analysis using the scenario technique [8–10]. With the sce-nario technique possible future situations are developed to assess the potential changes of conditions for the system. From these condition changes disruptions of the system, which could be either external disturbances or internal damages, can occur [11]. After-wards, based on a system analysis the sensitivity of the system towards the disruptions is determined considering implemented measures or inherent abilities of the system to deal with certain disruptions leaving the critical disruptions [8, 12]. This identification of the system's weaknesses is requisite for resilience design because it provides information about the critical quantities, the system needs to be resilient towards.

2.3 Functional Structure Modelling

During the systematic product development process for robust systems one of the first steps is modelling the functional structure. In the functional structure all functions and the flows of energy, material and signals interconnecting them, which are required to realise the overall function of a system, are depicted. The use of functional structures improves the understanding of the system and its aspired operating principles and enables to derive the required subfunctions and flows, regardless of possible function carriers, before choosing the appropriate components and modules for their realisation [13].

3 Research Question

The concept of resilience in load-carrying systems is understood as an extension of robust system design. As the extension by resilient functionality requires a different mindset compared to robust design, for a systematic system development it is crucial to also extend the development methodologies [1]. For resiliently mastering extreme situations resilience characteristics and resilience behaviour have to be combined appropriately, which requires determining strategies that exceed design principles. The resilient reaction of a system to a disruption is described as *coping*. Hence, the basic characteristic of the system's reaction is defined as a *coping strategy*, here [14].

The identification of the critical system conditions caused by unknown influences or unexpected component failures are determinable using the vulnerability analysis. As soon as the crucial vulnerabilities are identified a *basic coping strategy* is required to deal with the disruption in case it occurs. The static and dynamic resilience properties, depicted in the resilience application model, are a first concretisation of the coping strategy. It can be further concretised based on the functional resilience characteristics as the *system coping strategy*. The obvious coping strategies known from natural organisms are mainly characterised by utilising signals, material and energy resources. Hence, to realise the coping strategy during the systematic product development process, modelling the system coping strategy in combination with the functional structure is essential, because the signal, material and energy flows are first described within the functional structure during the development process. As the coping strategies always require the functional resilience characteristic responding, they implicitly require a purposive system adaptivity [3, 15]. Thus, for the development of resilient load-carrying systems this adaptivity has to be modelled within the functional structure, which is not possible using the conventional functional structure model for robust design [3]. In this context the objective of this contribution is to answer the following research question.

- How to define basic resilient coping strategies as an extension of robust design and model the mandatory system adaptivity in functional structures for methodological development of resilient systems?

4 Modelling Resilient System Structures

During the product development process, including resilient system behaviour, first resilient requirements are deduced using the vulnerability analysis as described in

Sect. 2.2. The current as well as the aspired system properties are depictable in the resilience application model, which also enables to quantify the resilience requirements as shown in [3]. Afterwards, the deduction and formulation of the resilient coping strategy follow in two steps. First a suitable *basic coping strategy* for the whole system is identified. Thereafter, the realisation of the coping strategy is concretised as a *system coping strategy*. The system coping strategy is modelled in the functional system structure as an extension using *adaption functions* based on the functional resilience characteristics. The adaption functions are depicted outside the robust system boundary and connected to the robust subfunctions and flow parameters.

4.1 General and Basic Resilient Coping Strategies

Defining coping strategies different consideration horizons have to be taken into account. The robust system structure is assumed to be predetermined and located in the central position. It is describable by the conventional functional structure model. The resilient consideration horizon exceeds the robust system boundary and takes the superordinate system into account, as well, because in case of extreme disruptions resilient systems may rely on external resources as well as flexible functions. For some approaches of realising resilient design an even wider, so called extended, consideration horizon is applied. The extended consideration horizon additionally comprises the system environment and enables to identify threats and utilise resources from beyond the superordinate system boundary.

Three suitable *general coping strategies* based on nature have been identified for load-carrying systems, as a first exemplary result: internal/external degradation, usage of alternative internal/external resources and purposeful overload. The internal degradation looks at the robust system, wherein less important functions are switched off or reduced in performance to maintain full performance of essential functions. External degradation uses an extended consideration horizon, which also regards the superordinate system. The degradation then is executed outside the robust system boundary but safes, e.g., resources for the regarded subsystem. The usage of alternative internal or external resources allows the system to draw on resources, which are not originally intended to be used by this subsystem or function. Alternative external resources are available in the superordinate system or the environment and taken into account by the extended consideration horizon. The strategy of purposeful overload uses a certain subsystem or function, of which the demand increased overly, excessively, accepting a possible damage. The choice of a coping strategy depends on the system's requirements and properties, as well as the disrupted quantities. Exemplary assignments of general coping strategies for typical vulnerabilities are given in Table 1.

After choosing a suitable *general coping strategy*, it has to be substantiated with the characteristics required by the system to a *basic coping strategy*, which means, e.g. the used external resource is specified as energy. Afterwards, the *system coping strategy* has to be derived and modelled as a combination of the resilience functions. The simplest system coping strategy only includes the resilience function responding. Thus, the system reacts upon occurrence of a disruption. Enhanced resilience functionality can be attained by including the resilience functions monitoring and anticipating.

Table 1. Exemplary assignment of general coping strategies for typical vulnerabilities

Vulnerability	Applicable general coping strategies
Lack of resources	Usage of alternative resources Internal/external degradation: reduction of resource consumption within the system boundary/in the system's direct environment
Increased power/performance demand	Purposeful overload: partly accepting system damage Increased functionality: booster, change of function characteristic Usage of alternative resources

4.2 Modelling of Adaptivity for Resilient System Coping Strategies

After the system coping strategy is defined it has to be modelled in combination with the functional system structure. Therefore, new elements to depict the adaptivity of the system structure and function elements to model the coping strategy are required and have been developed exemplary [3]. This contribution shows how the interface between the system coping strategy, modelled by resilient and adaption functions, and the robust functional structure is complemented. The adaptivity is modelled using the enablers and disablers for functions, signals and flows shown in Table 2. The enablers are connected to the adaption functions. The adaption functions are based on the functional resilience characteristics and defined as 'execute adaption' as the functional element for responding, 'gather data' for monitoring and 'interpret data' for anticipating. In Table 3 their general functionality and exemplary design principles used for their realisation are listed. The adaption functions are depicted by a greyed out rhomboid and additionally required robust functions for the realisation of resilience (resilient functions) are represented by a greyed out cuboid as depicted in Fig. 1.

Table 2. Enabler elements

Designation	Representing element	
function enabler/disabler	enable function	disable function
signal enabler/disabler	enable signal	disable signal
flow enabler/disabler	enable additional flow, enable flow	disable additional flow, disable flow

Table 3. Adaption functions

Adaption function	Description	Design principles
Execute adaption	Adjusts system according to a new situation/condition	Physical/functional redundancy, fail-safe functionality
Gather data	Detects influencing parameters' values	Sensory functionality
Interpret data	Analyses gathered data and concludes which adaption to execute	Algorithms

a)   adaption function   b)   resilient function

Fig. 1. Depiction of a) adaption functions and b) resilient functions

5 Example of by-wire Car Brake System

The application of the extended functional structure model is discussed looking at a by-wire car brake system. The by-wire car brake system consists of an electronic unit, including the brake control system and the brake force amplifier, and a hydraulic unit, including the functions of building up and reducing the hydraulic braking pressure, and consequently the brake force to decelerate the wheel. The parking brake is included within the brake system's robust system boundary, too. The brake system is powered by the central board net, which is also connected to other consumers within the superordinate system, the car. The functional structure of the by-wire brake system is shown in Fig. 2 with the engine starter as one exemplary external consumer of the board net energy [3, 16].

Due to the many consumers connected to the board net, which, depending on their particular application, require high currents, and disturbances of the vehicle battery that arise, e.g., from low battery temperatures, a decrease of the power supply voltage can occur. The brake system is based on the electronic unit, which breaks down eventually as the power supply voltage decreases. Thus, the brake system's functionality is reduced to the hydraulic unit. The hydraulic unit then is controllable by the human operator via a hydraulic crackdown activated by muscular power [3]. This solution follows the fail-safe principle as it keeps up an option to brake, but the measure only applies when the disruption already occurred, no defined minimum functionality is guaranteed and the recovery time until full braking functionality is available again after the disruption's decline, is determined by the duration of the reboot of the electronic unit [3, 16].

As the full functionality of the brake system is aspired to be available at any time the relatively long recovery time of over two seconds is considered as the crucial vulnerability. A significant improvement of the resilient system behaviour shall be achieved by reducing the recovery time. The reboot of the whole control unit is decisive for the recovery time. If the CPU of the control unit can be kept running the recovery time is reduced to less than one second. To realise this, a minimum power supply voltage is required, that consequently defines the minimum functionality of the brake system [3].

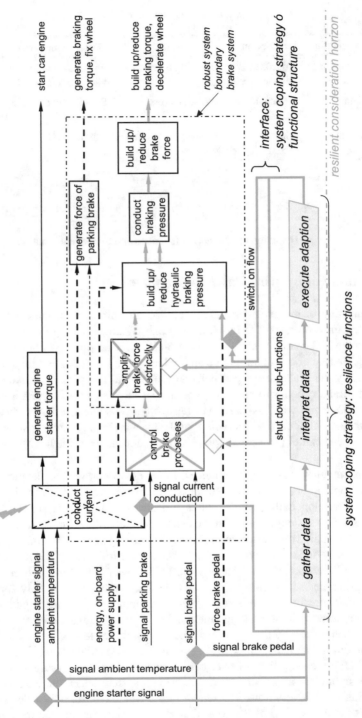

Fig. 2. Functional structure for a combined robust and resilient car brake system, following [3]

The described disruption falls in the category 'lack of resources'. According to Table 1, i.a., a degradation strategy is suitable for this disruption and is exemplary chosen, here. For the basic coping strategy the lacking resource is specified as energy, whose consumption shall be reduced using degradation. Applying the degradation means switching off expendable functions, which consume energy, in case of a decreasing power supply voltage. Looking at the brake system expendable functions are assistant systems like ABS for an internal degradation.

The system in the resilient configuration using internal degradation as a coping strategy is depicted in Fig. 2 according to the introduced additional elements for integrating the system adaptivity in the functional structure. The resilient consideration horizon is given below the robust system and the coping strategy consists of the three possible adaption functions, whereas no additional resilient function is required. The gather data function is connected to signal enablers for detecting the ambient temperature, the engine starter signal and the current conduction within the power supply unit. The collected data are processed by the interpret data function, which enables the system to anticipate a potential disruption of the power supply voltage and, in case of the occurrence of the disruption, activates the execute adaption function, which switches off expendable assistant functions within the electronic unit and enables the alternative load path of the hydraulic crackdown, depicted by the dis- and enablers connected to these functions and flows. The disrupted function here is the energy supply, which is crossed with dashed lines. The disrupted signal, material and energy flows, which arise from the disrupted energy supply, are denoted with dashed arrows and the flows manipulated by the resilient adaption functions are marked with thick grey arrows. These are also either solid for enabled flows or dashed for disabled flows.

6 Summary and Conclusions

Resilience design for load-carrying systems in addition to robust design offers a high potential, especially for highly safety relevant systems, like the shown car brake system. For the systematic development of resilient load-carrying systems the system needs to be analysed with regard to its weaknesses and inherent resilient behaviour using, e.g., the vulnerability analysis taking into account an extended consideration horizon, including the robust system itself as well as the superordinate system and the environment.

For the system description and the deduction of required resilient properties the resilience application model has been developed. Based on the analysis an appropriate basic coping strategy can be derived according to the disruption's character, the system properties and the environmental conditions. The definition of an appropriate basic coping strategy can be supported by a catalogue of general coping strategies derived from, e.g. successful resilient natural organisms as exemplary presented. Afterwards, the basic coping strategy is transferred into a system coping strategy.

For systematically realising the system coping strategy the static robust functional structure has to be extended to a dynamic resilient functional structure. Within the resilient functional structure flows and functions are controlled via a newly developed interface. The interface comprises mainly enablers and disablers controlled by the adaption functions. Having developed the resilient functional structure the conventional systematic product development process can be followed using resilient solution principles, in addition [4]. The depicted functional structure of a by-wire car brake system shows that resilient approaches in load-carrying systems are already realised, like the introduced degradation strategy.

Using a comprehensive resilience design approach as aspired with the introduced new or extended models and methods holistic resilience concepts for technical systems including all identified vulnerabilities are achievable.

Acknowledgement. Funded by the Deutsche Forschungsgemeinschaft (DFG, German Research Foundation) – Projektnummer 57157498 – SFB 805.

References

1. Hollnagel, E., Woods, D.D.: Prologue: resilience engineering concepts. In: Hollnagel, E. et al. (eds.) Resilience Engineering   Concepts and Precepts. Ashgate Publishing Ltd, Farnham (2006)
2. Altherr, L.C., et al.: Resilience in mechanical engineering - a concept for controlling uncertainty during design, production and usage phase of load-carrying structures. In: Pelz, P.F., Groche, P. (eds.) Applied Mechanics and Materials, vol. 885, pp. 187–198. Trans Tech Publication Ltd, Zurich (2018)
3. Schulte, F., et al.: Beitrag zur Entwicklungsmethodik für resiliente Systeme des Maschinenbaus. In: Proceedings of the 30th Symposium Design for X (DFX 2019), pp. 1–12. Jesteburg, Germany, 18-19 September 2019, The Design Society (2019)
4. Schulte, F., et al.: Analysis and synthesis of resilient load-carrying systems. In: Proceedings of the 22nd International Conference on Engineering Design (ICED19), pp. 1403–1412. Delft, Netherlands, 5-8 August 2019, Cambridge University Press (2019)
5. Woods, D.D.: Essential characteristics of resilience. In: Hollnagel, E. et al. (eds.) Resilience Engineering - Concepts and Precepts. Ashgate Publishing Ltd, Farnham (2006)
6. Hollnagel, E.: RAG-The resilience analysis grid. In: Hollnagel, E. et al. (eds.) Resilience Engineering in Practice: A Guidebook, pp. 275–296. Ashgate Publishing Ltd., Farnham (2011)
7. Turner, B.L., et al.: A framework for vulnerability analysis in sustainability science. In: Proceedings of the National Academy of Sciences of the United States of America (PNAS), vol. 100, no. 14, pp. 8074–8079. National Academy of Sciences, Washington (2003)
8. von Gleich, A., et al.: Resilienz als Leitkonzept - Vulnerabilität als analytische Kategorie. In: Fichter, K. et al. (eds.) Theoretische Grundlagen für erfolgreiche Klimaanpassungsstrategien, Nordwest 2050, pp. 13–51. Oldenburg, Bremen (2010)
9. Ratter, B.M.W.: Island vulnerability and resilience. In: Ratter, B.M.W. (ed.) Geography of Small Islands - Outposts of Globalisation, pp. 173–200. Springer International Publishing, Cham (2018)
10. Rowan, E., et al.: Indicator approach for assessing climate change vulnerability in transportation infrastructure. Transportation Research Record: Journal of the Transportation Research Board No. 2459 (1), pp. 18–28. Transportation Research Board of the National Academies, Washington (2014)
11. Döniz, E.J.: Was ist die Szenariotechnik? In: Döniz, E.J. (ed.) Effizientere Szenariotechnik durch teilautomatische Generierung von Konsistenzmatrizen - Empirie, Konzeption, pp. 6–44. Fuzzy- und Neuro-Fuzzy-Ansätze, Gabler, Wiesbaden (2009)

12. Gallopin, G.C.: Linkage between vulnerability, resilience, and adaptive capacity. Global Environ. Change **16**(3), 293–303 (2006)
13. Feldhusen, J., Grote, K.-H.: Pahl/Beitz Konstruktionslehre – Methoden und Anwendung erfolgreicher Produktentwicklung". 8th completely revised edition, pp. 237 ff. Springer Vieweg Verlag, Berlin (2013)
14. Cutter, S.L., et al.: A place-based model for understanding community resilience to natural disasters. Global Environ. Change **18**(4), 598–606 (2008)
15. Schlemmer, P.D., et al.: Adaptivity as a property to achieve resilience of load-carrying systems. In: Pelz, P.F., Groche, P. (eds.) Applied Mechanics and Materials, vol. 885, pp. 77–87. Trans Tech Publication Ltd, Zurich (2018)
16. Breuer, B., Bill, K.H.: Bremsenhandbuch – Grundlagen, Komponenten, Systeme, Fahrdynamik. Springer, Wiesbaden (2017)

Validation of an Optimized Resilient Water Supply System

Tim M. Müller[1]([⊠]), Andreas Schmitt[2], Philipp Leise[1], Tobias Meck[1],
Lena C. Altherr[3], Peter F. Pelz[1], and Marc E. Pfetsch[2]

[1] Chair of Fluid Systems, Department of Mechanical Engineering, TU Darmstadt,
Darmstadt, Germany
`tim.mueller@fst.tu-darmstadt.de`
[2] Discrete Optimization Group, Department of Mathematics, TU Darmstadt,
Darmstadt, Germany
[3] Faculty of Energy, Building Services and Environmental Engineering,
Münster University of Applied Sciences, Münster, Germany

Abstract. Component failures within water supply systems can lead
to significant performance losses. One way to address these losses is the
explicit anticipation of failures within the design process. We consider a
water supply system for high-rise buildings, where pump failures are the
most likely failure scenarios. We explicitly consider these failures within
an early design stage which leads to a more resilient system, i.e., a system
which is able to operate under a predefined number of arbitrary pump
failures. We use a mathematical optimization approach to compute such
a resilient design. This is based on a multi-stage model for topology optimization, which can be described by a system of nonlinear inequalities
and integrality constraints. Such a model has to be both computationally
tractable and to represent the real-world system accurately. We therefore
validate the algorithmic solutions using experiments on a scaled test rig
for high-rise buildings. The test rig allows for an arbitrary connection of
pumps to reproduce scaled versions of booster station designs for high-
rise buildings. We experimentally verify the applicability of the presented
optimization model and that the proposed resilience properties are also
fulfilled in real systems.

Keywords: Optimization · Validation · Resilience · Mixed-integer
nonlinear programming · Water distribution system · High-rise

1 Introduction

The design and usage of technical systems is subject to uncertainty, which can even
lead to a failure of a part or of the complete system. One way to anticipate this
uncertainty is to explicitly consider resilience within the design process. A technical system is resilient if it is able to fulfill a predefined functional level even if
failures occur. One particular approach to measure resilience with respect to failures is given by the so-called buffering capacity. A technical system has a buffering

P. F. Pelz and P. Groche (Eds.): ICUME 2021, LNME, pp. 70–80, 2021.
https://doi.org/10.1007/978-3-030-77256-7_7

capacity k, if up to k arbitrary components can fail or be manually deactivated for maintenance and the disturbed system still reaches a predefined level of functionality, see [2]. Using mathematical optimization, the buffering capacity and thus resilience can be ensured in the design process. This leads to a multi-level problem, e.g., a min-max-min problem, since the system may react to failures.

Such multi-level problems are notoriously hard to solve. It is therefore crucial to choose a model of the system that is both computationally tractable and adequately represents the considered system. This trade-off introduces another source of uncertainty, namely that of the model. Thus, a validation of the model is needed. However, how to do this is not obvious, since the model could be valid for a reference solution that can be tested experimentally, but might be inaccurate if failures occur.

In this paper, we consider this issue for the particular example of water supply of high-rise buildings. In such systems, booster stations consisting of one or more pumps are necessary to increase the water pressure to supply all floors of the building. Overall, multiple system layouts are possible. In [3] and [9] it was shown that a decentralized arrangement of pumps allows to achieve significant energy savings due to a reduction of throttling losses. The design and control of such sustainable systems, however, is highly complex and requires the usage of algorithmic approaches. Following [3], we use a Mixed-Integer Nonlinear Programming (MINLP) approach. As an objective, we use a linear-combination of the pump investment cost and the operational cost, which approximates the true life-cycle costs of the system.

The integration of resilience considerations via the buffering capacity yields complex models. In principle, each possible failure scenario resulting from the combination of failures of single pumps must be considered within the constraints of the optimization program, leading to very large models. In order to reduce complexity, usually problem-specific approaches are used. For example, [8] takes one arbitrary component failure in the optimization of an energy system design into account. Considering resilience in layout optimization is also prominent in electric grid planning and commonly known as the N-K property – out of N components K may fail, see e.g., [1,4,11,12]. For water distribution systems in high-rise buildings a method to optimize the buffering capacity with regard to pump failures is presented in [3]. In this paper, we apply a more general algorithm, described in [10], which produces according to the model correct results in acceptable time for small systems.

As mentioned above, it is also important to use mathematical models that represent the considered technical system accurately. For models which describe complex physical phenomena, experiments are the ultimate tool for validation in addition to simulation. Validation is a common step in Operations Research, as mentioned for instance by S. I. Gass in 1983 in [6] and as part of standard references in Operations Research, cf. [5] and [7].

The main contribution of this paper is the experimental validation of resilience properties for topologies generated by the above mentioned algorithmic approach. For this, we use a modular test rig which was presented in [9] to validate the correctness of the underlying MINLP to model the physics of a high-rise water distribution system. The main point is that the computed optimal solution is not only valid for standard situations, but also if failures occurs.

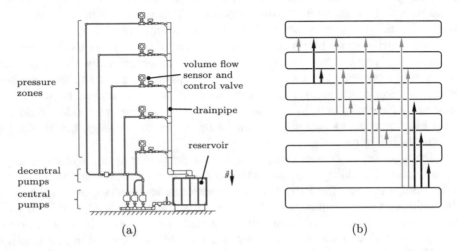

Fig. 1. (a) Sketch of the test rig to validate the solutions of the optimization program model. (b) Graph of the possible configurations which are considered in the optimization program. The black connections represent the configuration shown in Fig. 1a.

The paper is organized as follows. Section 2 contains the description of the test rig and Sect. 3 the corresponding optimization model. The experimental validation is presented in Sect. 4. Afterwards, we give a short summary and address future research directions.

2   Test Rig

The test rig presented in [9], and shown schematically in Fig. 1a, represents a downscaled high rise building with five pressure zones on different height levels. Its purpose is to supply each zone with a predefined volume flow and minimum pressure approximating the behavior a building with the same number of pressure zones. In [9] cost and energy optimal solutions have been computed based on different modeling and solution approaches, and the obtained results were validated on the test rig. These experiments do not consider resilience as it is done in this contribution.

In each pressure zone of the test rig, the volume flow is measured and the required demand is set by a control valve. The water is pumped from a reservoir under ambient pressure via various (decentralized) pumps into the pressure zones. In addition to the central pumps, which connect the reservoir and the pressure zones directly, further decentralized pumps may be used. The configuration (pump types, placement, rotational speed of the pumps) can be adjusted according to the optimization results. The possible pipe topologies considered within the optimization model and realizable in validation are shown in Fig. 1b.

In total there are 13 pumps available, cf. [9]. Besides the volume flow and valve-position, the power consumption can be measured at the test rig which enables a validation of the obtained optimized results.

We use five different demand scenarios, which differ in their probabilities of occurrence, volume flow demands (up to $q^{\mathrm{nom}} = 4.28 \ \mathrm{m}^3\mathrm{h}^{-1}$) and pressure losses in accordance to [9]. The demand of the different pressure zones is assumed to be equal for the same scenario. Note that the pressure loss is a function of the geodetic height, the volume flow as well as the friction in the system. Due to the various influences, the pressure loss is subject to considerable uncertainty.

As described in the introduction, a failure or deactivation of up to k pumps should be tolerated in the derived system topology and a minimum fulfillment of a predefined function performance has to be guaranteed, cf. [2]. We define that in each failure scenario, at least $\tilde{q}^{\mathrm{fail}} = q^{\mathrm{fail}}/q^{\mathrm{nom}} = 70\%$ of the maximum required volume flow q^{nom} has to be supplied.

3 Mathematical Optimization Model

In this section we present a Mixed-Integer Nonlinear Program (MINLP) to find a cost optimal test rig design. Afterwards, we describe the consideration of failures.

A general water network design problem is specified by a directed graph (V, A), for which the vertices V denote in-/outputs of the network and transition points between components. The arcs $A = A^p \cup A^a$ are divided in passive and active arcs and represent possibilities to place pipes and pumps, respectively. Further, the set of demand scenarios S specifies, for each node $v \in V$ and each scenario $s \in S$, lower/upper bounds $\underline{q}_{v,s}/\overline{q}_{v,s}$ on the volume flow demand (negative if v is a sink) and $\underline{p}_{v,s}/\overline{p}_{v,s}$ on the pressure-head. Each arc $a \in A$ also has lower/upper bounds $\underline{q}_a/\overline{q}_a$ on the volume flow. For passive arcs, pressure along the pipe does not change, i.e., we assume friction does not depend on the flow and is included in the pressure bounds. An active arc $a \in A^a$ can increase the pressure by an amount Δp_a, which is bounded above and below by a quadratic polynomial in the flow q_a over the arc:

$$\underline{\alpha}_a q_a^2 + \underline{\beta}_a q_a + \underline{\gamma}_a \leq \Delta p_a \leq \overline{\alpha}_a q_a^2 + \overline{\beta}_a q_a + \overline{\gamma}_a.$$

This, however, consumes an energy e_a according to a cubic polynomial in q_a and Δp_a

$$e_a = \sum_{0 \leq i+j \leq 3} \alpha_a^{i,j} q_a^i \Delta p_a^j.$$

Note that this differs from the pump model used in [9], where we obtain the power consumption and pressure increase in two approximations depending on the volume flow and the pump operating speed.

Altogether, we obtain the following optimization problem, which searches for a network specified by binary variables x_a and its operation such that the arc costs given by C_a and the total energy cost under the demands of each scenario $s \in S$, weighted by C_s, are minimized. Here, the usage of the active arcs is represented by binary variables $y_{a,s}$. For each scenario the model further-more contains volume flow variables q on each arc, pressure variables p for each node and lastly variables Δp for the pressure differential on active arcs. The notation $\delta^-(v)$ and $\delta^+(v)$ is used for the incoming respectively outgoing arcs of node v. We refer to [3] for an in-depth explanation of the constraints.

$$
\begin{aligned}
\min \quad & \sum_{a \in A} C_a x_a + \sum_{a \in A^a} \sum_{s \in S} C_s \Big(\sum_{1 \le i+j \le 3} \alpha_a^{i,j} q_{a,s}^i \, \Delta p_{a,s}^j + \alpha_a^{0,0} y_{a,s} \Big) \\
\text{s.t.} \quad & \underline{q}_{v,s} \le \sum_{a \in \delta^-(v)} q_{a,s} - \sum_{a \in \delta^+(v)} q_{a,s} \le \overline{q}_{v,s}, && v \in V, s \in S, \\
& \underline{q}_{a,s} x_a \le q_{a,s} \le \overline{q}_{a,s} x_a, && a \in A^p, s \in S, \\
& \underline{q}_{a,s} y_{a,s} \le q_{a,s} \le \overline{q}_{a,s} y_{a,s}, && a \in A^a, s \in S, \\
& (p_{v,s} - p_{u,s}) x_a = 0, && a = (u,v) \in A^p, s \in S, \\
& (p_{v,s} - p_{u,s}) y_{a,s} = \Delta p_{a,s}, && a = (u,v) \in A^a, s \in S, \\
& y_{a,s} \le x_a, && a \in A^a, s \in S, \\
& \underline{\alpha}_a q_{a,s}^2 + \underline{\beta}_a q_{a,s} + \underline{\gamma}_a y_{a,s} \le \Delta p_{a,s}, && a \in A^a, s \in S, \\
& \Delta p_{a,s} \le \overline{\alpha}_a q_{a,s}^2 + \overline{\beta}_a q_{a,s} + \overline{\gamma}_a y_{a,s}, && a \in A^a, s \in S, \\
& q \in \mathbb{R}^{A \times S}, \; p \in [\underline{p}, \overline{p}], \; \Delta p \in \mathbb{R}_+^{A^a \times S}, \; x \in \{0,1\}^A, \; y \in \{0,1\}^{A^a \times S}.
\end{aligned}
\tag{1}
$$

The possible test rig layouts are modeled by the following graph (V, A). There exists a node in V for the basement. For each of the five pressure zones two nodes v_i^{in} and v_i^{out} are introduced. The input nodes have a flow demand of zero and no restrictions on the pressure. The output nodes have a flow demand and pressure requirements according to the scenarios in S. The set of arcs contains, for each pump and each pressure zone, an active arc from v_i^{in} to v_i^{out} and another active arc, which models a bypass without costs or friction (all coefficients in the pump approximations set to zero). Furthermore, there are arcs from the basement to each input node v_i^{in} and from each output node v_i^{out} to the input nodes above v_j^{in}, $i < j$. To model the test rig accurately, cardinality constraints are added to Problem (1), which restrict the number of possible arcs corresponding to a given pump to be at most one. Furthermore, for each v_i^{in} there may be at most one incoming arc.

A solution topology x most likely does not have a buffering capacity k. Thus, there exists a failure scenario of the active arcs such that there exists no operation of the remaining pumps to supply the network, even with the reduced demand q^{fail} and the corresponding node bounds like $\underline{q}^{\text{fail}}$ and $\underline{p}^{\text{fail}}$. The solution topology x would be resilient, if for each failure scenario, encoded in a binary vector $z \in \{0,1\}^{A^a}$ with $\sum_{a \in A^a} z_a \le k$, there exists an operation for the remaining pumps. This can be ensured, if for each z the following system in variables y, q, p and Δp has a solution:

$$
\begin{aligned}
& y_a \leq 1 - z_a, && a \in A^a, \\
& \underline{q}_v^{\text{fail}} \leq \sum_{a \in \delta^-(v)} q_a - \sum_{a \in \delta^+(v)} q_a \leq \overline{q}_v^{\text{fail}}, && v \in V, \\
& \underline{q}_a x_a \leq q_a \leq \overline{q}_a x_a, && a \in A^p, \\
& \underline{q}_a y_{a,s} \leq q_a \leq \overline{q}_a y_a, && a \in A^a, \\
& (p_v - p_u) x_a = 0, && a = (u,v) \in A^p, \\
& (p_v - p_u) y_a = \Delta p_a, && a = (u,v) \in A^a, && (2) \\
& y_a \leq x_a, && a \in A^a, \\
& \underline{\alpha}_a q_a^2 + \underline{\beta}_a q_a + \underline{\gamma}_a y_a \leq \Delta p_a, && a \in A^a, \\
& \Delta p_a \leq \overline{\alpha}_a q_a^2 + \overline{\beta}_a q_a + \overline{\gamma}_a y_a, && a \in A^a, \\
& q \in \mathbb{R}^A, \ p \in [\underline{p}^{\text{fail}}, \overline{p}^{\text{fail}}], \ \Delta p \in \mathbb{R}_+^{A^a}, \ y \in \{0,1\}^{A^a}.
\end{aligned}
$$

One theoretical possibility to obtain optimal resilient solutions is to integrate System (2) for each considered failure scenario z into Problem 1 and solve this enlarged MINLP. However, due to the problem size of our instances, this is unsolvable in a tolerable amount of time. To circumvent this, we use the algorithm proposed in [10]. Here, the restriction to be resilient is integrated into the branch and cut algorithm used to solve Problem 1. For solution candidates an auxiliary optimization problem is solved to check whether there exists a violated failure scenario. If this is the case, a linear inequality is derived to cut off this infeasible solution. The approach presented in [3] is not applicable, since it utilizes the structure of the auxiliary problem and requires that only pumps of the same type can be build in parallel.

4   Results and Validation

Using the above model, we computed three optimal solutions, which have a guaranteed buffering capacity of $k \in \{0, 1, 2\}$ for a minimal relative volume flow of $\tilde{q}^{\text{fail}} = 70\%$, respectively. This means that – according to the model – for a solution with a specified buffering capacity of k, at least k pumps may fail and the system will still achieve a minimum volume flow of $\tilde{q}^{\text{fail}}$. Together with a reference solution, which consists of only parallel pumps of the same type, these solutions are shown in Fig. 2. Note that the reference solution has a buffering capacity of $k = 1$.

All of the optimized solutions contain one or several parallel central pumps and a smaller decentralized pump for the highest pressure zones. The predicted power consumption of the optimized solutions is roughly equal and saves about 22% compared to the reference solution, cf. Fig. 2. The required resilience is ensured by an increased number of central pumps, leading to higher investment and thus higher total cost. However, not just redundant pumps are used, but different pumps are combined.

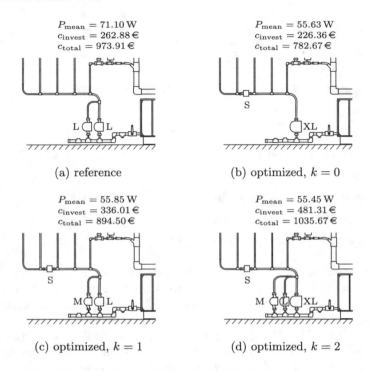

$$P_{\text{mean}} = 71.10\,\text{W}$$
$$c_{\text{invest}} = 262.88\,\text{€}$$
$$c_{\text{total}} = 973.91\,\text{€}$$

(a) reference

$$P_{\text{mean}} = 55.63\,\text{W}$$
$$c_{\text{invest}} = 226.36\,\text{€}$$
$$c_{\text{total}} = 782.67\,\text{€}$$

(b) optimized, $k = 0$

$$P_{\text{mean}} = 55.85\,\text{W}$$
$$c_{\text{invest}} = 336.01\,\text{€}$$
$$c_{\text{total}} = 894.50\,\text{€}$$

(c) optimized, $k = 1$

$$P_{\text{mean}} = 55.45\,\text{W}$$
$$c_{\text{invest}} = 481.31\,\text{€}$$
$$c_{\text{total}} = 1035.67\,\text{€}$$

(d) optimized, $k = 2$

Fig. 2. Solution of optimization and illustration of the set-up on the test rig. The letters S, M, L and XL indicate the pump type and refer to their maximum hydraulic power.

In our experiment we validated all four solutions by setting up the topologies shown in Fig. 2. For the reference system, we solve the optimization problem with fixed topology variables. The input for the test rig in each demand scenario is the pump operation (rotational speeds of the pumps) according to the optimization results. The valves are set such that the volume flow coincides for each zone. The output of the experiment is the measured total power consumption of the pumps and the measured total volume flow for each demand scenario.

Figure 3 compares the theoretical results of the optimization (squares) with the results of the measurements (circles). Associated points of a demand scenario are connected by a line. The measurement errors are rather small ($\Delta q \leq 0.024\,\text{m}^3/\text{h}$; $\Delta p \leq 2.61\,\text{W}$). Thus, the error bars of the experimental results would vanish behind the markers and are therefore not shown in the figure.

When comparing optimization and experiment, it is noticeable that there are deviations due to inaccuracies in the used model: Due to uncertainty in the pressure loss of the test rig and in the characteristic curves of the pumps, the predicted volume flow and power consumption at a given pump rotational speed differ from the measured values. Note that in real systems, such volume flow deviations could be compensated by using a volume flow control rather than a speed control, as assumed here for modeling reasons to validate the computed

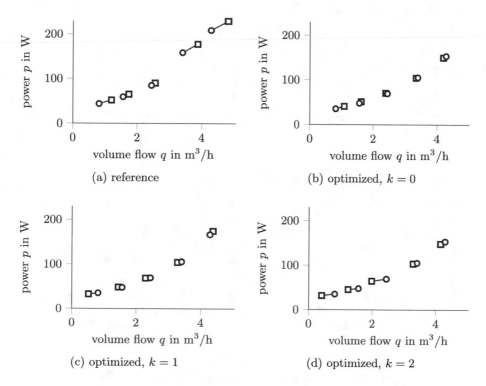

Fig. 3. Comparison of the optimization results (circles) and measured (squares) power consumption and volume flow of the solutions for different load scenarios without failures. Associated points of a demand scenario are connected by a line.

optimization results. The magnitudes of the deviations depend on the pumps installed and the optimized system topology, as this influences the pressure loss of the system. Overall, the decisive trend between power consumption and volume flow is well correlated, which is crucial for the expected energy consumption. Thus, the experiment confirms the reduced energy consumption for the optimized and decentralized systems and thus the benefit of the optimization. This is consistent with the results of [9].

To validate the buffering capacity of the design, the experimental setup is as follows: For each solution, we configure the remaining system for every possible combination of one up to three failing pumps and measure the maximal achievable volume flow. Thus, we also check the cases in which there are more or less failures than anticipated in the optimization, leading to a total of 28 experimental setups. To simulate a pump failure, the respective pump is replaced by a pipe, which corresponds to a bypass around the pump. In the failure scenario, the remaining pumps are operated at maximum speed. Again, the valves are used to balance the volume flow on different zones. If one zone can not be supplied, the measured total volume flow is set to zero.

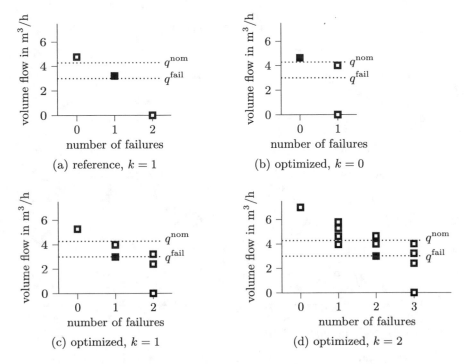

Fig. 4. Measured maximal volume flow for the configurations derivable from the combination of the solutions and the failure scenarios with 0 up to 3 failures. Note that some markers overlap and cover each other. The worst-case failure for the anticipated number of failures is indicated by filled markers.

These measurements are shown in Fig. 4, in which each marker represents the measured volume flow for a configuration. One can see, that the required minimum functionality q^{fail} is always achieved. This means that for a specified buffering capacity of k in the optimization and less than k arbitrary pump failures in the experiment, the minimum volume flow of q^{fail} is fulfilled for all cases. The worst-case failure of all possible failure combinations, shown as a filled marker, is decisive here, as all failure scenarios must be covered. This worst-case volume flow coincides with the minimum functional level (q^{fail}) for the optimized resilient solutions and thus, there is no buffer in the case of failures. This is characteristic for optimization algorithms, which tend to produce solutions close to the border of the feasible solution space.

If more pumps than expected fail, the functional level can not be satisfied. A special case is if all central pumps are affected since the lower pressure zones are not supplied anymore. The results show that a higher volume flow can be achieved if there are less failures than expected. For example, if $k = 2$ is specified and only one of the pumps fails, a volume flow of $\tilde{q}^{\text{fail}} \geq 92.26\% > 70\%$ can be achieved for any failure combination (Fig. 4d).

Also, if no pump fails, a higher volume flow than required can be achieved, which increases with higher k, $\overline{q}^{\max}_{k=2} \geq \overline{q}^{\max}_{k=1} \geq \overline{q}^{\max}_{k=0} \geq 100\%$, cf. Fig. 4. Since the pressure losses in the system increase quadratically with the volume flow, a significantly higher pressure than originally planned can be achieved as well. These two facts show a desirable feature of the system and confirm the concept of resilience: the system is able to react even to unforeseen events. This can be, for example, a higher volume flow demand than expected, but also covers deviations in the pressure loss of the system (e.g. due to uncertainty during the design phase or due to wear of the components).

5   Conclusion

In this paper we have validated the resilience of solutions given by an optimization method to design resilient water distribution systems. This was done by examining the system for each possible combination of missing pumps. Even for the relatively small system sizes this leads to a high number of costly measurements. Future research could address this by consideration of only those failure scenarios, which are predicted to be critical given some further measure. In our case, these could be all failures for which the maximal volume flow is below 80% of q^{nom}, i.e., the minimum required performance plus an additional safety offset of 10%, assuming the model error is smaller. For the validation of the $k = 2$ solution this could have reduced the number of measurements from 15 to 4, which would significantly reduce the effort of validation. To efficiently compute all critical failure scenarios, the adaptive algorithms given in [10] and [3] could be used.

Acknowledgments. This research was partly funded by Deutsche Forschungsgemeinschaft (DFG, German Research Foundation) – Project Number 57157498 – SFB 805, subproject A9. Furthermore, the presented results were partly obtained within the research project "Exact Global Topology-Optimization for Pumping Systems", project No. 19725 N/1, funded by the program for promoting the Industrial Collective Research (IGF) of the German Ministry of Economic Affairs and Energy (BMWi), approved by the Arbeitsgemeinschaft industrieller Forschungsvereinigungen "Otto von Guericke" e.V. (AiF).

References

1. Alguacil, N., Delgadillo, A., Arroyo, J.M.: A trilevel programming approach for electric grid defense planning. Comput. Oper. Res. **41**, 282–290 (2014)
2. Altherr, L.C., Brötz, N., Dietrich, I., Gally, T., Geßner, F., Kloberdanz, H., Leise, P., Pelz, P.F., Schlemmer, P.D., Schmitt, A.: Resilience in mechanical engineering-a concept for controlling uncertainty during design, production and usage phase of load-carrying structures. In: Applied Mechanics and Materials, vol. 885, pp. 187–198. Trans Tech Publications (2018)
3. Altherr, L.C., Leise, P., Pfetsch, M.E., Schmitt, A.: Resilient layout, design and operation of energy-efficient water distribution networks for high-rise buildings using MINLP. Optim. Eng. **20**(2), 605–645 (2019)

4. Bienstock, D.: Electrical Transmission System Cascades and Vulnerability: An Operations Research Viewpoint. SIAM-MOS Series on Optimization (2015)
5. Eiselt, H., Sandblom, C.L.: Operations Research - A Model-Based Approach. Springer, Heidelberg (2012)
6. Gass, S.I.: Decision-aiding models: validation, assessment, and related issues for policy analysis. Oper. Res. **31**(4), 603–631 (1983)
7. Hillier, F., Lieberman, G.: Introduction to Operations Research. McGraw-Hill Education, New York (2015)
8. Hollermann, D.E., Hoffrogge, D.F., Mayer, F., Hennen, M., Bardow, A.: Optimal $(n-1)$-reliable design of distributed energy supply systems. Comput. Chem. Eng. **121**, 317–326 (2019)
9. Müller, T.M., Leise, P., Lorenz, I.-S., Altherr, L.C., Pelz, P.F.: Optimization and validation of pumping system design and operation for water supply in high-rise buildings. Optim. Eng. (2020). https://doi.org/10.1007/s11081-020-09553-4
10. Pfetsch, M.E., Schmitt, A.: A framework for optimizing resilient layouts. Manuscript in preparation (2021)
11. Wang, Q., Watson, J., Guan, Y.: Two-stage robust optimization for $n-k$ contingency-constrained unit commitment. IEEE Trans. Power Syst. **28**(3), 2366–2375 (2013)
12. Xiong, P., Jirutitijaroen, P.: An adjustable robust optimization approach for unit commitment under outage contingencies. In: 2012 IEEE Power and Energy Society General Meeting, pp. 1–8 (2012)

Comparability of Water Infrastructure Resilience of Different Urban Structures

Imke-Sophie Lorenz[(⊠)], Kevin Pouls, and Peter F. Pelz

Chair of Fluid Systems, Technische Universität Darmstadt, Darmstadt, Germany
imke.lorenz@fst.tu-darmstadt.de

Abstract. Urban water distribution systems (WDS) ensure the demand-driven supply of a city at multiple ends. Well-being of the population as well as multiple economic sectors depend on its viability and thereby classify it as a critical infrastructure. Therefore, its behavior when exposed to changes is of interest to water suppliers as well as local decision-makers. It can be determined by resilience metrics, assessing the capability to meet and recover its functioning when exposed to disturbances. These disturbances can occur in form of changes in the water availability, the WDS topology, or the water demand pattern. Since networks as WDS are studied by graph theory, also different graph-theoretical resilience metrics were derived. In this work a well-established topology-based resilience metric is adapted and deployed to assess the present resilience of the urban main-line WDS of the German city of Darmstadt as well as of a suburb in the Rhine-Main region. Thereby, the intercomparability of the resilience for the different urban structures were of interest. Based on this analysis the comparability of different urban main-line WDS regarding their resilience is facilitated. Additionally, the conducted approach to allow for the comparability of absolute resilience values of urban structures of varying size can be applied to different resilience metrics as well as technical systems.

Keywords: Water distribution systems · Resilience · Urban design

1 Introduction

An integral part of urban structures are their infrastructures – forming its design and furthermore significantly influencing its livability. One of the main infrastructures is the water distribution system (WDS) meeting multiple needs, starting at the well-being of the population and reaching into the economic prosperity of an urban region. Its layout greatly depends on the present urban structure, especially the urban transportation infrastructure as these two infrastructures are highly in parallel [1].

Urban structures themselves form by different means, studied in the field of urban morphology. Naturally they form when a rural area grows into a metropole or several close rural areas grow into a single urban structure, as it is the case for Dublin and on smaller scale as well the city of Darmstadt [2]. Additionally, there exist planned urban areas. These were often characterized by geometric and homogeneous areal urban design planned and rolled out in a short period of time, even though this approach is

© The Author(s) 2021
P. F. Pelz and P. Groche (Eds.): ICUME 2021, LNME, pp. 81–90, 2021.
https://doi.org/10.1007/978-3-030-77256-7_8

challenged today [3]. Both methods of developing urban structures do not have a well-functioning WDS as main objective in mind. *Form follows function,* a guiding principle defining product design as well as architecture doesn't seem to apply for urban design considering the WDS. While on small scale this principle is followed by when designing components as pipes, valves, and pumps of the WDS, the overall layout of the WDS then follows the already existing or planned form of the urban structure. This can as well be interpreted as its function, yet it is an interesting question to pose, whether the urban structure shouldn't be planned or reconstructed to allow for the best WDS design.

To answer this question, one needs to define the function of the WDS as well as a criterion to measure its fulfillment. In this case not only the fulfillment of the water demand itself should be considered but also the ever changing environment it is embedded in. Therefore, it is of interest what changes apply to urban structures and its WDS. On the one hand, there is a trend of constant urbanization [4], which leads to changing water demands. On the other hand, WDS are subject to hazards leading to different disturbances in the availability of water, the WDS's components as well as changes of the demand pattern. The functioning (to a minimum degree) under changing conditions and furthermore the recovery or adaptation of a system to fully function again is a given for resilient systems [5]. This aim is also often shared in asset management [6]. Therefore, different resilience metrics for infrastructures were defined, also specifically for WDS.

These metrics are often applied, for example in asset management, to improve one system's resilience. This led to the finding that the present urban structure determines a saturation like maximum resilience for an optimized WDS [7]. Yet, the comparison of different systems towards their resilience, especially for WDSs, is, to the best of the authors' knowledge, dealt with solely to compare different system adaptations. Therefore, the resilience metric's informative value to compare different systems not achieving the same goal, as feeding a given urban structure with water, is not given within these metrics. This issue is also addressed in present urban design research [8, 9]. Relative comparison of WDSs' resilience depends not solely on the present urban design but, depending on the resilience metric considered, furthermore on its size, overall demand resulting from its size as well as demand density, and redundancy as well as robustness. As a result, these additional aspects influencing the resilience have to be considered non-dimensioned within a resilience metric, to compare the possible maximum resilience of different urban designs.

In this research an existing resilience metric is adapted to allow for a relative comparison of urban structures. The derived non-dimensioned resilience metric is tested for the comparison of the main line WDSs of two different real-world cities varying in their urban structure, area size and demand density.

2 Related Work

The resilience of infrastructures, especially of WDSs is subject to present research [10]. There have been established different approaches to measure the resilience of WDSs, most based on graph theory to describe the underlying infrastructure network and its functioning. Therefore, the WDS is represented by a planar graph $G = (N, E)$ in which the pipes are represented by its edges E and the sources, tanks, junction and consumer nodes by nodes N. There are different approaches of analyzing the resilience of WDSs, which can be categorized into three types.

First, there are resilience metrics which assess the WDS's resilience based on its topology. While there are simple graph-theoretical metrics as spectral and statistical metrics which can already count as resilience metrics [11], there are also metrics on a higher scale of sophistication. Analyzing the simpler metrics in context of resilience, resilient networks can be characterized as, at the same time, redundant and robust networks [11, 12]. An established metric to assess the resilience of single consumer nodes considering both of these aspects is the resilience metric first introduced by Herrera et al. [13]. Its physical derivation and its application to optimize the resilience of WDSs has been studied by Lorenz et al. [7, 14]. The topological resilience assessment has the advantage that it is independent of a disruption scenario.

Second and in turn, there are resilience metrics assessing the demand satisfaction considering specific disruptive events. For these metrics special attention has to be given whether they assess robustness instead of resilience. The latter allows for a minimum functionality independent of the disruption scenario. This is considered and realized by the resilience index derived by Todini, applicable independent of the disruption scenario [15] and extended variously [10]. Studying on the one hand increased demand and on the other hand reduced water availability, Amarasinghe et al. [16] introduced a further disruption scenario independent resilience metric considering the demand satisfaction.

Last, the whole dynamic process following a disruptive event and, in the case of a resilient system, recovery is studied by different resilience metrics. Again, attention is to be paid to distinguish resilience and robustness assessment depending on the definition of the disruptive event. A well-known resilience metric for the assessment of the recovery process is the resilience triangle introduced by Bruneau et al. [17]. This metric quantifies the integral time span during which the functionality of a system is impaired after a disruptive event. Apart from that, Meng et al. apply and compare multiple criteria to assess the recovery process following a disruptive event [18].

These metrics have in common that applied to different networks, their absolute values do not allow for a clear distinction of the better network, as the network size and demand are not considered non-dimensioned. Yet, this is of interest when comparing infrastructure layouts towards their resilience against disruptive events.

The infrastructure layout and adaptations are highly dependent on the existing urban structure as infrastructures as the WDS, the power grid and information and communication technology lay highly in parallel to the urban transportation infrastructure, forming the visual structure of a city [19]. Three main types of urban structures can be distinguished. While linear and concentric cities form generically, homogeneous areal cities often are planned [20]. These predefined structures as well as their hybrid forms limit the infrastructure layout significantly. Therefore, Lynch proposed to define performance dimensions of cities to compare cities qualitatively and quantitatively [8]. In present work, the urban structure's influence on the resilience, which can be understood as a performance dimension, is studied. Fischer et al. [9] analyze the resilience of different urban structures based on the development. Buildings are distinguished based on their construction type as well as their use type. A further topological study conducted by Giudicianni et al. considers the WDS's functioning depending on the underlying urban structure by analyzing multiple spectral and statistical graph metrics [21]. It was found that the random structure has better values for most graph metrics compared to a homogenous areal urban structure.

Derived from the presented work, the interest emerged to define a resilience metric which allows for a comparison of different urban structures instead of the increased resilience within a system. This resilience metric then allows for infrastructure comparison independent of the present scales.

3 Materials and Methods

This work considers a topology based resilience metric to study the influence of urban structures on the maximum possible resilience of the realizable WDS. Therefore, the topological resilience metric on which this work is based is introduced. The dimensional dependencies and the subsequent non-dimensioning of the resilience metric is studied. Finally, the derived relative resilience metric is tested for two sample real-world WDSs of differing urban structures as well as area and overall demand.

3.1 Resilience Assessment

The graph theoretical resilience metric, first introduced by Herrera et al. [13], considers robustness and redundancy by taking into account, on the one hand, the hydraulic resistance of the feeding path $r_{s,c}$ between source node s and consumer node c and, on the other hand, the alternative k-shortest feeding paths from any source s of the WDS. The latter is indexed by k in the three-dimensional matrix of the hydraulic resistance of feeding paths $r_{s,c,k}$. Furthermore, as introduced by Lorenz and Pelz [7], to assess the resilience of the overall WDS given in Eq. (1), the resilience of each consumer node c is weighted with its relative demand q_c/Q to find an overall resilience measure for the present WDS. In this q_c is the demand at the consumer node c while Q is the overall demand in the WDS given as $Q = \sum_{c \in C} q_c$. Thereby, the resilience of single consumer nodes is prioritized by their relative demand.

$$I_{GT} = \sum_{c \in C} (q_c/Q \sum_{s \in S} (1/K_{\max} \sum_{k=1,...,K_{s,c} \leq K_{\max}} 1/r_{s,c,k})) \tag{1}$$

The hydraulic resistance of each feeding path results as the sum of the hydraulic resistance of the pipes making up the feeding path. The hydraulic resistance of a pipe is a non-dimensioned measure of the pressure losses along the considered pipe. In the case of WDSs, it can be derived that the pipes can be considered to be hydraulic rough in which turbulent flow takes place. The physical derivation of the hydraulic resistance is given by Lorenz et al. [14].

The number of alternative paths to consider for the resilience assessment is given by $K_{\max}$, which allows for a significant computational facilitation compared to considering all possible paths. At the same time, alternative feeding paths with a much higher resistance than the shorter feeding paths do not increase the overall resilience significantly. The sum of the hydraulic conductance of single feeding paths, i.e. the inverse of its resistance, is either limited by the number of alternative paths to be considered or if there do not exist as many alternative feeding paths by the number of existing alternative feeding paths for the specific pair of consumer node c and source s, given in the two-dimensional matrix $K_{s,c}$.

3.2 Comparability of Resilience

The previously introduced resilience index allows for the assessment of a relative resilience change within one network when adapted. At the same time the absolute value does not allow for a clear comparison of two different WDSs serving two different urban structures. Therefore, the presented resilience index is analyzed towards its non-dimensional characteristics and further is adapted to allow for a relative comparability of WDS within the adapted metric.

As stated before besides topological differences, WDSs can differ in areal size and overall demand or demand density. To assess which topological structure, dictated by the underlying urban structure, allows for the highest resilient WDS, these factors have to be considered in a non-dimensional form. The introduced definition of the WDS's resilience takes into account the non-dimensioned demand by the weighted sum of each consumer node as well as a non-dimensioned length through the consideration of the hydraulic resistance, which includes the length of the pipes making up a feeding path.

However, the introduced definition of the resilience for a WDS does not differentiate between one node having multiple feeding paths of a certain resistance and single nodes having solely one feeding path of the same resistance as to all nodes having the same number of feeding paths as well as the same total number of feeding paths compared to the first scenario. If consumer nodes are considered as equally important, the second scenario should yield a higher resilience than the first scenario. Therefore, the influence of second, third and k^{th} feeding path on the overall resilience has to be weighted so that a homogenous supply by alternative paths is evaluated superior. This can be realized by adding the hydraulic resistance of the shorter paths to the considered alternative path $(\sum_{i=1,...,k} r_{s,c,i})$. Therefore, the resilience automatically increases less for the $k + 1$-shortest path than it does for the k-shortest path. This also allows for the drop of the division by the maximum number of alternative paths considered, K_{max}, compared to Eq. (1). As some consumer nodes have higher demand than others, the importance of high demand consumer nodes is still factored by the weighted sum of the resilience of single consumer nodes. This is mathematically formulated by Eq. (2).

$$I_{GT} = \sum_{c \in C} \left(q_c / Q \sum_{s \in S} \left(\sum_{k=1,...,K_{s,c} \leq K_{\text{max}}} \left(1 / \sum_{i=1,...,k} r_{s,c,i} \right) \right) \right) \quad (2)$$

3.3 Real-World Sample WDSs

The present work studies the application of the former derived resilience metric for two different real-world main WDSs. The studied main WDSs differ in areal size, demand density as well as in their urban structure. While the first main WDS provides for the German city of Darmstadt, the second feeds a suburban area in the Rhine-Main region. To obtain the main WDS, pipes with diameters smaller than 100 mm are neglected while their demand is added to the nearest node still considered. The remaining connected graph including the source is considered as the main WDS.

The considered urban forms differ greatly in their urban structure. While the city of Darmstadt can be classified as a concentric city, the suburb is mostly of homogeneous areal structure. Additionally, the development in both urban forms leads to a high discrepancy in the demand density. In the city of Darmstadt the demand density is at least 2.5 times larger than in the suburb. Moreover, for the city of Darmstadt there exist two supply lines functioning as sources. In turn, the suburb is solely fed by one source.

To exploit the full potential of the respective urban structures, each existing main WDS is enhanced to the maximum possible main WDS. This is done by adding pipes with the minimum diameter of 100 mm to represent the characteristics of the urban transportation infrastructure. The respective existing and maximum main WDS for each urban structure are presented in Fig. 1. Thereby, the existing main WDS is represented by the blue lines and the additional pipes in the maximum main WDS by the green lines. The grey lines represent the urban transportation system, which is highly parallel to the WDS and therefore, the lines of the WDS often hide the lines representing the urban transportation system.

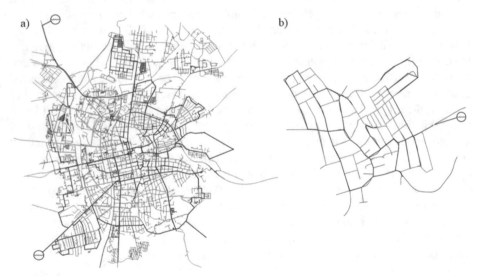

Fig. 1. Representation of the existing main WDSs marked by blue lines, the additional pipes of the maximum main WDSs marked by green lines, and the underlying urban transportation infrastructure marked by grey lines respectively for a) the city of Darmstadt and b) a suburb in the Rhine-Main region.

Each main WDS is represented by a mathematical graph made up of nodes and edges. Edges represent the pipes connecting the different nodes. The nodes can be distinguished into source nodes with unlimited water availability, junction nodes, consumer nodes with a specific water demand and tanks which can store a defined amount of water. The studied WDSs differ greatly in their number of nodes, specifically consumer nodes, for which the resilience index is determined, and their number of edges. Therefore, the computational expenses when assessing the resilience of both WDSs differ.

4 Results

The present data of the WDSs was altered as described in Sect. 3.3 in python using the python package WNTR [22] to receive the main WDSs. To further reduce the complexity of the resulting main WDSs, consumer nodes with lower than the median demand are clustered into the closest consumer node of demand larger than the median demand. In a next step, all nodes which neither have a demand nor serve as significant junction nodes, connecting three or more pipes, are removed while the two pipes connected are replaced by a single pipe of the respectively summed length. Thereby, the forthcoming computational expenses to assess the main WDS's resilience can be reduced significantly while the possible paths remain unchanged.

As discussed earlier, the existing main WDS is enhanced by adding pipes to better represent the underlying urban structure and leading to a maximum main WDS. This is done with the help of the tool QGIS as well as the openstreetmap data for the existing urban transportation infrastructure [23, 24]. Thereby, solely new pipes and junction nodes are added while the consumer nodes stay the same. This allows for an increased number of alternative paths and at the same time lower resistance paths. The main WDS for the city of Darmstadt consists of 180 consumer nodes while the suburb in the Rhine-Main region has 53 consumer nodes.

The resilience assessment is implemented and conducted in python making use of the packages WNTR as well as NetworkX [25]. In this study both resilience metrics are considered. On the one hand, the alternative paths are considered as parallel hydraulic resistance paths for which their conductance adds up, see Eq. (1). On the other hand, the alternative feeding paths are prioritized so that WDSs with homogeneously multiple feeding paths have a higher resilience value than those with single consumer nodes fed by a large number of feeding paths, see Eq. (2). These resilience definitions are applied to both, the existing and the maximum main WDSs of each urban structure. The absolute values for each resilience metric as well as both WDSs and each urban structure are listed in Table 1.

Table 1. Resilience measure of the urban structures for the existing and maximum WDSs considering prioritized alternative feeding paths or as parallel hydraulic resistances.

WDS	Existing main WDS		Max. possible main WDS	
	Prioritized alt. paths	Parallel alt. paths	Prioritized alt. paths	Parallel alt. paths
Homogeneous city	0.0098	0.0027	0.0189	0.0042
Concentric city	0.0263	0.0039	0.0306	0.0065

The comparison of the two real-world main line WDSs of different urban structure, area size, demand density, as well as number of sources shows that for both resilience metrics the concentric urban structure is superior to the homogeneous city. As there are several differences between the two WDSs for both urban structures, a definite claim is not possible. The influence of a second source allows for a resilience increase in the same order of magnitude as the resilience of the same WDS considering solely one source. This impact given by the factor of 2 between the resilience of the two urban structures is solely exceeded when considering the existing main WDSs assessed with the prioritized alternative path resilience index.

5 Discussion

In this work the question of intercomparability of urban structures towards their WDS's resilience is addressed. Therefore, the non-dimensional form of the resilience metric is of importance, which was derived and tested for two real-world WDSs within this work.

The herein proposed resilience metric takes up an interesting point, whether alternative paths are of higher value when they are spread within the considered WDS. This idea results from the redundancy and robustness aspect of resilient infrastructures. Thereby, the resilience increases when both, redundancy and robustness increase. Prioritizing alternative paths which have less k shorter preceding paths allows to consider the redundant aspect of alternative paths and at the same time sets a focus on the overall network robustness.

This resilience metric is studied for the existing as well as maximum WDSs of each urban structure. By considering the latter, the full potential of an urban structure can be studied. As the difference in the number of sources has a high impact on the WDS's resilience, there is no definite claim possible towards which urban structure allows for higher resilience. The resilience assessment suggests that a concentric urban structure allows for a higher resilient main WDS compared with a homogeneous areal urban structure. This hypothesis seems to be in line with previous research conducted by Giudicianni et al. [21], yet has to be validated by further studies of different main WDSs for the considered urban structures to apply as a guiding principal for future urban design.

The introduced resilience metric considers a resistance of paths, which can also be found for different transportation infrastructures, such as the gas network, the electrical network or even the urban transportation system. Therefore, the adapted resilience metric can assess the resilience of different urban infrastructures and allow for a comparison of the suitability of the underlying urban structures for the studied infrastructure.

Acknowledgements. The authors thank the KSB Stiftung Stuttgart, Germany for funding this research within the project "Antizipation von Wasserbedarfsszenarien für die Städte der Zukunft". Moreover, we thank the LOEWE center of Hesse State Ministry for Higher Education, Research and the Arts for partly funding this work within the project emergenCITY.

References

1. Mair, M., Zischg, J., Rauch, W., Sitzenfrei, R.: Where to find water pipes and sewers?—On the correlation of infrastructure networks in the urban environment. Water **9**(2), 146 (2017)
2. Berry, B.J.L.: Urbanization. In: Marzluff, J.M., Shulenberger, E., Endlicher, W., Alberti, M., Bradley, G., Ryan, C., ZumBrunnen, C., Simon, U. (eds.) Urban Ecology - An International Perspective on the Interaction Between Humans and Nature, pp. 103–119. Springer, US (2008)
3. Watson, V.: 'The planned city sweeps the poor away...': Urban planning and 21st century urbanisation. Prog. Plann. **72**(3), 151–193 (2009)
4. United Nations: World Urbanization Prospects: The 2018 Revision - Key Facts (2018)
5. Altherr, L.C., et al.: Resilience in mechanical engineering - a concept for controlling uncertainty during design, production and usage phase of load-carrying structures. Appl. Mech. Mater. **885**, 187–198 (2018)
6. Ugarelli, R., Venkatesh, G., Brattebø, H., Di Federico, V., Sægrov, S.: Asset management for urban wastewater pipeline networks. J. Infrastruct. Syst. **16**(2), 112–121 (2010)
7. Lorenz, I.-S., Pelz, P.: Optimal resilience enhancement of water distribution systems. Water **12**(9), 2602 (2020)
8. Lynch, K.A.: What is the form of a city, and how is it made? In: Marzluff, J.M., Shulenberger, E., Endlicher, W., Alberti, M., Bradley, G., Ryan, C., ZumBrunnen, C., Simon, U. (eds.) Urban Ecology. Springer, Boston, pp. 677–690 (2008)
9. Fischer, K., Hiermaier, S., Riedel, W., Häring, I.: Morphology dependent assessment of resilience for urban areas. Sustainability **10**(6), 1800 (2018)
10. Shin, S., et al.: A systematic review of quantitative resilience measures for water infrastructure systems. Water **10**(2), 164 (2018)
11. Yazdani, A., Otoo, R.A., Jeffrey, P.: Resilience enhancing expansion strategies for water distribution systems: a network theory approach. Environ. Model Softw. **26**(12), 1574–1582 (2011)
12. Yazdani, A., Jeffrey, P.: Complex network analysis of water distribution systems. Chaos Interdiscip. J. Nonlinear Sci. **21**(1), (2011)
13. Herrera, M., Abraham, E., Stoianov, I.: A graph-theoretic framework for assessing the resilience of sectorised water distribution networks. Water Resour. Manag. **30**(5), 1685–1699 (2016)
14. Lorenz, I.-S., Altherr, L.C., Pelz, P.F.: Resilience enhancement of critical infrastructure – graph-theoretical resilience analysis of the water distribution system in the German City of Darmstadt. In: 14th WCEAM Proceedings Book (2019)
15. Todini, E.: Looped water distribution networks design using a resilience index based heuristic approach. Urban Water **2**(2), 115–122 (2000)
16. Amarasinghe, P.: Resilience of Water Supply Systems in Meeting the Challenges Posed by Climate Change and Population Growth. Queensland University of Technology, Brisbane (2014)
17. Bruneau, M., et al.: A framework to quantitatively assess and enhance the seismic resilience of communities. Earthq. Spectra **19**(4), 733–752 (2003)
18. Meng, F., Fu, G., Farmani, R., Sweetapple, C., Butler, D.: Topological attributes of network resilience: a study in water distribution systems. Water Res. **143**, 376–386 (2018)
19. DIN-Normausschuss Bauwesen: DIN 1998 - Placement of service conduits in public circulation areas – Guideline for planning, p. ICS 93.025 (2018)
20. Reicher, C.: Städtebauliches Entwerfen. Springer Fachmedien Wiesbaden, Wiesbaden (2017)
21. Giudicianni, C., Di Nardo, A., Di Natale, M., Greco, R., Santonastaso, G.F., Scala, A.: Topological taxonomy of water distribution networks. Water (Switzerland) **10**(4), 1–19 (2018)

22. Klise, K.A., Murray, R., Haxton, T., Luther King, M.: An overview of the water network tool for resilience (WNTR). In: 1st International WDSA/CCWI 2018 Joint Conference (2018)
23. QGIS Development Team: QGIS Geographic Information System. Open Source Geospatial Foundation Project (2020)
24. OpenStreetMap contributors: "Planet dump" (2015)
25. Hagberg, A.A., Schult, D.A., Swart, P.J.: Exploring network structure, dynamics, and function using NetworkX. In: Proceedings of the 7th Python in Science Conference (SciPy), vol. 836, pp. 11–15 (2008)

Uncertainty in Production

Dealing with Uncertainties in Fatigue Strength Using Deep Rolling

Berkay Yüksel[1,2](✉) (iD) and Mehmet Okan Görtan[1,2] (iD)

[1] Hacettepe University, 06800 Ankara, Turkey
berkay.yuksel@hacettepe.edu.tr
[2] National Nanotechnology Research Center (UNAM), Bilkent University,
06800 Ankara, Turkey

Abstract. Mechanical properties inherently possess uncertainties. Among these properties, fatigue behavior data generally shows significant scatter which introduces a challenge in the safe design of dynamically loaded components. These uncertainties in fatigue behavior are mainly results of factors related to surface state including: Roughness, tensile residual stresses, scratches and notches at surface. Therefore, controlling these parameters allows one to increase fatigue strength and reduce scatter and uncertainties in fatigue behavior. Mechanical surface treatments are applied on parts to increase fatigue strength via introducing compressive residual stresses and work-hardening at surface. Two of the most common among these treatments are shot peening and deep rolling. Shot peening has found many applications in industry because of its flexibility. However, it introduces irregularities at the surface and may increase roughness which causes uncertainties in the fatigue behavior data; especially for low-medium strength materials. Unlike shot peening, deep rolling reduces surface roughness. Therefore, it has the capability to reduce uncertainty in the fatigue behavior. To this date, rolling direction of deep rolling was selected as tangential direction to turning direction for axisymmetric parts. Nonetheless, the authors believe that the rolling direction has an apparent effect on the fatigue behavior. In this study, longitudinal direction was also applied for deep rolling operation and the results of these two direction applications on the EN-AW-6082 aluminum alloy were investigated. It was shown that, longitudinal rolling had yielded less scatter and uncertainty in the fatigue behavior than the tangential rolling together with the higher fatigue strength.

Keywords: Deep rolling · Fatigue behavior · Aluminum alloy

1 Introduction

Due to the tightening in regulations on emission standards [1], usage of lightweight materials like aluminum alloys has increased over the years in the automotive industry. One of the most commonly used aluminum alloys are the heat-treatable 6xxx series owing to their strength-to-weight ratio, formability characteristics and corrosion resistance [2, 3]. 6xxx series alloys are commonly used for the design of critical parts which are subjected to dynamic loading like suspension parts [4].

© The Author(s) 2021
P. F. Pelz and P. Groche (Eds.): ICUME 2021, LNME, pp. 93–103, 2021.
https://doi.org/10.1007/978-3-030-77256-7_9

Mechanical surface treatments are commonly applied on the parts which are dynamically loaded in order to increase their fatigue strength properties. Among these treatments shot peening and deep rolling are the most commonly applied ones [5]. These treatments are applied to induce local plastic deformation on the parts which causes favorable compressive residual stresses and a work-hardened layer at the surface region. Shot peening is applied via peening the part using steel or ceramic balls with a pre-determined intensity and time whereas deep rolling is applied using hydraulically or mechanically supported roller or a ball to deform the material surface with a well-defined force. Although the compressive residual stresses induced by deep rolling generally reaches higher depths than shot peening [6], shot peening process has been a more investigated subject compared to deep rolling process to this date owing to its flexibility and ease of application. However, shot peening inherently has a negative effect on the surface roughness parameters due to its principles [5]. Low-to-medium strength materials are especially prone to these negative effects and if the severity of the shot peening is not chosen correctly, process may induce crack-like irregularities on the material surface [5, 7, 8]. On the other hand, deep rolling reduces surface roughness if applied with correct forces [5, 9]. Reduced surface roughness is known to be beneficial for fatigue strength properties. Because of these reasons deep rolling stands out to be an attractive option for the critical parts which are to be manufactured from 6xxx series alloys in automotive industry.

Although improvement can be made via surface treatments, fatigue behavior data generally shows significant scatter and uncertainty. Therefore, it should be subjected to statistical methods in order to overcome these uncertainties to allow safe design of the parts. Within engineering community, lower 2-sigma or 3-sigma design curves commonly employed. However, these methods fail to account for providing the definite confidence intervals and reliability levels [10]. On the other hand, Owen one-side tolerance limit method is a viable option in order to determine the confidence and reliability levels [11]. In both of these methods, standard error of the specimen set about the median curve is used and assumed to be uniform for the entire range of the data set. For lower 3-sigma approach, standard error is multiplied by three and algebraically subtracted from the least-squares curve to obtain the lower limit curve. For the Owen one-side tolerance limit, coefficient K_{owen} is found by employing empirical coefficients and multiplied with the standard error [10, 11]. Different K_{owen} values corresponding to various confidence and reliability levels were tabulated in [12].

In this study, commercial EN-AW 6082 aluminum alloy specimens were subjected to deep rolling to improve fatigue strength properties. Although deep rolling is conventionally applied in the tangential direction for axisymmetric parts, rolling in the longitudinal direction was also applied. Effects of rolling direction on fatigue strength and its uncertainty were investigated and the results were compared with untreated specimens. Statistical approaches were used to evaluate safe design ranges for fatigue strength. Both lower-3 sigma design curves and Owen one-side tolerance limits were considered and compared.

2 Materials and Methods

For this study, EN-AW 6082 aluminum bars of 1 m length and 15 mm diameter were supplied. Chemical composition of the material was shown in Table 1. From these bars specimens of 120 mm length were cut and then solution heat-treated at 550 °C for 1.5 h, quenched and artificially aged at 180 °C for 8 h. After the heat treatment procedure, both tensile and fatigue test specimens were machined to their respective geometries using Spinner TC600-35 CNC turning machine with 0.4 mm depth of cut and 0.105 mm/revolution feed rate. Finishing cut was done using 0.1 mm depth of cut.

Table 1. Chemical composition of 6082 aluminum alloy

Al	Si	Fe	Cu	Mn	Mg	Cr
97.2	0.93	0.295	0.015	0.53	1.00	0.0055

Mechanical properties of the alloy were determined using tensile tests. In these tests, specimens of Type A in accordance with DIN 50125: 2009-07 standard were used. Specimens had 40 mm gauge length and 8 mm diameter and the strain rate was selected as 1×10^{-3}. Yield point was determined using 0.2% offset rule. Mechanical properties were averaged from the data of 3 specimens. Apart from these tests, Vickers micro-hardness measurements were done on the section perpendicular to the longitudinal axis of the specimens using 500 g force and 10 s dwell time. Mechanical properties of the alloy were summarized in Table 2.

Table 2. Mechanical properties of 6082 aluminum alloy

Yield strength [MPa]	Tensile strength [MPa]	Elongation at fracture [%]	HV0.5
280.8 ± 5.9	312.0 ± 9.8	22.3 ± 1.4	123.1 ± 3.1

Deep rolling process was applied on the specimens in two different directions: Tangential (conventional) and longitudinal directions. These directions can be seen in Fig. 1. For both directions, 250 N of rolling force was used and feed was selected as 0.1 mm/pass. Rolling speed was selected as 10 mm/s for both processes. Roller radius and tip radius of the deep rolling apparatus were 42.5 mm and 2.5 mm, respectively. Using Hertz Theory; maximum contact pressures were calculated as 2610 MPa and 1750 MPa for tangentially rolled (TR) and longitudinally rolled (LR) cases, respectively.

Fig. 1. Deep rolling directions

Surface roughness parameters of Ra and Rz were determined using Mitutoyo SJ-210 profilometer with skidded stylus tip. Average of 10 measurements were used for each set in accordance with ISO 4287-1994 standard and in the direction parallel to the longitudinal direction of the specimen. Since the apparatus was skidded, roughness measurement in the tangential direction was not possible because of the measurement errors associated with the specimen curvature. Measurement section can be seen in Fig. 2.

Deep rolling induced work-hardening state of the material surface was qualitatively determined via Vickers micro hardness measurements close to the surface with 10 g force and 10 s dwell time. Measurements were started from the near-surface region and proceeded towards the bulk of the material in order to be able to observe the differences between near-surface region and bulk of the material.

Residual stresses at the surface were measured by using X'Pert PANalytical multi-purpose x-ray diffractometer at National Nanotechnology Research Center (UNAM). Peak location for CuKα radiation of $2\theta = 138°$ were used and 9 different tilt angles were measured between $-39.25° < \psi < 39.25°$. Residual stress measurements were done on fatigue specimens. Diffraction area on the specimen surface was a circle with an approximate diameter of 2 mm in order to minimize the errors associated with the curvature of the specimens. Measurement section can be seen in Fig. 2.

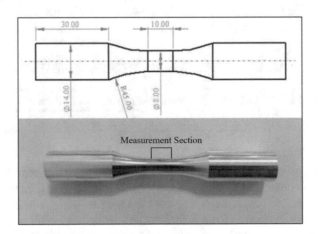

Fig. 2. Fatigue test specimen geometry

Fatigue tests were conducted as force-controlled constant amplitude tests by employing servo-hydraulic axial fatigue testing machine with 4 Hz frequency and load ratio of $R = -1$. In these tests specimens with gage length of 10 mm and diameter of 8 mm were used. Specimen can be seen in Fig. 2. Since high cycle fatigue is technologically more important in automotive industry, tests were done at stress levels that corresponds to cycle counts that are above 10^4. 12 specimens for each case were tested for untreated (UT) tangentially rolled (TR) and longitudinally rolled (LR) cases with the total of 36 specimens. After the tests were finished, obtained fatigue behavior data were subjected to lower 3-sigma method and Owen one-side tolerance limit method with 90% reliability and 90% confidence (R90C90). This way, practical limits which allow for the safe design were determined.

3   Results and Discussions

Roughness alteration after deep rolling process was shown in Table 3. Both TR and LR specimen sets showed similar improvements for roughness parameters Ra and Rz. Ra values dropped from approximately 0.7 μm to 0.1 μm; whereas Rz values dropped from approximately 2.8 μm to 0.5 μm. These improvements on the roughness values has a beneficial effect on the fatigue behavior since surface irregularities act as stress concentrators and crack nucleation sites. However, it is important to note that roughness after deep rolling is strongly dependent on the roughness prior to deep rolling [9].

Table 3. Roughness parameters

	Ra [μm]	Rz [μm]
UT	0.720 ± 0.010	2.797 ± 0.128
TR	0.092 ± 0.007	0.542 ± 0.103
LR	0.107 ± 0.018	0.535 ± 0.072

Figure 3 demonstrates the roughness profiles of LR and TR specimen. Even though the obtained numerical values for Ra and Rz were similar, there were significant differences in the roughness profiles. The regions marked with grey ellipses for TR specimen's roughness profile exhibited significantly higher valley depths compared to the other regions which may act as crack nucleation sites. Therefore, these regions may

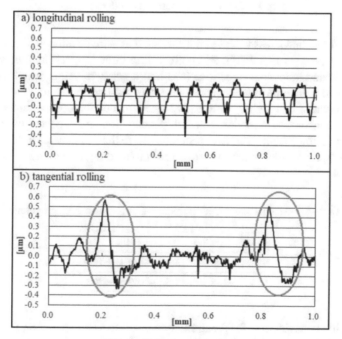

Fig. 3. Roughness profiles

increase probability of the crack nucleation and increase the uncertainties in the fatigue behavior data.

Hardness distribution within the material in the depth direction can be seen in Fig. 4. Solid line on the figure represents the nominal untreated material hardness value of 123.1 HV. Increase in the hardness is associated with the work-hardening. Therefore, plastic deformation depth induced by deep rolling process can be assessed. It can be seen that, for both rolling directions, plastic deformation depth reached up to nearly 0.5–0.6 mm. Hardness increase usually retards crack initiation; therefore, is expected to have a positive effect on the fatigue behavior. Near surface, hardness increase was approximately 10%.

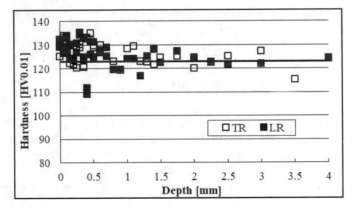

Fig. 4. Hardness distribution after deep rolling

Measured residual stress values at the surface that were developed after deep rolling operation can be seen in Table 4. For both rolling direction, residual stresses in longitudinal (σz) and tangential directions (σx) had a negative value; which is beneficial for fatigue strength improvement. It can be observed that the residual stresses in the tangential direction had higher compressive stress values for LR specimens compared to TR. This is favorable to hinder the crack propagation and expected to contribute to fatigue strength improvement of LR specimens more than TR ones. On the other hand, difference between residual stresses in longitudinal (loading) direction is relatively small and negligible.

Table 4. Residual stress measurements after the deep rolling operation

	σz [MPa]	σx [MPa]
TR	-189.9 ± 13.3	-98.1 ± 15.8
LR	-178.3 ± 18.5	-190.4 ± 20.0

Fatigue life behavior of UT, TR and LR specimens can be seen in Fig. 5. Note that the drawn curves are least-squares regression curves in semi-log coordinates and represents the median curves for the fatigue life. It can be clearly seen that both deep rolling operations improved high cycle fatigue behavior. Although tangential rolling resulted in only a slight improvement, longitudinal rolling had a significantly pronounced positive effect on the fatigue behavior. For the materials which do not exhibit a well-defined endurance limit, stress amplitude level at 10^6 cycles is a commonly adopted value in order to be able to quantify the fatigue strength of the material. Median curve of the UT specimen set indicated the fatigue strength of approximately 136 MPa at 10^6 cycles; whereas for the TR and LR specimen sets, this value increased to 145 MPa and 167 MPa, respectively.

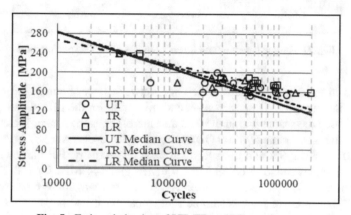

Fig. 5. Fatigue behavior of UT, TR and LR specimen sets

Although median fatigue curves are useful in order to make comparison between different sets, they often possess significant amount of scatter and uncertainty associated with this scatter as seen from the Fig. 5. Figure 6, 7 and 8 shows the lower 3-sigma and the R90C90 Owen one-sided limits together with the median curves for UT, TR and LR sets, respectively.

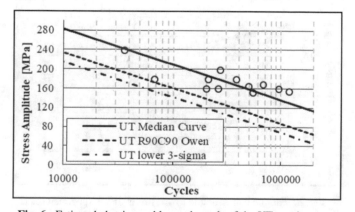

Fig. 6. Fatigue behavior and lower bounds of the UT specimen set

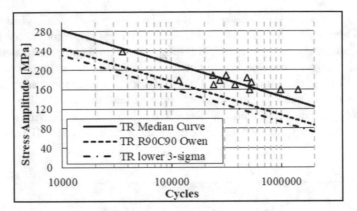

Fig. 7. Fatigue behavior and lower bounds of the TR specimen set

As seen in the Fig. 6, 7 and 8, difference between lower bounds and the median curve was maximum for UT specimen set and minimum for the LR specimen set. This difference is due the scatter within the data and increases as the scatter increases.

Fatigue strength for the R90C90 Owen curves at 10^6 cycles were 87 MPa, 107 MPa and 150 MPa for UT, TR and LR specimen sets, respectively. For the lower 3-sigma curves, these values found to be 68 MPa, 93 MPa and 144 MPa for UT, TR and LR sets, respectively. As can be seen, lower-3 sigma curve was more conservative than the R90C90 Owen curve for this sample size; meaning it had higher confidence and reliability level. However, care should be taken to make a conclusion as the K_{owen} values change with the sample size. Therefore, if the lower sigma approaches are to be used, it can be a better practice to use lower-2 sigma rather than lower-3 sigma for the large sample sizes (above 30 samples). Because, K_{owen} values decreases as the sample size increases due to the reduction in the uncertainties. In that case, use of the lower-3 sigma curve may cause excessive material usage and increase the weight of the structure.

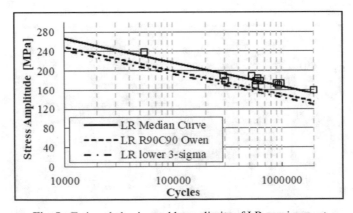

Fig. 8. Fatigue behavior and lower limits of LR specimen set

Improvement of fatigue strength at 10^6 cycles in percentage were given in the Table 5. Since the UT median curve possessed higher level of uncertainties compared to the TR and LR sets, improvement in the fatigue strength after the deep rolling became more pronounced with the statistical treatment applied on the experimental data set. It can also be seen that LR yielded significantly higher fatigue strength and less scatter than the TR. This behavioral difference between rolling direction is believed to be mostly related to geometrical alterations at the surface. Because, residual stresses were expected to have similar values for all specimens within each set. Therefore, uncertainty difference between TR and LR set could not be explained by residual stresses only.

Table 5. Fatigue strength improvements [%]

	Median curve	R90C90 owen	Lower 3-Sigma
TR	6.6	23.0	36.8
LR	22.8	72.4	111.8

Differences between TR and LR sets in terms of experimental fatigue data were mainly attributed to the direction of the grooves formed by deep-rolling operation and the differences in the roughness profiles. Authors acknowledge the fact that for LR, roughness profile would look similar to the profile shown in Fig. 3b if the measurement were done in the tangential direction. However, it was stated in [13] that polishing in which the fine scratches oriented parallel to the loading direction had yielded better results than perpendicular direction in terms of fatigue strength. Analogous to that phenomenon it can be expected that the grooves/ridges parallel to the loading direction would result in consistently higher fatigue strength than the perpendicular one as the latter case may promote crack nucleation. In addition to that, the aforementioned differences in the roughness profiles in the longitudinal (loading) direction are believed to be one of the major reasons of the lower fatigue strength and higher uncertainty of the TR specimen set compared to LR. Since the effects of the longitudinal rolling direction compared to tangential direction were not investigated in the literature before; differences between LR and TR sets mentioned above are of technological importance.

4 Conclusions and Recommendations

It can be seen in the above results that the fatigue strength determination is prone to uncertainties. Because of these uncertainties, usage of lower-bound curves are essential for safe design. In this study, beneficial effects of the deep rolling in different rolling directions on surface properties and fatigue behavior of the EN-AW 6082-T6 aluminum alloy were shown. These effects can be listed as below;

- Although surface roughness values were significantly reduced to similar values for both rolling directions, there were significant differences between profiles. TR specimen set had more inconsistent roughness profile and deeper valleys than LR set.

- Hardness values showed increased values near-surface region after deep rolling because of the work-hardening behavior without a significant difference between TR and LR specimen sets.
- Favorable compressive residual stresses formed at the surface after deep rolling.
- Fatigue strength properties were improved after deep-rolling compared to the untreated parts. This improvement was significantly more pronounced for the longitudinal rolling than the tangential rolling.
- For the untreated specimen set, scatter within the data and the uncertainties associated with this scatter was found to be highest among the specimen sets. Because of this, fatigue strength improvements due to deep rolling operation became even more pronounced after the statistical treatment of the median curves.
- Longitudinal rolling yielded less scatter and uncertainties than the tangential rolling.

In the future, reach of this study can be extended to include different deep rolling parameters to acquire more comprehensive knowledge on the material and the process.

Acknowledgments. The authors thank the Scientific and Technological Research Council of Turkey (TÜBİTAK) for financial support (Project Number: 217M962)

References

1. Advanced High-Strength Steel Applications Guidelines 6.0. World Steel Association. http://www.worldautosteel.org/projects/advanced-high-strength-steel-application-gui delines/. Accessed 14 Aug 2017
2. Aruga, Y., Kozuka, M., Takaki, Y., Sato, T.: Effects of natural aging after pre-aging on clustering and bake-hardening behavior in an Al-Mg-Si alloy. Scripta Mater. **116**(1), 82–86 (2016)
3. Abdulstaar, M., Mhaede, M., Wagner, L.: Pre-corrosion and surface treatments effects on the fatigue life of AA6082 T6. Adv. Eng. Mater. **15**(10), 1002–1006 (2013)
4. Jeswiet, J., Geiger, M., Engel, U., Kleiner, M., Schikorra, M., Duflou, J., Neugebauer, R., Bariani, P., Bruschi, S.: Metal forming progress since 2000. CIRP J. Manufact. Sci. Technol. **1**, 2–17 (2008)
5. Schulze, V.: Modern Mechanical Surface Treatments, 1st edn. WILEY-VCH Verlag Gmbh & Co. KGaA, Weinheim (2006)
6. Sticchi, M., Schnubel, D., Kashaev, N., Huber, N.: Review of residual stress modification techniques for extending the fatigue life of metallic components. Appl. Mech. Rev. **67**(1), 1–9 (2015)
7. Trsko, L., Guagliano, M., Bokuvka, O., Novy, F.: Fatigue life of AW 7075 aluminum alloy after severe shot peening treatment with different intensities. Procedia Eng. **74**(1), 246–252 (2014)
8. Gonzalez, J., Bagherifard, S., Guagliano, M., Pariente, I.F.: Influence of different shot peening treatments on the surface state and fatigue behaviour of Al 6063 alloy. Eng. Fract. Mech. **185**(1), 72–81 (2017)
9. Beghini, M., Bertini, L., Monelli, B.D., Santus, C., Bandini, M.: Experimental parameter sensitivity analysis of residual stresses induced by deep rolling on 7075-T6 aluminium alloy. Surf. Coat. Technol. **254**(1), 175–186 (2014)

10. Lee, Y., Pan, J., Hathaway, R., Barkey., M.: Fatigue Testing and Analysis: Theory and Practice. 1st edn. Elsevier Butterworth-Heinemann, Burlington (2004)
11. Shen, C.L., Wirshing, P.H., Cashman, G.T.: Design curve to characterize fatigue strength. J. Eng. Mater. Technol. **118**, 535–541 (1996)
12. Williams, C.R., Lee, Y., Rilly, Y.T.: A practical method for statistical analysis of strain-life fatigue data. Int. J. Fatigue **25**(5), 427–436 (2003)
13. Dieter, G.E., Bacon, D.: Mechanical Metallurgy. 3rd edn. McGraw-Hill Book Co., Singapore (1988)

Investigation on Tool Deflection During Tapping

Felix Geßner[(⊠)], Matthias Weigold, and Eberhard Abele

Institute of Production Management, Technology and Machine Tools (PTW),
Technische Universität Darmstadt, Otto-Berndt-Str. 2, 64287 Darmstadt, Germany
f.gessner@ptw.tu-darmstadt.de

Abstract. Tapping is a challenging process at the end of the value chain. Hence, tool failure is associated with rejected components or expensive rework. For modelling the tapping process we choose a mechanistic approach. In the present work, we focus on the tool model, which describes the deflection and inclination of the tool as a result of the radial forces during tapping. Since radial forces always occur during tapping due to the uneven load distribution on the individual teeth, the tool model represents an essential part of the entire closed-loop model. Especially in the entry phase of the tap, when the guidance within the already cut thread is not yet given, radial forces can lead to deflection of the tool. Therefore, the effects of geometric uncertainty in the thread geometry are experimentally investigated, using optical surface measurement to evaluate the position of the thread relative to the pre-drilled bore. Based on the findings, the tool deflection during tapping is mapped using a cylindrical cantilever beam model, which is calibrated using experimental data. The model is then validated and the implementation within an existing model framework is described.

Keywords: Tapping · Deflection · Model

1 Introduction

In industrial applications and in everyday live, threaded joints are widely used as detachable connecting elements. The most used process for machining internal threads is tapping [1]. During tapping the thread is cut successively into the wall of an existing bore by the tooth of the tapping tool. Since the process is located at the end of the value chain, tool failure is associated with rejected components or expensive rework [2]. The industry is therefore constantly striving to increase process stability, to reduce failures or the need for human intervention [3]. To achieve this, it is necessary to increase the understanding of the tapping process especially regarding the effect of uncertainty.

Like all machining processes, drilling–tapping process chains are generally affected by uncertainty, since not all properties and process characteristics can be fully determined. Forms of geometrical uncertainty, which occur in tapping process chains, are shown in Fig. 1. In the process chain uncertainty can arise from the preceding or the current process step itself, or from the interlinking of both steps. Deviations of the pre-drill geometry, like variations in diameter, straightness or cylindrical shape may occur in the

© The Author(s) 2021
P. F. Pelz and P. Groche (Eds.): ICUME 2021, LNME, pp. 104–114, 2021.
https://doi.org/10.1007/978-3-030-77256-7_10

preceding process. Runout and synchronization errors are allocated in the current process step. Uncertainty arising from the interlinking of both process steps are positioning errors, like axis offset and inclined pre-drill bores [4]. Positioning errors, can be caused by misplacement, inaccuracy of the machine tool or reclamping of the part between the process steps [5].

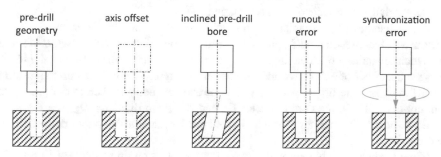

Fig. 1. Forms of uncertainty in a drilling-tapping process chain [4]

For the present work, the focus is mainly on axis offset and inclined pre-drill bores, that are relevant for tapping and other subsequent processes of drilling. Investigations on the tool deflection in reaming illustrate for example, that the axis offset leads to unbalanced radial forces during the entry phase of the reaming tool, which causes tool deflection and a center offset of the bore [6]. Just like the reaming tool, the tapping tool is not yet guided when entering the pre-drilled bore. In addition, when tapping, the cutting load is not homogeneously distributed over the cutting section and over the lands of the tap due to the individual tooth geometry [7] (see Fig. 2).

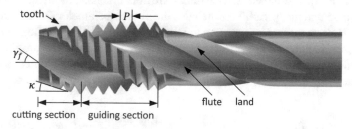

Fig. 2. General geometry of a blind-hole tapping tool

Although general descriptions of the tapping process can be found in numerous manuals, it has received less attention for research activities, than other processes like turning, milling or drilling [8]. This is also evident in studies on tool deflection since many studies exist for the milling process [9–12]. Kim et al. [9] describe the tool deflection during ball-end milling. They calculate the cutter deflection due to the cutting force using a two-step cylindrical cantilever beam and estimate the form error of the finished part. An extension of this approach is presented by Smaoui et al. [10] who compare the results of the analytical beam model to numeric and experimental model approaches. Zeroudi and Fontaine [11] use the analytical tool deflection model for tool path compensation. Another use of beam theory is presented by Moges et al. [12] to describe tool deflection in micro milling. A consideration of the effect of uncertainty on tool deflection are described by Hauer [6] for the reaming process, by Bölling [13] for valve guide and seat machining and by Hasenfratz [14] for milling of deep cavities in TiAl6V4. All three works use beam theory and Jeffcott rotor theory to reproduce the tool displacement as well as the dynamic effects of the fast-rotating tools.

The influences of process faults on tapping are described by Dogra et al. [2, 15]. They use a mechanistic model approach to predict the effect of axis offset, tool runout and synchronization errors on the resulting torque and forces. Mezentsev et al. [16, 17] describe the effect of axis offset and tool runout on tapping, with focus on the resulting radial forces and the thread geometry. The thread geometry is therein defined by the nominal diameter D, the core diameter D_1 and the pitch diameter D_2 (see Fig. 3). It should be noted, that neither Mezentsev et al. nor Dogra et al. take tool deflection into account.

Many studies already deal with the displacement of cutting tools due to resulting forces. The aim of this article is to extend the previous findings and to transfer them to tapping. For this purpose, the phenomenon of tool deflection during tapping is first described based on experimentally recorded data. This is followed by the derivation of a model that can be used to describe the phenomenon. Subsequently, the model created is calibrated and validated in experimental tests. Finally, the integration of the tool deflection model into an existing framework to simulatively investigate the effect of uncertainty on the tool deflection during tapping is described.

2   Effect of Tapping Tool Deflection on Thread Geometry

As described in [5], the diameters in the thread are not influenced by an offset between the pre-drilled bore and the thread. This includes the nominal diameter D, the core diameter D_1 and the flank diameter D_2 according to DIN 2244 [18]. Therefore, the detection of an axis offset is a challenging task. However, there is an influence on the profile height H_4, which is influenced by the position of the centers of the described diameters in relation to each other [5]. Therefore, conclusions about the offset between the center axis of

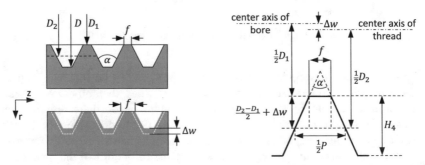

Fig. 3. Measuring the axis offset Δw between the center axis of the pre-drilled bore and thread based on the width of flat f

the pre-drilled bore and the thread can be drawn based on the width of flat f, which is influenced by the profile height.

The offset Δw between the thread and the pre-drilled bore at a specific position z_i can be calculated using the following formula, where P is the pitch of the thread, and α is the thread profile angle. In case of the M8 thread the values are $P = 1.25$ mm and $\alpha = 60°$.

$$\Delta w(z_i) = \frac{\frac{P}{2} - f(z_i)}{2\tan\left(\frac{\alpha}{2}\right)} - \frac{D_2 - D_1}{2} \tag{1}$$

To measure the width of flat f, the drilled threads are cut open perpendicular to the axis offset and the thread profile is recorded with the optical surface measuring device Alicona Infinite Focus G5. As the diameters can no longer be measured after cutting, the calculation of the axial offset of the thread relative to the pre-drilled bore is based on the width of flat f_{ref} of a reference thread produced with the same tools.

$$\Delta w(z_i) = \frac{f_{ref} - f(z_i)}{2\tan\left(\frac{\alpha}{2}\right)} \tag{2}$$

If the position of the pilot bore w_{bore} is known, the position of the thread w_{thread} can be calculated based on the position of the thread relative to the pilot bore as follows:

$$w_{thread}(z_i) = w_{bore}(z_i) + \Delta w(z_i) = w_{bore}(z_i) + \frac{f_{ref} - f(z_i)}{2\tan\left(\frac{\alpha}{2}\right)} \tag{3}$$

The deflection is evaluated for M8x1.25 threads (see Fig. 4). The threads were machined on a GROB G350 machining center, using a tool holder with minimum length compensation and internal cooling with an emulsion containing 7% oil. The used M8 machine tap has a helix angle γ_f of $45°$ and chamfer form C. The tolerance class is ISO2/6H. The tap is made of HSS-E and has a GLT-1 coating. The cutting depth is 19.75 mm, the cutting speed is 15 m/min and the workpiece material is 42CrMo4. The pre-drilled bore is a blind hole with a nominal diameter of 6.8 mm. During the pre-drilling, uncertainty was applied in the form of an offset of the pilot bore of $\pm 280\,\mu$m and an inclination with an angle of $\pm 1°$.

The deflection of the thread for a pre-drilled bore with axis offset shows that most of the deflection of the tool happens directly when entering the bore. The resulting inclination of the tapping tool then leads to an increasing deviation of the center axis of the thread over the drilling depth. The effect can be seen for both positive and negative axis offsets. With an inclined pre-drilled bore, there is only a slight deflection when first entering the pre-drilled bore, thus the tapping tool is guided in a central position. However, the tool deflection increases with increasing drilling depth. This can be attributed to unbalanced radial forces due to the greater material thickness on one side. With the combination of axial offset and inclined pilot hole, the inclination of the tapping tool, as it enters the pre-drilled bore, causes the tap to follow the bore center axis, which in turn leads to uniform chip removal. Thus, the combination of uncertainty may reduce the negative effects.

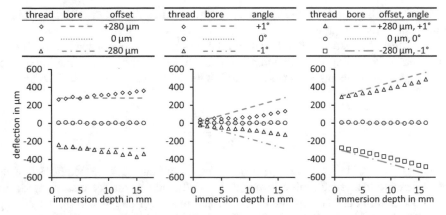

Fig. 4. Calculated thread deflection based on the experimentally measured width of flat

3 Model for Determining Tool Deflection During Tapping

3.1 Model Description

Based on the findings of Sect. 2, the system of tool holder and tapping tool is modelled using Euler-Bernoulli-Beam theory. First, a substitute model is set up to map the tool deflection due to lateral forces. Therefore, the tapping process is divided into two stages. In the first stage, until the time t_e, the tool is assumed to be unguided. Here, only a radial force F_R acts on the tool, which corresponds to the resulting force due to the chip removal of the tap F_{res}.

$$F_R = F_{res}, \text{ for } t < t_e \qquad (4)$$

Based on the resulting lateral force F_{res}, the deflection w and the inclination w' of the tool can be calculated during this period. After the time t_e the tool is assumed to be guided. The guidance of the tap in the already cut thread is mapped with a sliding sleeve, which in addition to the force F_R also takes up the moment M_{AO} (see Fig. 5).

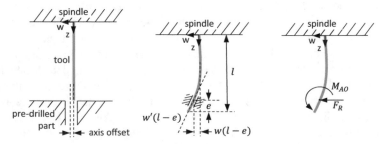

Fig. 5. Mechanical substitute model of the tapping tool before entering the pre-drilled bore (left), the guidance of the tool in the thread modelled with a sliding sleeve (middle) and the free body diagram of the deflected tool (right)

Based on the findings from Sect. 2, the axial offset and the inclination of the sliding sleeve are assumed to be constant for the rest of the tapping process. Only the position relative to the spindle changes depending on the current cutting depth $e(t)$.

$$w(l - e(t)) = \text{const., for } t \geq t_e \tag{5}$$

$$w'(l - e(t)) = \text{const., for } t \geq t_e \tag{6}$$

To accurately map the tool holder and the tool, a cylindrical cantilever beam with three segments is used (see Fig. 6). Regarding boundary conditions it is assumed that the cantilever beam has a fixed end on the left side ($z = 0$) and a free end on the right side ($z = a+b+c$). Furthermore, the inclination $w'(z)$ is assumed to be a continuous function over all segments of the beam. The first segment represents the tool holder with a length a, a Young's modulus E_C and an area moment of inertia I_1. The second two segments represent the tool. The tool used is a M8x1.25 tap, as described in Sect. 2, with a central cooling channel of diameter 0.5 mm and reinforced shank according to DIN 371 [19]. Therefore, the tool is represented by two beam segments with the length b and c, which share the same Young's modulus E_T but differ in their diameter and consequently in the area moment of inertia I_2 and I_3.

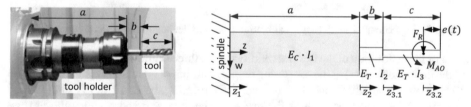

Fig. 6. Euler-Bernoulli-Beam model of the tool holder and tool system

The deflection of the beam shown in Fig. 6 can be determined from the respective curvature w'' of the individual segments due to the prevailing bending moment $M(z)$.

$$w''(z) = -\frac{M(z)}{EI} \tag{7}$$

Double integration of the equations thus established and determination of the integration constants using the boundary conditions described yields the following formulas for the deflection for the respective segments.

$$w_1(z_1) = \frac{1}{E_C I_1}\left[F_R\left(-\frac{1}{6}z_1^3 + \frac{1}{2}(a+b+d)z_1^2\right) - \frac{1}{2}M_{AO}z_1^2\right] \tag{8}$$

$$w_2(z_2) = \frac{1}{E_T I_2}\left[F_R\left(-\frac{1}{6}z_2^3 + \frac{1}{2}(b+d)z_2^2\right) - \frac{1}{2}M_{AO}z_2^2\right] + w_1'(a)z_2 + w_1(a) \tag{9}$$

$$w_{3.1}(z_{3.1}) = \frac{1}{E_T I_3}\left[F_R\left(-\frac{1}{6}z_{3.1}^3 + \frac{1}{2}dz_{3.1}^2\right) - \frac{1}{2}M_{AO}z_{3.1}^2\right] + w_2'(b)z_{3.1} + w_2(b) \tag{10}$$

$$w_{3.2}(z_{3.2}) = w_{3.1}'(d)z_{3.2} + w_{3.1}(d) \tag{11}$$

3.2 Model Calibration

To calibrate the beam model, the deflection is measured at several points on the tool and the tool holder while a lateral force is applied to the tool by means of a pneumatic piston. To apply the force in a targeted manner, a nut is screwed and glued onto the tool. In steps of 50 N, a lateral force F_R of 250 N is applied with a distance of $e = 3.4$ mm to the tip of the tool and the deflection is measured at six points on the tool and the tool holder. The test setup and the deflection resulting from a lateral force of 50 N are shown in Fig. 7.

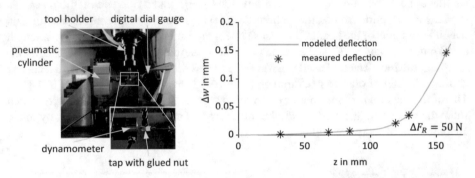

Fig. 7. Experimental setup for the calibration tests and deflection for a lateral force of 50 N

For the calibration of the flexural rigidity for the three segments the averaged gradient of the deflection $\Delta w(z)$ at the measuring points resulting of an increasing load of $\Delta F_R = 50$ N (in the range of 50 to 200 N) is used. Transforming the Eqs. (8), (9) and (10) and inserting ΔF_R, $\Delta w(z)$ as well as the measured lengths a, b and c yields the flexural rigidity of the segments listed in Table 1.

Table 1. Results of the calibration

Parameter	Value	Parameter	Value
a	102.5 mm	$E_C\, I_1$	3392.1 Nm^2
b	21.1 mm	$E_T\, I_2$	40.49 Nm^2
c	37.6 mm	$E_T\, I_3$	8.97 Nm^2

3.3 Experimental Model Validation

The tool displacement model is validated using experimental tests performed on a DMC 75 V linear machining center. Tool holder, tapping tool, cutting speed and workpiece material are chosen the same as in Sect. 2. The radial forces are measured with a Kistler 9272 dynamometer. The tool displacement is measured using two eddy current sensors pointing perpendicular on the shaft of the tool with a position of $z = 128.4$ mm. The sensors are attached to the spindle housing via a mounting system as shown in Fig. 8. To protect the measuring equipment from chips a protection shield was applied, and the tests were carried out under dry conditions. The cutting depth of the tap is reduced to 16 mm, to prevent the sensors from colliding with the component. Subtracting the length of the cutting section this results in a usable thread length of 12.25 mm, which equals approximately 1.5xD.

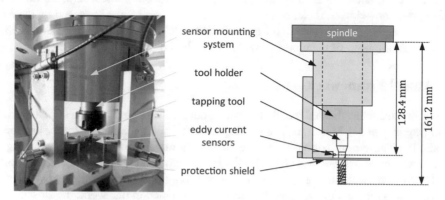

Fig. 8. Experimental setup to measure the tool deflection during tapping

The measured lateral force and tool deflection at the sensor position are shown in Fig. 9. In addition, the tool deflection at the sensor position is calculated using the calibrated model from Sect. 3.2, the measured lateral force and the current cutting depth $e(t)$. The measured and the calculated deflection show good accordance.

The start of the spindle rotation is set as time $t = 0$ s. The deflection before this is constant and is set as the zero point. With the start of the spindle rotation, a periodic signal with the frequency of the spindle rotation is visible. This can be interpreted as the combined runout error of the tool and the tool holder. When the tool cuts into the pre-drilled bore, a reduction in the amplitude of the runout error can be seen. This indicates that the tool is guided in the already cut thread. For the tests with specifically applied axis offset, a step in the measured lateral force and the measured tool deflection can be observed when entering the pre-drilled bore at $t = 0.67$ s.

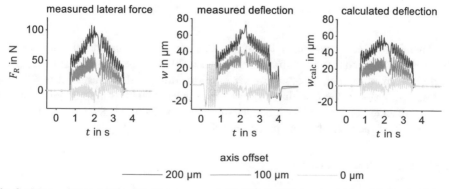

Fig. 9. Measured lateral force (left), measures deflection (middle) and calculated deflection based on the measured lateral forces (right) at the sensor position of $z = 128.4$ mm

The subsequent increase in force and deflection is due to the increasing cutting depth e, which has the effect of shortening the distance between the sensor and the guided part of the tool, as described in Sect. 3.1. After approx. 2 s, the rotation of the tool is stopped and reversed to unscrew the tool from the thread. During the unscrewing process, forces and deflection decrease analogously to the cutting process. However, since no more chips are removed here, it can be concluded that the lateral force is caused by the tool deflection due to the guidance in the already cut thread. It can therefore be summarized that the assumptions of the model are also shown in the experiment.

4   Model Framework

The calibrated and validated tool deflection model can be implemented within a mechanistic tapping model. Mechanistic modelling is a widely used method for torque and force prediction. The approach is based on the chip load-cutting force relationship according to Koenigsberger and Sabberwal [20]. The general structure of the mechanistic framework, as described by Kapoor et al. [21], combines this relationship with other submodels to a closed loop model. For the tapping process, as presented in this work, the closed loop model is shown in Fig. 10.

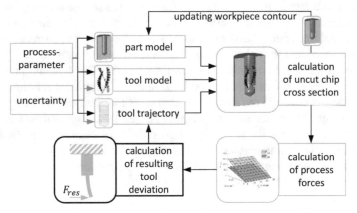

Fig. 10. Framework of the mechanistic process model

Therein the local time-dependent chip sizes considering uncertainty, as shown in Fig. 1, are calculated using a chip cross-section model [5]. The geometrically calculated chip sizes are then used as input for the empiric force model, presented in [7]. The tool deviation due to the resulting process forces is calculated using the tool deflection model, presented in Sect. 3. The influence of the tool deflection is fed back into the chip cross-section model via the manipulation of the tool trajectory.

5   Summary and Outlook

In this article it was shown how uncertainty in form of positioning errors can affect the tapping process and the thread geometry. With the measurement and evaluation of the width of flat, a method was described to determine the axial offset between the pre-drilled bore and the thread. Based on the knowledge gained from this, a substitute beam model was developed, which represents the tool deflection in the unguided and guided state of the tapping process. The model was calibrated and validated experimentally. Finally, the integration of the model into an existing framework was described, which can be used in further works to predict and describe the effect of uncertainty on the tapping process.

Acknowledgement. The authors would like to thank the German Research Foundation (DFG) for funding the research activities at the Collaborative Research Centre (CRC) 805 – Control of Uncertainty in Load-Carrying Structures in Mechanical Engineering.

References

1. Brandão, G.L., Silva, P.M.d.C., de Freitas, S.A., et al.: State of the art on internal thread manufacturing: a review. Int. J. Adv. Manufact. Technol. **110**, 3445–3465 (2020). https://doi.org/10.1007/s00170-020-06107-x
2. Dogra, A.P.S., Kapoor, S.G., DeVor, R.E.: Mechanistic model for tapping process with emphasis on process faults and hole geometry. J. Manufact. Sci. Eng. **124**, 18–25 (2002). https://doi.org/10.1115/1.1430237
3. Patel, H.J., Patel, B.P., Patel, S.M.: A review on thread tapping operation and parametric study. Int. J. Eng. Res. Appl. (IJERA) **2**, 109–113 (2012)
4. Pelz, P.F., et al.: Mastering uncertainty in mechanical engineering. Springer (2021)
5. Abele, E., Geßner, F.: Spanungsquerschnittmodell zum Gewindebohren: Modellierung der Auswirkung von Unsicherheit auf den Spanungsquerschnitt beim Gewindebohren. wt Werkstattstechnik online, vol. 108, pp. 2–6 (2018)
6. Hauer, T.: Modellierung der Werkzeugabdrängung beim Reiben – Ableitung von Empfehlungen für die Gestaltung von Mehrschneidenreibahlen. Dissertation, TU Darmstadt (2012)
7. Geßner, F., Weigold, M., Abele, E.: Measuring and modelling of process forces during tapping using single tooth analogy process. Prod. Eng. Res. Dev. (2020). https://doi.org/10.1007/s11740-020-01004-4
8. Özkaya, E.: FEM-basiertes Softwaresystem für die effiziente 3D-Gewindebohrsimulation und Werkzeugoptimierung mittels CFD-Simulation, 1. Auflage. Schriftenreihe des ISF, vol 89. Vulkan, Essen (2017)
9. Kim, G.M., Kim, B.H., Chu, C.N.: Estimation of cutter deflection and form error in ball-end milling processes. Int. J. Mach. Tools Manufact. **43**, 917–924 (2003). https://doi.org/10.1016/S0890-6955(03)00056-7
10. Smaoui, M., Bouaziz, Z., Zghal, A., et al.: Simulation of the deflected cutting tool trajectory in complex surface milling. Int. J. Adv. Manufact. Technol. **56**, 463–474 (2011). https://doi.org/10.1007/s00170-011-3213-x
11. Zeroudi, N., Fontaine, M.: Prediction of tool deflection and tool path compensation in ball-end milling. J. Intell. Manufact. **26**, 425–445 (2013). https://doi.org/10.1007/s10845-013-0800-8
12. Moges, T.M., Desai, K.A., Rao, P.V.M.: Modeling of cutting force, tool deflection, and surface error in micro-milling operation. Int. J. Adv. Manufact. Technol. **98**, 2865–2881 (2018). https://doi.org/10.1007/s00170-018-2415-x
13. Bölling, C.: Simulationsbasierte Auslegung mehrstufiger Werkzeugsysteme zur Bohrungsfeinbearbeitung am Beispiel der Ventilführungs- und Ventilsitzbearbeitung. Dissertation, TU Darmstadt (2018)

14. Hasenfratz, C.: Modellgestützte Prozessauslegung zum linearen Schaftfräsen von tiefen Kavitäten in TiAl6V4 - Verdichterscheiben. Dissertation, TU Darmstadt (2018)
15. Dogra, A.P.S., DeVor, R.E., Kapoor, S.G.: Analysis of feed errors in tapping by contact stress model. J. Manufact. Sci. Eng. **124**, 248–257 (2002). https://doi.org/10.1115/1.1454107
16. Mezentsev, O.A., DeVor, R.E., Kapoor, S.G.: Prediction of thread quality by detection and estimation of tapping faults. J. Manufact. Sci. Eng. **124**, 643–650 (2002). https://doi.org/10.1115/1.1475319
17. Mezentsev, O.A., Zhu, R., DeVor, R.E., et al.: Use of radial forces for fault detection in tapping. Int. J. Mach. Tools Manufact. **42**, 479–488 (2002). https://doi.org/10.1016/S0890-6955(01)00139-0
18. DIN Deutsches Institut für Normung e.V.: Gewinde - Begriffe und Bestimmungen für zylindrische Gewinde (DIN 2244:2002-05) (2002)
19. DIN Deutsches Institut für Normung e.V.: Maschinen-Gewindebohrer mit verstärktem Schaft für Metrisches ISO-Regelgewinde M1 bis M10 und Metrisches ISO-Feingewinde M1 × 0,2 bis M10 × 1,25 (DIN 371:2016-01) (2016)
20. Koenigsberger, F., Sabberwal, A.J.P.: An investigation into the cutting force pulsations during milling operations. Int. J. Mach. Tool Des. Res. **1**, 15–33 (1961). https://doi.org/10.1016/0020-7357(61)90041-5
21. Kapoor, S.G., DeVor, R.E., Zhu, R., et al.: Development of mechanistic models for the prediction of machining performance: model building methodology. Mach. Sci. Technol. **2**, 213–238 (1998). https://doi.org/10.1080/10940349808945669

How to Predict the Product Reliability Confidently and Fast with a Minimum Number of Samples in the Wöhler Test

Jens Mischko[1]([✉]), Stefan Einbock[1], and Rainer Wagener[2]

[1] Robert Bosch GmbH, Schwieberdingen, Germany
Jens-Frederik.Mischko@de.bosch.com
[2] Fraunhoferinstitut LBF, Darmstadt, Germany

Abstract. To accurately estimate and predict the (product) lifetime, a large sample size is mandatory, especially for new and unknown materials. The realization of such a sample size is rarely feasible for reasons of cost and capacity. The prior knowledge must be systematically and consistently used to be able to predict the lifetime accurately. By using the example of Wöhler test, it will be shown that the lifetime prediction with a minimum number of specimen and test time can be successful, when taking the prior knowledge into account.

Keywords: Increasing efficiency · Wöhler test · Lifetime model · Prior knowledge · Test planning

1 Introduction

The reliable design of components or design elements is an essential part of the development areas in the product development process. This is particularly the case if oversizing, e.g. lightweight engineering, must be avoided. In order to be able to accurately estimate the strength, lifetime models are used. Lifetime models show the correlation between the stress or load and the lifetime of a failure mechanism. The challenge for test engineers is to determine the parameters of the lifetime models as accurately as possible based on lifetime tests in order to obtain the best possible estimation of the predicted lifetime. However, this can only be achieved with a very large sample size, as the test results are statistically distributed. In contrast, shorter development times and increasing competitive pressure lead to a lack of time and costs to determine the parameters of the lifetime models in a confident manner.

2 Results

This paper aims to show a method that can efficiently determine the parameters of lifetime models. By estimating the parameters based on prior knowledge, the

© The Author(s) 2021
P. F. Pelz and P. Groche (Eds.): ICUME 2021, LNME, pp. 115–128, 2021.
https://doi.org/10.1007/978-3-030-77256-7_11

test conduction can be optimised. First results show that the parameters can be reasonably estimated even with a very small sample size of $n = 5$ samples.

A high accurate knowledge regarding the parameters for lifetime models, ensure better results of the method. The prior knowledge can for example be determined through prior tests, based on the results of large statistical data collection or derived from mathematical models of guidelines by starting from chemical and physical material properties. However, the method can also be successfully applied to new or unknown materials or material qualities where only a limited amount of prior knowledge is available, e.g. in the form of previous test results.

The increased efficiency of this method is demonstrated by using a real Wöhler test. It turns out that a good estimation of the slope k in particular leads to a good estimation of the Wöhler curve. In the example, the number of samples can be reduced by 75% from $n = 20$ to $n = 5$.

In order to examine the validity of the method, further statistical investigations must be carried out. Furthermore, the extension of the method to other lifetime models is to be examined.

3   General Structure of Lifetime Models

In general, lifetime models specify the correlation between the sustainable stress or load and the expected lifetime. This correlation is described by a mathematical model. Characteristic for lifetime models is that a lower lifetime can be expected at a higher load. However, there are limits to this correlation in most cases. The load is limited by the change of failure mechanism upwards and downwards. On the one hand, a static limit value can be reached upwards and on the other hand, no degradation occurs if the value is below a certain threshold value.

Examples of lifetime models are the Wöhler model or the Arrhenius model. The Wöhler model describes the fatigue of varying cyclic mechanical loads on the lifetime of materials. Upwards, the model is limited by the static failure of plasticisation. In contrast, the Arrhenius model describes the dependence of the reaction rate of a chemical reaction with the absolute temperature. A prerequisite for the reaction is that the specific activation energy is exceeded [1]. Table 1 shows both models. Further lifetime models refer to [2] for example.

Table 1. Overview of different lifetime models

Name	Model
Wöhler	$N = N_A \cdot \left(\dfrac{S}{S_A} \right)^{-k}$
Arrhenius	$t = t_0 \cdot \exp\left(\dfrac{E_a}{k \cdot T} \right)$

A characteristic feature of lifetime models is that they usually only provide a simple, linear relationship between the applied load and the expected life-time. Mostly, the individual models differ only in the scaling of the axes. The Wöhler curve for example uses a double logarithmic scaling and results in a linear decrease of the high cycle fatigue.

4   Proposal for an Efficient Test Procedure for the Determination of Wöhler Curves

In this chapter, a method will be introduced to conduct fast and efficient lifetime tests. The focus of this method will be on conducting Wöhler test. A transfer to other lifetime models is possible due to the similarities between the lifetime models as described in Sect. 3.

Efficient Wöhler tests are limited to the high cycle fatigue, since a discrete value is determined in the experiments. The aim is to define a procedure in which the Wöhler curve can be determined efficiently with a sample size of $n = 5$ test specimens.

To make this possible, research on the test conduction of Wöhler experiments will first be analysed and the results will be derived from it. In addition, prior knowledge will be used for tests with such a small sample size. Due to the small sample size, the test will also be conducted according to the pearl chain method, where the experiment with one sample is carried out at different load levels.

4.1   Research on the Test Conduction of Wöhler Tests

Essential investigations on Wöhler experiments were presented in MÜLLER [3]. With the help of Monte Carlo simulations, Wöhler tests were simulated and conclusions were obtained. The following descriptions are based on this work results.

The mean load level of the Wöhler curve at a specific number of load cycles can be estimated unbiased. The greater the spread of the number of load cycles, the fewer samples have to be rejected due to an increasing slope. The higher the spread and the smaller the scatter of the Wöhler curve, the better the estimated accuracy of the Wöhler curve.

The situation is similar when estimating the slope based on test results. The slope is estimated independently of the predetermined slope. Small spreads $N_{max}/N_{min} < 10$ partly lead to the rejection of the estimation, since a increasing Wöhler curve is predicted. The best prediction is achieved when the test results are distributed over the entire range of the high cycle fatigue. The smaller the scatter of the Wöhler curve, the better the slope can be estimated.

An estimation for the scatter is asymptotically unbiased. With a very small sample size, e.g. $n < 7$, the scatter minor underestimated. The estimation is independent of the spread, the slope and the scatter range. The uncertainty of the estimated scatter range is too large for a sample size of $n \leq 5$. Therefore, the parameter cannot be reliably estimated with this method.

According to an efficient test conduction the following aspects should therefore be noted and implemented:

- The mean load level and the slope of the Wöhler curve can be estimated without bias.
- A wide spread $N_{\max}/N_{\min} > 10$ is to be aimed for a good estimation of the slope.
- The scatter range $T_N = N_{90\%}/N_{10\%}$ of the Wöhler curve cannot be reliably estimated with a sample size of $n = 5$.

4.2   Prior Knowledge of the Wöhler Curve Parameters

Prior knowledge of the Wöhler curve parameters is required for an efficient test conduction in determining the Wöhler curve. The prior knowledge can be obtained from different sources. The following section will present some methods which provide substantial elements for this prior knowledge.

In general, an estimate of the characteristic values can always be made from previous tests results. Alternatively, similar tests can be researched in databases and these values can be used as prior knowledge.

The fatigue limit or mean load can be calculated using mathematical estimations. A well-known example for the calculation of the fatigue limit depending on various influencing parameters, such as the mean load, the size influence, the notch condition or surface influences can be found in the FKM guideline [4]. The fatigue limit is estimated on the basis of easily determined material properties. Calculations can also be conducted using the finite element method.

Calculated procedures for determining the slope of the Wöhler curve can be found, for example, in [5,7]. Proposals for an estimated slope can also be found in the FKM guideline [4]. Another possibility for predicting the slope is the concept of the standardised Wöhler curve, as described in [6].

With standardised Wöhler curves, the scatter range of the Wöhler curve can also be reliably predicted. Evaluations of the scatter ranges of Wöhler curves were published in detail in ADENSTEDT [8]. These results can be used for efficient test conduction.

A suggestion for the parameter of the Wöhler curve slope k can be taken from the following Table 2 and for the scatter range from Table 3. If no scatter ranges are known, a scatter range in the direction of load cycles of $T_N \approx 3.25$ can be assumed according to [8,9].

Table 2. Proposal for the slope k of steel [10]

Notched	Welded	Not welded
Not		$k = 15$
Mildly	$k = 3...5$	$k = 5$
Sharply		$k = 3$

Table 3. Proposal for the scatter range T_S of steel [6,8]

Notched	Welded	Not welded
Not		$T_S = 1.10...1.15$
Mildly	$T_S = 1.25...1.45$	$T_S = 1.20...1.30$
Sharply		$T_S = 1.25...1.30$

4.3   Derivation of an Efficient Test Conduction

An efficient method for conducting Wöhler tests will be introduced, in this section. The results regarding the test conduction described in Sect. 4.1 and take prior knowledge into account as presented in Sect. 4.2. The procedure is based on the following basic ideas:

- the test results should confirm the existing prior knowledge,
- prior knowledge is valid until it has been rejected by the experimental test results.

For this purpose, the high cycle fatigue area of the Wöhler curve is divided into five sectors. In general, the high cycle fatigue area starts at $N = 1 \cdot 10^4$ numbers of load cycles [9] at the latest and ends at about $N = 1 \cdot 10^6$ numbers of load cycles. The aim is to have a test result in each of the five sectors. In an experiment, the load S must always be specified and the number of load cycles N is the statistically dependent parameter. Therefore, it is not trivial to construct a test where the number of load cycles covers the specified range. The experimental results should ideally be conducted according to the order shown in Fig. 1. The order was chosen considering the following aspects:

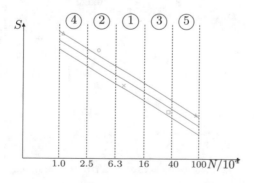

Fig. 1. Dividing the high cycle fatigue area from $N = 1 \cdot 10^4$ to $N = 1 \cdot 10^6$ into five sections

1. The test point is located in the centre of the high cycle fatigue area. This increases the probability that even if the Wöhler curve is poorly estimated, the test point will still be within the high cycle fatigue area.

2. The second test level is above the load level of the first and has a shorter test time than the first test point. This test should be in the second of five sections and thus continue to be as safe as possible in the area of fatigue strength.
3. The third sample is in the fourth section and thus further within a safe section of the high cycle fatigue area.
4. In order to achieve the largest possible spread of the Wöhler curve, the fourth test point is located in the first section and thus as close as possible to the low cycle fatigue area. This sample is chosen in a way that it has a short test time.
5. The fifth sample is close to the transition area, to further maximise the spread. As this is the test point with the longest test time, it should also be estimated best. Therefore, the fifth sample is conducted at the end of the test series.

After each of the conducted tests it will be assessed, whether the result of the experiment applies to the prior knowledge. If exactly $n = 1$ test result is known, then, as shown in the Fig. 2, the test result is compared to the Wöhler curve by checking whether the result lies within or outside the 95% interval of the Wöhler curve:

$$N_1 \in \left(N_{1,2.5\%}, N_{1,97.5\%}\right)$$

If the result is within the given range, the Wöhler curve from the prior knowledge is assumed. The second test level is determined with the prior knowledge. However, if the test result is outside the defined limits, the Wöhler curve is moved parallel through the test result and the corrected prior knowledge is used to determine the next test level, see Fig. 2.

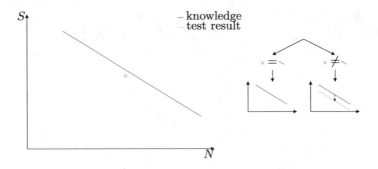

Fig. 2. Testing strategy for $n = 1$ samples

If at least two tests have been conducted at two different load levels, then both the slope k and the mean load level from the test results $\bar{S}$ can be compared with the prior knowledge. The procedure is shown in Fig. 3:

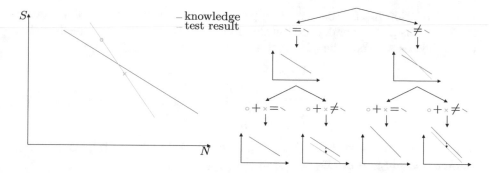

Fig. 3. Testing strategy for $n > 1$ samples

1. Check the slope from the prior knowledge k_0 with the slope k determined from the test results.
2. Verify the mean load level from prior knowledge S_0 with the mean load level $\bar{S}$ determined from the test results. Depending on the prior test decision, the slope from the prior knowledge k_0 or the slope k determined from the regression calculation is chosen for the verification.

Since there are now at least two samples, the parameters determined from the experimental results and the prior knowledge can be tested using a two-sided one-sample t-test. Within the method it is specified that the results are conducted at the significance level of $\alpha = 5\%$. The sample test is divided into three steps [11]:

1. Formulate null hypothesis:
   $H_0 : \mu = \mu_0$
   $H_1 : \mu \neq \mu_0$
2. Calculate test value:
   $t = \sqrt{n} \cdot \frac{\bar{x} - \mu_0}{s}$
3. Test decision:
   Rejection area for H_0: $\left(-\infty, -t_{1-\frac{\alpha}{2};n-1}\right) \cup \left(t_{1-\frac{\alpha}{2};n-1}, \infty\right)$

In the first sample test, the slope is tested. Therfore, the slope k determined from the regression calculation is tested against the slope from the prior knowledge k_0. In order to be able to conduct the sample test, a scatter of the slope s_k is required. This depends on the following factors:

- Sample size n
- Spread $N_{\max}/N_{\min}$
- Slope k
- Scatter range in direction of the load cycles $s_N = \dfrac{\log(T_N)}{2.56}$

Monte Carlo simulations with different parameter variations were conducted to determine the scatter of the slope as a function of the mentioned parameters. The general relationship for the scattering parameter of the slope s_k was determined in Eq. 1. This parameter depends on the slope k itself, the scatter in the direction of load cycles s_N and a further scatter parameter $s(n, N_{max}/N_{min})$:

$$s_k = k \cdot \frac{\dfrac{1 + s_N \cdot s(n, N_{max}/N_{min})}{1 - s_N \cdot s(n, N_{max}/N_{min})} - 1}{\dfrac{1 + s_N \cdot s(n, N_{max}/N_{min})}{1 - s_N \cdot s(n, N_{max}/N_{min})} + 1} \tag{1}$$

The scattering parameter $s(n, N_{max}/N_{min})$ can be tabulated or calculated using the relationship given in Eq. 2:

$$s(n, N_{max}/N_{min}) = \frac{a(n)}{\ln\left(N_{max}/N_{min}\right)} \tag{2}$$

The constant $a(n)$ is only dependent on the sample size n. The characteristic values for this parameter can be taken from the Table 4.

Table 4. Characteristic values for the calculation of the parameter $s(n, N_{max}/N_{min})$, see Eq. 2

$n =$	2	3	4	5	7	10
$a(n) =$	3.256	3.255	3.089	2.911	2.610	2.282

The mean load level $\bar{S}$ is to be tested with the mean load level from the prior knowledge S_0 in a second one-sample test. The load level is calculated depending on the determined slope from the first test decision, see Fig. 3. The individual test points are transformed via the slope k to the level of the mean number of load cycles N_0 from the prior knowledge. The following characteristic values are then used for the test:

$$\mu = \log\left(\bar{S}\right) = \frac{1}{n} \sum_{i=1}^{n} (\log S_i)$$

$$\mu_0 = \log\left(S_0\right)$$

Using prior knowledge, the scatter range in load direction T_S is already defined. For the one sample test this can be used by the following calculation:

$$s_S = \frac{\log(T_S)}{2.56}$$

5   Example of Increasing the Efficiency of Costs and Time When Conducting Wöhler Tests

5.1   Data Basis

The procedure for efficient test conduction is demonstrated by using a concrete test result. The test and its results are shown in the Table 5.

Table 5. Results of a Wöhler test

i	S_i	N_i	i	S_i	N_i
1	114	$3.61 \cdot 10^5$	11	117	$3.41 \cdot 10^5$
2	128	$1.23 \cdot 10^5$	12	128	$7.51 \cdot 10^4$
3	128	$7.45 \cdot 10^4$	13	101	$4.29 \cdot 10^5$
4	114	$4.73 \cdot 10^5$	14	101	$3.55 \cdot 10^5$
5	117	$3.87 \cdot 10^5$	15	101	$6.86 \cdot 10^5$
6	133	$1.75 \cdot 10^4$	16	133	$4.67 \cdot 10^4$
7	133	$3.49 \cdot 10^4$	17	128	$1.28 \cdot 10^5$
8	133	$3.76 \cdot 10^4$	18	117	$3.94 \cdot 10^5$
9	117	$3.17 \cdot 10^5$	19	101	$1.98 \cdot 10^6$
10	107	$5.13 \cdot 10^5$	20	99	$6.94 \cdot 10^5$

A total of $n = 20$ samples were conducted at seven test levels. All test results are shown normalised, where the value $S = 100$ corresponds to the load level at $N_0 = 1 \cdot 10^6$ which in this case is 36% of the tensile strength R_m. The test series shown originates from a $K_t = 1$ sample of a material. The evaluation of the entire test series from Table 5 resulted in the following:

$$k = 10 \tag{3}$$
$$T_S = 1.15 \tag{4}$$
$$S_0 \left(N_0 = 1 \cdot 10^6\right) = 0.36 \cdot R_m = 100 \tag{5}$$

5.2   Evaluation Taking Prior Knowledge into Account

In this test, a material with comparatively few preliminary tests had been examined. Characteristic strength values and Wöhler curve parameters can only be taken from the literature to a limited extent and not from standards. However, the following characteristic values could be derived from the existing preliminary tests:

$$k = 10 \tag{6}$$
$$T_S = 1.15 \tag{7}$$
$$S_0 \left(N_0 = 1 \cdot 10^6\right) = 0.27 \cdot R_m = 75 \tag{8}$$

Procedure for Evaluating the Test Results. The test results from Table 5 should be taken to apply the method described in Sect. 4.3. The following principle is applied in each case:

1. The next test level is calculated by using the prior knowledge or the prior knowledge corrected by the test results.

2. The test level closest to the calculated test level is then selected.
3. If several tests were conducted at the test level, then the test result is taken into account in the order of the tests conducted. This also applies if a test level is selected several times.
4. If the second test level corresponds to the first test level or if no further test results are available, the test level that is second closest to the calculated test level is used as an exception.

In order to better classify the results, different variants were chosen for the evaluation of the test results. In the first part, the correct slope was used and the mean load level was varied. In a second evaluation, the correct mean load level was specified and the slopes varied.

Specification of the Correct Slope and Variation of the Mean Load Level. The prior knowledge used for the tests is shown in Table 6. For variant A, the mean load level was underestimated by 25% which corresponds to the estimation as it would have been conducted in the experiment, refer to Eq. 8. Variant C corresponds to the result as it turned out in the evaluation of all test results.

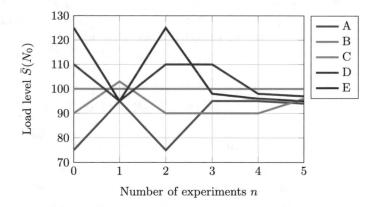

Fig. 4. Evaluated mean load level $\bar{S}$ at $N_0 = 1 \cdot 10^6$ for the prior knowledge ($n = 0$) and after $n = 1...5$ experiments. The prior knowledge for the variants is shown in Table 6.

The results of these evaluations are shown in Fig. 4 and Fig. 5 as a function of the conducted samples. The slope was estimated correctly in each case and was not corrected in any of the cases. In contrast, the estimated mean load level was already corrected after the first test in all cases, with exception of variant C. After the tests were conducted, all test results were between $S = 94$ and $S = 100$. The deviation from the evaluation of all test results was therefore $<10\%$.

Specification of the Correct Mean Load Level and Variation of the Slope. In the second evaluation of the experiment, the slope of the prior

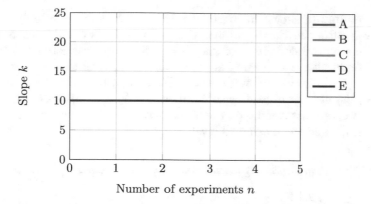

Fig. 5. Evaluated slope k for the prior knowledge ($n = 0$) and after $n = 1...5$ experiments. The prior knowledge for the variants is shown in Table 6.

Table 6. Parameters for the different prior knowledge with a correct slope and a variation of the mean load level

Variant	T_S	N_0	S_0	k_0
A	1.15	$1 \cdot 10^6$	75	10
B	1.15	$1 \cdot 10^6$	90	10
C	1.15	$1 \cdot 10^6$	100	10
D	1.15	$1 \cdot 10^6$	110	10
E	1.15	$1 \cdot 10^6$	125	10

knowledge is varied. The scatter range T_S was chosen so that the scatter range in the load direction $T_N = T_S{}^k$ was kept approximately constant. All variants are shown in Table 7. With variant c, the correct test result was given as prior knowledge as a reference.

Table 7. Parameters for the different prior knowledege with a correct mean load level and a variation of the slope

Variant	T_S	N_0	S_0	k_0
a	1.30	$1 \cdot 10^6$	100	5
b	1.20	$1 \cdot 10^6$	100	8
c	1.15	$1 \cdot 10^6$	100	10
d	1.10	$1 \cdot 10^6$	100	15

The results are shown in Fig. 6 and Fig. 7. The slope was only corrected respectively after the fourth and fifth tests. For variants a and b, each of which underestimated the slope in prior knowledge, the slope was estimated too large in after conducting $n = 5$ samples. The mean load level at the reference load cycle number of $N_0 = 1 \cdot 10^6$ was estimated too high. In contrast to that, due to the slope, the number of load cycle numbers for $N = 1 \cdot 10^4$ was estimated too low. The two variants c and d on the other hand returned the correct results for both the slope and the mean load level.

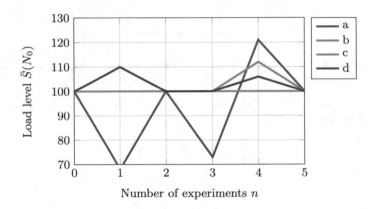

Fig. 6. Evaluated mean load level $\bar{S}$ at $N_0 = 1 \cdot 10^6$ for the prior knowledge ($n = 0$) and after $n = 1...5$ experiments. The prior knowledge for the variants is shown in Table 7.

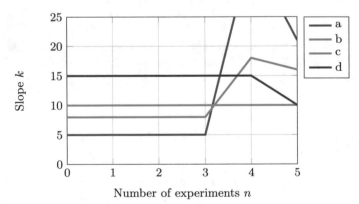

Fig. 7. Evaluated slope k for the prior knowledge ($n = 0$) and after $n = 1...5$ experiments. The prior knowledge for the variants is shown in Table 7.

5.3 Concluions

Using this example, good results can already be achieved for the experiment with a reduced sample size. Even with incorrectly assumed prior knowledge, it was possible to correct this circumstance and reliable results were already achieved after $n = 5$ samples, which came close to the test results with $n = 20$ samples. It is obvious that the mean load level is corrected much earlier with a wrong estimation than it is the case with a wrong estimated slope.

In this first example, the saving on specimens was 75%. For the number of load cycles, the saving was even between 77% and 92%. Good results could be achieved with the method when the slope was estimated properly. If the slope is estimated inaccurately, a larger sample size is needed.

The method focuses on the determination of the parameters for lifetime models. There are further uncertainties in the determination of product reliability that have not been addressed so far. This is the case when for example several failure mechanisms of a system are critical or influences of tolerances (geometry, surface quality, ...) are present.

6 Outlook

In a first example, it was shown that with the method for efficient test conduction taking prior knowledge into account, it is quite possible to significantly reduce the sample size. Good prior knowledge about the parameters of the lifetime model is an advantage. Further research will target the following three focus points:

- Further statistical studies on the method to demonstrate validity,
- Procedure for further validation of the lifetime model for $n > 5$ samples,
- Transferability of the method to other failure mechanisms.

References

1. Bargel, H., Schulze, G.: Werkstoffkunde. Springer, Heidelberg (2013)
2. Jakob, F.: Nutzung von Vorkenntnissen und Raffungsmodellen für die Zuverlässigkeitsbestimmung. In: Berichte aus dem Institut für Maschinenelemente. PhD thesis (2017)
3. Müller, C.: Zur statistischen Auswertung experimenteller Wöhlerlinien. PhD thesis (2015)
4. FKM-Richtlinie Rechnerischer Festigkeitsnachweis für Maschinenbauteile aus Stahl, Eisenguss und Aluminiumwerkstoffen, 6. überarbeitete Auflage (2012)
5. Bergmann, J., Thumser, R.: Synthetische Wöhlerlinien für Eisenwerkstoffe. In: Forschungsbericht, p. 249 (1999)
6. Haibach, E.: Betriebsfestigkeit: Verfahren und Daten zur Bauteilberechnung. Springer, Berlin (2006)
7. Hück, M., Thrainer, L., Schütz, W.: Berechnungen von Wöhlerlinien aus Stahl. In: Bericht der Arbeitsgemeinschaft Betriebsfestigkeit, Stahlguß und Grauguß (1981)
8. Adenstedt, R.: Streuung der Schwingfestigkeit. PhD thesis (2001)

9. DIN 50100: Schwingfestigkeitsversuch – Durchführung und Auswertung von zyklischen Versuchen mit konstanter Lastamplitude für metallische Werkstoffproben und Bauteile (2015)
10. Radaj, D., Vormwald, M.: Ermüdungsfestigkeit: Grundlagen für Ingenieure. Springer, Heidelberg (2010)
11. Einbock, S.: Statistik für Ingenieure und Naturwissenschaftler (mit Excel): endlich verständlich! (2018)

Tuning and Emulation of Mechanical Characteristics – Tunable Mounts and a Mechanical Hardware-in-the-Loop Approach for More Efficient Research and Testing

Jonathan Millitzer[✉], Jan Hansmann, Giovanni Lapiccirella, Christoph Tamm, and Sven Herold

Fraunhofer Institute for Structural Durability and System Reliability LBF, 64289 Darmstadt, Germany
Jonathan.Millitzer@lbf.fraunhofer.de

Abstract. Numerical simulations offer a wide range of benefits, therefore they are widely used in research and development. One of the biggest benefits is the possibility of automated parameter variation. This allow testing different scenarios in a very short period of time. Nevertheless, physical experiments in the laboratory or on a test rig are still necessary and will still be necessary in the future. The physical experiments offer benefits e.g. for very complex and/or nonlinear systems and are needed for the validation of numerical models.

Fraunhofer LBF has developed hardware solutions to bring the benefit of rapid and automated parameter variation to experimental environments. These solutions allow the tuning and emulation of the mechanical properties, like stiffness, damping and eigenfrequencies of structures.

The work presents two approaches: First a stiffness tunable mount, which has been used in laboratory tests in the field of semi-active load path redistribution. It allowed the researcher to test the semi-active system under different mechanical boundary conditions in a short period of time. Second, a mechanical Hardware-in-the-loop (mHIL) approach for the NVH development of vehicles components is presented. Here a mHIL-system is used to emulate the mechanical characteristics of a vehicle's body in white in a wide frequency range. This allows the experimental NVH optimization of vehicle components under realistic boundary conditions, without actually needing a (prototype) body in white.

Keywords: Uncertainty · Smart dynamic testing · Tunable stiffness · Mechanical Hardware-in-the-loop

1 Introduction

1.1 Uncertainties in Early Phases of the Product Development

Uncertainty is considered a potential deficiency in any development phase of a technical system that has arisen due to a lack of information and/or knowledge. The behavior of a

P. F. Pelz and P. Groche (Eds.): ICUME 2021, LNME, pp. 129–144, 2021.
https://doi.org/10.1007/978-3-030-77256-7_12

mechanical system is not deterministic due to uncertainty in the system and its environment; i.e. the system behavior cannot be clearly determined [1]. *Uncertainty Quantification* (UQ) deals with the identification, quantification and reduction of uncertainty in models, experiments and their effects on selected targets of a technical system [2]. Using UQ errors are usually classified as being either random error (precision) or systematic error (bias). There are three kinds of uncertainties. Parameter uncertainty describes an uncertainty associated with the parameters of a numerical model. An error describes a recognizable deficit in the numerical simulation of a system. Model uncertainty describes the accuracy with which a numerical model depicts reality.

In many studies, the variations in the properties of a technical system are explored according to the procedure of the *Uncertainty Quantification*. Research focuses on the optimization of a technical system taking parameter uncertainty into account, and the determination of variations in system behavior due to parameter uncertainty in the system properties [3]. Variations in geometric, mechanical, electrical and material properties such as the length and thickness of a beam, the fuselage length and width of an aircraft, the mass and the damping coefficient of a vehicle body etc. are described either with intervals or with distribution functions such as normal and gamma distributions. Probabilistic simulation methods such as Monte-Carlo-Simulations are frequently used to determine the influences of parameter uncertainty on a system property. In most studies, the intervals and distribution functions used to describe the variations in system properties and parameters are based on the assumptions of the respective authors.

The numerical models of the investigated systems are often analytically well known, which allows a comparison between a system optimization with probabilistic and non-probabilistic simulation methods. However, the validation of these models with experimental data is usually highly time-consuming or even impossible and a criterion for adequate and sufficient prediction is not defined or proposed. When it comes to experimental validation of numerical models for vibroacoustic applications especially two tasks within the experiment are often time-consuming:

1) The variation of mechanical characteristics like stiffness, damping or elastomer-like characteristics. Often different parts (e.g. rubber mounts with different stiffness and loss-factor) are needed to realize different characteristics and they need to be exchanged in the test setup.
2) The realization of adequate mechanical boundary conditions in case single components or subsystems are investigated. Therefore, often auxiliary constructions have to be designed and manufactured. E.g. auxiliary constructions are used to realize a desired dynamic stiffness as boundary conditions e.g. for active vibration reduction systems in a marine [4] and an automotive application [5].

Smart dynamic testing is an approach that can be used to make these experimental tests more efficient.

1.2 Smart Dynamic Testing

The basic idea behind *smart dynamic testing* is to make physical experiments in research and development as straightforward as possible for the researcher. This means the focus should be on the device under test (DUT) and the respective research objective. Additional efforts, e.g. auxiliary constructions as mechanical boundary conditions or the exchange of parts of the test setup to realize different configurations should be minimized. To make dynamic tests smarter and reduce time-consuming tasks there are two main approaches:

Rapid Parameter Variation of Mechanical Characteristics. Whether it comes to the mechanical boundary conditions of the DUT or mechanical characteristics like stiffness or damping values from parts of the DUT it is often beneficial if there is the opportunity to do a parameter variation on these values during experiments. Therefore, tools like tunable mounts, tunable vibration absorbers (TVA) or the later presented mechanical Hardware-in-the-loop system can be used. A brief overview is given in [6, 7] gives an application example of tunable mounts for laboratory tests in the field of uncertainty research.

Active Emulation of Mechanical Boundary Conditions. Typically, subsystems of a vehicle, e.g. suspension systems or drive trains, are developed and tested by suppliers. In the vehicle, the body in white defines the mechanical boundary conditions of these subsystems. Especially in early stages of the development process, the body in white of the vehicle is often not available. Nevertheless, the correct boundary conditions are crucial for the vibroacoustic development [8] as well as for durability testing [9]. Instead of designing and building auxiliary constructions with a desired dynamic characteristic, these characteristic can be emulated by an active system. This emulation is often referred to (mechanical) Hardware-in-the-loop testing. Different application examples can be found in [10–16].

Figure 1 shows an exemplary test setup for noise, vibration and harshness (NVH) investigations on an electric drivetrain, which implements these two approaches. The drivetrain with its subframe (1) and the rubber mounts is the DUT.

Goal of this setup is to test the DUT under mechanical boundary conditions close to the actual installation situation in the vehicle in order to reduce uncertainties in the NVH development process. From a vibroacoustic point of view, the installation setup is characterized by its mechanical boundary conditions i.e. the dynamic stiffness of the body in white at the mounting points (reference characteristic). The shown setup allows the active emulation of this reference characteristic through four mechanical Hardware-in-the-loop interfaces (4).

Further, a tunable vibration absorber (3) is shown as an example for rapid parameter variation in physical test setups. It allows to test different vibration absorber configurations (tuning frequency and damping) without actually exchanging the vibration absorber in the test setup. Nevertheless the focus of this paper is the mHIL-system for the emulation of the mechanical boundary conditions.

Focusing on the reduction of uncertainties during early stage development and hence the correct emulation of the boundary conditions, the requirements depend on the actual research objective. Core requirements are the number of mechanical degrees of freedom (DOF) and the frequency range in which the emulated characteristics should be close to the reference characteristics.

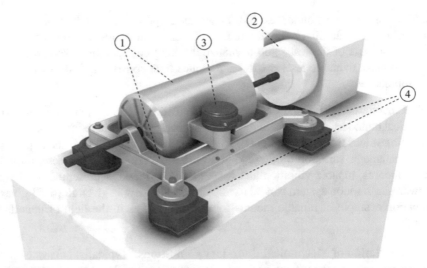

Fig. 1. Schematic test setup for NVH investigations on an electric drivetrain with: (1) drivetrain with subframe as DUT, (2) driven machine, (3) tunable vibration absorber as an example for rapid parameter variation and (4) four mHIL-interfaces for active emulation of the DUTs mechanical boundary conditions.

In this paper, the results of a 1-DOF active emulation of the dynamic stiffness are shown and discussed. The requirements were chosen according to typical issues in the field of automotive NVH development. The dynamic stiffness range reaches from 500 N/mm (rubber mount) up to 10.000 N/mm (body in white) and the considered frequency range reaches from 0 Hz to 1 kHz.

2   Active Dynamic Stiffness Emulation by the Mechanical Hardware-in-the-Loop Approach

2.1   The Mechanical Hardware-in-the-Loop (mHIL) System

Figure 2 gives an overview of the main components of the mHIL-system used to emulate the dynamic stiffness which defines the mechanical boundary condition for the DUT. The dynamic stiffness is defined by the user in a numerical simulation model, e.g. a finite element model. This numerical simulation model is converted into a numerical real-time capable model, the target model. Based on this target model the mHIL-interface is controlled using an adaptive controller which minimizes the difference between the target behavior and the actual behavior measured between the mHIL-interface and the DUT.

Numerical target models of the full system or of individual components of the system are set up with common analytical or numerical tools. Usually the Finite Element method in combination with suitable model order reduction methods are used in simulation [17]. In the dynamic testing environment, the simulation models have to be solved in real-time, i.e. the simulation has to meet requirements regarding timeliness,

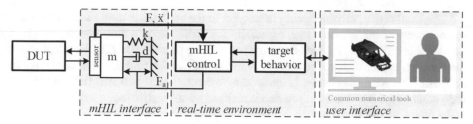

Fig. 2. Overview of the mHIL-system.

simultaneousness and responsiveness. The mandatory model properties are dependent on the specific characteristics of the mHIL-system, e.g. frequency range, computational power of the real-time simulator.

The mHIL-interface demonstrator was designed to allow 1-DOF dynamic stiffness emulation for typical scenarios in automotive testing. [6] shows an example where this interface was used for the characterization of automotive shock absorbers for different installation scenarios. Figure 3 shows the basic topology of the interface, which is mounted to a surface and can be connected to the DUT at its moved mass.

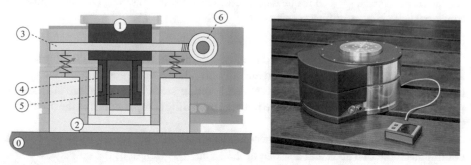

Fig. 3. Topology of the mHIL-interface with: The mounting surface (0), the moved mass (1), the housing (2), the tunable spring (3), the voice-coil actuators coil (4) and magnet (5) and the worm gear and electric motor for stiffness tuning (6) (left). Picture of the interface (right).

When it comes to the actual mechanical design of the mHIL-interface it is beneficial to have a low moved mass, a small installation space and no mechanical resonances in the frequency range in which the target behavior should be emulated. The focus in the presented work was on the emulation capability of the whole mHIL-system, the requirements "installation space" and "low moved mass" of the interface had no priority for the presented design and are to be further optimized with respect to a distinct application.

Compared to [8] a voice coil actuator (VCA) is used instead of a piezo-actuator. Goals were to keep the costs low, have the possibility to use off the shelf components (VCA and HiFi amplifiers), and have a mechanical robust design for the use in the test field.

The semi-active, tunable spring in parallel to the actuator is used to keep the force requirements of the actuator low, especially for applications with higher static loads. The tunable spring was realized using the mechanism presented in [18]. Figure 4 shows the basic principle, which is to change to free length φ of a ring segment to change its stiffness. This stiffness change is realized by the rotation of two structures relative to each other. In the interface, this rotation is realized by an electric motor using a planetary and a worm gear in serial.

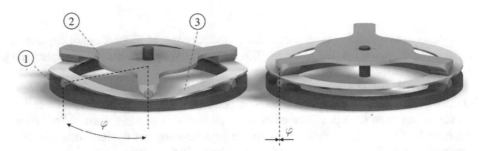

Fig. 4. Principle of semi-active stiffness tuning, with the first structure (1), the second structure (2), the spring element (3), and the tuning angle φ. Lowest stiffness setting (shown in deflected condition) (left). Highest stiffness setting (right).

The tunable stiffness was designed to have a minimum stiffness of 400 N/mm. This leads to an inner diameter of 160 mm and an outer diameter of 200 mm for the spring element, which was made out of high strength spring steel. The maximum stiffness is mainly defined through the compliance of the surrounding components like the interfaces housing.

The chosen VCA (Type "BEI Kimco, LA30-43-000A") has a peak force of 445 N, a continuous force of 185 N and a stroke of ±12 mm. The weight of the coil is 726 g and the weight of the magnet assembly is 1.9 kg. To keep the moved mass of the interface low the magnet is attached to the housing and the coil is part of the moved mass. The total mass of the interface is about 20 kg, whereas the moved mass is between 4.3 kg (highest stiffness setting) and 4.8 kg (lowest stiffness setting).

For a first mechanical characterization, the interface was excited with a white noise actuator current and the acceleration on the moved mass was measured. Figure 5 shows the magnitude response of the H1 transfer function estimate between the voice coil's current and the acceleration at the moved mass of mHIL-interface for different stiffness settings.

For the lowest stiffness setting the resonance of the interfaces moved mass is at 49.8 Hz and for the highest stiffness setting the resonance can be estimated around 590 Hz. Considering the mHIL-interface as a simplified spring-mass-damper system this corresponds to a tuning range from 470 N/mm to 59.100 N/mm for the stiffness element.

Further, there are effects which are considered to be caused by structural resonances of components of the interface itself, e.g. at 290 Hz, between 550 Hz and 730 Hz and

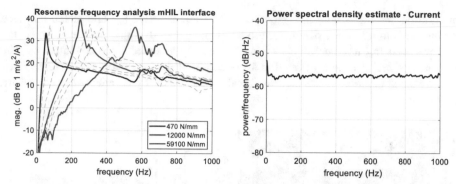

Fig. 5. Characterization results of the uncontrolled mHIL-interface for different settings of the tunable stiffness (left). Bandlimited white noise excitation of the voice coil's current used for characterization (right).

at 935 Hz. The cause of these effects is subject of ongoing research with the goal to eliminate these parasitic effects in a next iteration of the interface design.

An adaptive controller is used to adjust the mHIL-interface's movement based on the measured reaction force between the DUT and the mHIL-interface's mounting surface. Preferring an adaptive controller, i.e. a real-time estimator [19] over a fixed-parameter control approach, is motivated by two major facts: Firstly, the dynamics of the control path might be hard to model as they incorporate the conflated dynamics of the DUT, the mHIL-interface and the target behavior and thus an experimental modelling approach is highly advisable. Secondly, due to its iterative adaptation process the adaptive controller is able to minimize the controller's objective function even if slight deviations within the system occur. For an overview on possible mHIL control approaches, the reader is kindly referred to [20, 21].

2.2 Test Setup and Test Cases

Experimental investigations were carried out and focused on a preliminary study assessing synthetic test cases based on the emulation of the principle dynamic stiffness behavior present in an exemplary automotive application (c.f. Sect. 2.3). Envisioning a mechanical HIL test scenario, a substructure of a passenger car's chassis will be connected to the mHIL-interface, which in turn emulates the mechanical boundary condition the substructure would have experienced when installed into the car's chassis. Figure 6 shows the experimental test setup. The mHIL-interface demonstrator is mounted on a rigid supporting structure. An electrodynamic shaker is connected to the mHIL-interface through an impedance measurement head. The electrodynamic shaker introduces a broadband colored noise mechanical excitation into the interface, which in turn emulates the desired mechanical boundary condition by measuring the reaction force at the shaker's mounting position and by controlling the interface's movement (i.e. acceleration). Figure 6 also illustrates the test setup including the digital signal processing chain. The mHIL-interface is represented by the lumped parameter model including the tunable stiffness element k_i, a presumed viscous damping c_i, and the mHIL-interface's moved mass m_i.

In addition, the mHIL-interface incorporates a force actuator F_a. The mHIL-interface is connected to a lumped parameter model of the electrodynamic shaker given by the shaker's mass m_s, it's mechanical stiffness k_s and an assumed viscous damper c_s. The excitation signal is used for both, the introduction of the excitation shaker force F_s as well as the generation of the control signal F_a computed by an adaptive feedforward controller.

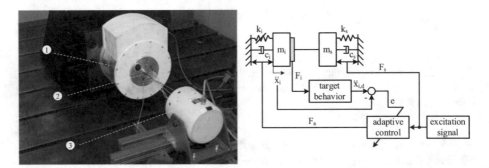

Fig. 6. Mechanical HIL test setup comprising the mHIL-interface demonstrator (1) and a primary mechanical excitation by means of an additional electrodynamic shaker (3). An impedance measurement head (2) captures interface forces and accelerations (left). Schematic illustration of the test setup including the adaptive controller and the numerical target model behavior (right).

Once an excitation force F_s has been introduced into the test rig by means of the electrodynamic shaker, the reaction force F_i is fed into a numerical model of the target mechanical behavior. The desired interface acceleration $\ddot{x}_{i,d}$ is derived from the measured interface force F_i and the numerical model of the target behavior. It is then compared to the actual, measured interface acceleration $\ddot{x}_i$. The deviation between the desired acceleration $\ddot{x}_{i,d}$ and the measured acceleration $\ddot{x}_i$ serves the computation of an error signal e. The adaptive control algorithm's objective is to minimize the Least-Mean-Squares error signal by means of an overlap-save frequency-domain Newton's algorithm [22]. Making use of feedforward topology, the controller is unconditional stable ensuring a bounded-input-bounded-output stability.

Table 1 shows the test cases that have been carried out with this setup. The motivation behind the selected test cases was to have a first proof for the mHIL-system to emulate different dynamic systems. Within this, the ability to vary the (static) stiffness, insert resonance effects and vary their damping are evaluated.

2.3   Test Results

Due to their high comparability, detailed experimental results are presented for the exemplary test case B-2 within the beginning of this section. A final assessment of all test cases is given at the end of this section (Fig. 10 and Fig. 11).

Experimental investigations are carried out for a stochastic excitation signal whereas bandlimited red noise with a maximum bandwidth of 1 kHz is introduced by the electrodynamic shaker. Figure 7 (left) shows the frequency response function of the numerical

Table 1. Synthetic test cases.

Test case	Quasistatic stiffness (N/mm)	1st resonance frequency (Hz)	Modal damping (%)	2nd resonance frequency (Hz)
A-1 to A-3	500 to 10,000	–	–	–
B-1 to B-3	2,250	300	1 to 100	–
C-1 to C-3	2,250	100 to 500	10	–
D-1 to D-3	2,250	190	Approx. 10	327 to 700

target behavior for the test case B-2. The frequency response function shows a single resonance frequency at 300 Hz with a modal damping of 10%. The computed desired acceleration $\ddot{x}_{i,d}$ and the measured acceleration signal $\ddot{x}_i$ serve the calculation of the error signal e.

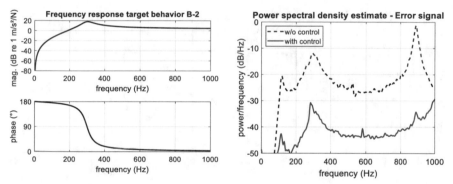

Fig. 7. Frequency response function of the numerical target behavior B-2 (left) and power spectral density estimate of the deviation between the actual mHIL-interface acceleration and the desired acceleration derived from the numerical model of the target behavior (right).

Figure 7 (right) illustrates a power spectral density (PSD) estimate of the error signal e. Deviations between the target and the desired acceleration with respect to the bandlimited red noise excitation are mainly observed at the three distinct resonance frequencies of the test setup. Two of the resonance frequencies originate from the mechanical test setup itself. The first resonance frequency at approx. 160 Hz originates from the mechanical resonance frequency of the mHIL-interfaces tunable stiffness k_i and the attached masses of the mHIL-interface m_i and the electrodynamic shaker m_s, respectively. The highest resonance frequency at approx. 890 Hz is observed due to a parasitic effect within the test setup and most likely results from a leak in stiffness of the used connection rod between the electrodynamic shaker and the mHIL-interface. This connection will be improved in further investigations. In addition, the third resonance frequency at 300 Hz within the measured PSD of the error signal e is caused by the numerical model of the target behavior. Hence, it should be noted that the adaptive controller has to deal with a control system whose dynamic properties originate from both the actual physical test rig setup as well as the dynamics given by the numerical model of the target behavior.

Once the adaptive controller is enabled and a steady state within the iterative controller adaptation process has been reached, the PSD of the error signal e is significantly lowered. A broadband reduction of -20 to -40 dB can be observed within the whole frequency band up to the target operational frequency of 1 kHz.

Reducing the deviation between the actually measured interface acceleration $\ddot{x}_i$ and the desired interface acceleration $\ddot{x}_{i,d}$ depicts that the mHIL-interface's mechanical behavior follows the target behavior given by the implemented numerical model. This also gets obvious considering the target and the measured acceleration signals in time domain. Figure 8 shows an exemplary section of the time series of the initial state (left). Disabling the control signal of the mHIL-interface, a significant deviation between the target acceleration and the actual measured interface acceleration can be observed. Once the adaptive controller is enabled only small deviations occur between the target behavior and the actual measured interface acceleration (c.f. Fig. 8, right).

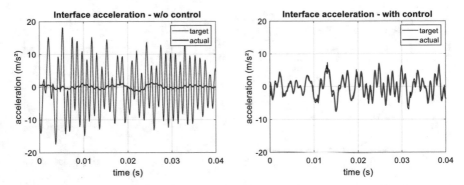

Fig. 8. Time series of the desired mHIL-interface's acceleration $\ddot{x}_{i,d}$ computed by the numerical model of the target behavior and the actual measured interface acceleration $\ddot{x}_i$.

Figure 9 shows the PSD estimates for both the mHIL-interface's acceleration $\ddot{x}_i$ (left) and the measured reaction force F_i (right). Again, significant changes can be observed, once the mHIL-interface's adaptive controller has been enabled and has reached steady state. For the uncontrolled case, the mHIL-interface's acceleration PSD estimate is mainly dominated by the resonance frequencies at approx. 160 Hz and at approx. 890 Hz resulting from the physical test setup. This behavior significantly changes once the mHIL-interface's controller is enabled. Here, the measured interface's acceleration PSD estimate is dominated by the resonance frequency at 300 Hz originating from the numerical model of the target behavior B-2. Changes within the PSD estimate can also be observed for the measured reaction force's PSD estimate whereas changes mainly occur within the frequency range above 100 Hz. A considerable reduction of the reaction force F_i for the controlled case is mainly observed at the parasitic resonance frequency at 890 Hz.

In order to assess the conflated behavior that results from both a change for the mHIL-interface's acceleration $\ddot{x}_i$ as well as the interface's reaction force F_i, the dynamic

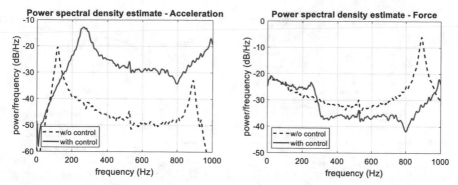

Fig. 9. Power spectral density estimate of the measured mHIL-interface's acceleration $\ddot{x}_i$ (left) and reaction force F_i (right). The black dashed line depicts the uncontrolled case (i.e. no control signal is fed to the mHIL-interface's actuator). The solid red line illustrates the measured steady state behavior.

stiffness frequency response has been calculated by means of both measured signals. Assuming the measurement noise to be uncorrelated to the input signals, the H1 method has been used for deriving the dynamic stiffness frequency response functions. Figure 10 and Fig. 11 show the estimated frequency response functions for the test cases A and B as well as for the test cases C and D, respectively. Due to a leak in coherence within the frequency range below 70 Hz, which is caused by an impropriate signal-to-quantization noise ratio of the analog-to-digital-converter for the acceleration measurement channel, the H1 transfer function estimate in the lower frequency range is shown only for the sake of completeness and has to be considered untrustworthy.

The test cases A-1 to A-3 (c.f. Fig. 10) show a varying quality in emulating the mechanical characteristics given by the numerical target behavior model. For the lowest stiffness setting (c.f. test case A-1, Table 1) a good emulation is achieved within the frequency range from approx. 100 Hz to 420 Hz. Deviations mainly occur in the higher frequency range above 420 Hz. For both test cases A-2 and A-3, the controlled mHIL-interface shows a sufficient performance, except for the frequency range at approx. 890 Hz. For the test cases B-1 to B-3, Fig. 10 illustrates the good performance of the controlled mHIL-interface up to the target frequency range of 1 kHz. Slight deviations occur for the test case B-1. Here, the amplitude response shows an error of a factor of five within the sharp resonance frequency (modal damping of only 1%) implemented by the numerical model of the target behavior B-1.

Figure 11 shows the exemplary results obtained for the test cases C and D (c.f. Table 1). The mHIL system shows a good performance for the test cases C. Hence, the controlled interface is able to emulate a shift in the resonance frequency for the considered test cases. Again, small deviations are observed in the frequency range at approx. 890 Hz. Introducing a second resonance frequency in the test cases D, the mHIL-interface is also able to emulate the desired mechanical behavior based on the numerical target model with slight restrictions. For the test case C-3 and D-2, slight deviations occur in the frequency range at approx. 890 Hz due to the aforementioned reasons.

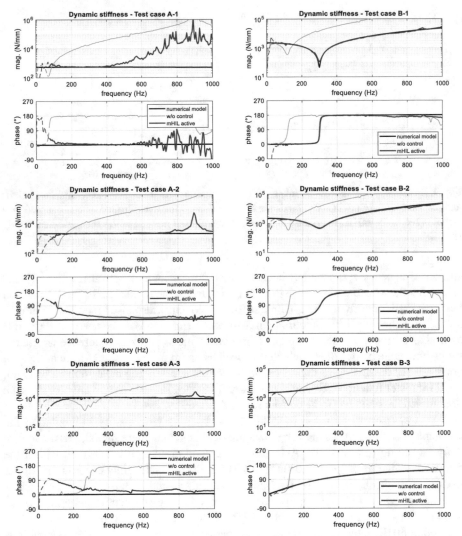

Fig. 10. Conflated behavior of the controlled mHIL-interface for test cases A (left, top to bottom) and B (right, top to bottom) evaluated by means of a H1 transfer function estimate of the dynamic stiffness computed from the measured interface's acceleration $\ddot{x}_i$ and reaction force F_i.

3 Discussion

The emulation of an ideal dynamic stiffness, the emulation of a single resonance frequency with varying modal damping and resonance frequency, and the emulation of a multi-resonant mechanical behavior (c.f. Table 1) have been demonstrated successfully. To further illustrate this, Fig. 12 illustrates the intended operating range of the controlled mHIL-interface with the shown test cases. The validations of the quasistatic stiffness ranges above 10.000 N/mm and the higher frequencies above 1 kHz are subject to current research activities.

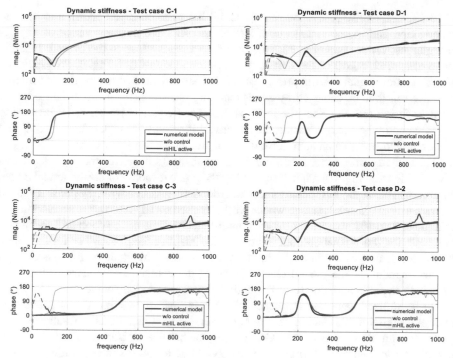

Fig. 11. Conflated behavior of the controlled mHIL-interface for exemplary test cases C (left) and D (right) evaluated by means of a H1 transfer function estimate of the dynamic stiffness computed from the measured interface's acceleration $\ddot{x}_i$ and reaction force F_i.

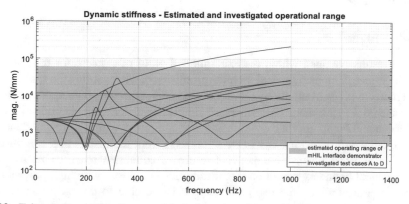

Fig. 12. Esitmated operational range (shaded area) and investigated operational range of the mHIL-interface demonstrator.

In the investigated operating range, limitations of the current setup mainly occur in the lower frequency range below 70 Hz. Here, performance limitations are mainly due to an inappropriate data acquisition setup of the digital signal processing chain. Even though the utilized dSPACE DS2004 analog-to-digital conversion hardware offers a resolution of 16 bit, an effective resolution of only 12 bit has been observed taking into account quantization and measurement noise. Hence, the effective dynamic range of analog-to-digital converter is approx. 72.2 dB, which is already close to the overall dynamic range of approx. 50 dB of the measured acceleration signal (c.f. Fig. 9, PSD estimate of acceleration signal). An improvement of the performance of the controlled mHIL-interface can thus be achieved by either reducing measurement noise, by introducing a sensor fusion technique for the lower and higher frequency range, or by increasing the dynamic range (i.e. increasing the effective bit resolution) of the analog-to-digital converter incorporating an alternative digital signal processing hardware.

The limited performance for the emulation of an ideal stiffness element (i.e. test cases A, c.f. Table 1 and Fig. 10) might be caused by the phase response of the numerical model for the target behavior. Introducing a second order derivative behavior, the phase response shows a +180° phase advantage whenever the desired mHIL-interface's acceleration is computed. Taking into account the minimum reaction time of the conflated secondary path behavior, the Wiener optimal solution of the adaptive feedforward control problem might thus be non-causal in the higher frequency range. This issue might be addressed by further reducing the overall reaction time of the conflated secondary path behavior, which requires an increase of the digital signal processing frequency fairly above the so far utilized sampling frequency of 10 kHz, the application of fast-conversion analog-to-digital and digital-to-analog converters as well as tailored anti-aliasing and reconstruction filters respectively. Considering the hardware effort required to achieve an appropriate control performance for the synthetic test cases incorporating an ideal stiffness element (i.e. test cases A), an end users survey has to reveal the importance of these test cases. Physical equivalent test cases are deemed to be occasional and found e.g. in the development of small optical instruments or within the semiconductor, MEMS, or sensor industry.

4 Conclusion

The test results obtained within the experimental proof of concept for the controlled mHIL-interface demonstrator depict a quite promising result regarding the application of the mHIL technology in automotive NVH development, especially if uncertainties should be considered. In accordance to the increased interest in uncertainty quantification, there exists a demand for appropriate experimental test equipment within structural dynamics. As uncertainties in structural dynamics often arise from the installation conditions of mechanical substructures and components, future test equipment must be able to emulate the mechanical boundary condition (i.e. installation condition) with high precision and bandwidth. Furthermore, the test equipment must be able to change its mechanical behavior ensuring ease of use.

Within this paper, a mHIL-system demonstrator was experimentally validated, demonstrating the highly automatable capability to change a mechanical boundary condition. It is based on both the application of a tunable mechanical stiffness element as well as the incorporation of a numerical model of a target mechanical behavior in an adaptive control loop. Investigations for different synthetic test cases illustrate the capabilities of emulating mechanical properties up to a frequency of 1 kHz and thus enabling experimental probabilistic analyses within the field of uncertainty quantification.

As a next step the mHIL-system demonstrator should be further developed towards industrial application, whereat a close cooperation with industrial partners is necessary.

Acknowledgements. Parts of the work have been funded by the Fraunhofer research project »Light Materials 4 Mobility LM4M« and the Fraunhofer technology transfer program »AHEAD« [23]. The financial support is gratefully acknowledge.

References

1. Oberkampf, W.L., Deland, S.M., Rutherford, B.M., Diegert, K.V., Alvin, K.F.: Error and uncertainty in modeling and simulation. Reliab. Eng. Syst. Saf. **75**(3), 333–357 (2002)
2. Smith, R.C.: Uncertainty Quantification: Theory, Implementation, and Applications. Bd. 12. SIAM (2013)
3. Platz, R., Ondoua, S., Enss, G.C., Melz, T.: Approach to evaluate uncertainty in passive and active vibration reduction. In: Model Validation and Uncertainty Quantification, Volume 3: Proceedings of the 32nd IMAC, A Conference and Exposition on Structural Dynamics, pp 345–352. Springer International Publishing (2014)
4. Matthias, M., Friedmann, A., Koch, T., Drögemüller, T.: Active mounts for marine application; the BMBF research project "active aggregate mounts to reduce structure vibrations and structure-borne sound transmission. In: Proceedings of SPIE (2006)
5. Kraus, R.: Auslegung und Umsetzung eines Labordemonstrators zur aktiven Lagerung im Fahrwerksbereich eines Kraftfahrzeugs. Diplomarbeit, Fachhochschule Gießen-Friedberg (2010)
6. Hansmann, J., Millitzer, J., Rieß, S., Balzer, L.: Symbiose virtueller und experimenteller Methoden für effizienteres Testen und Entwickeln. In: Experten-Forum Powertrain: Simulation und Test 2019. Springer Vieweg, Deutschland, Wiesbaden (2020). ISBN: 978-3-658-28706-1
7. Gehb, C.M., Platz, R., Melz, T.: Two control strategies for semi-active load path redistribution in a load-bearing structure. Mech. Syst. Sig. Process. **118**, 195–208 (2019). https://doi.org/10.1016/j.ymssp.2018.08.044
8. Mayer, D., Jungblut, T., Wolter, S., Millitzer, J.: Hardware-in-the-loop test environments for vibration control systems. In: NAFEMS Seminar Practical Aspects of Structural Dynamics, Deutschland, Wiesbaden (2015)
9. Landersheim, V., Möller, R., Dittmann, K.-J.: Ableitung Eines Vereinfachten Ersatzversuchs Für Pkw-Integralträger Mittels Numerischer Schädigungsbewertungen. In: Tagung des DVM Arbeitskreises Betriebsfestigkeit Potenziale im Zusammenspiel von Versuch und Berechnung in der Betriebsfestigkeit Volume: 43, Steyr, Österreich (2016)
10. Plummer, A.R.: Model-in-the-loop testing. Proc. Inst. Mech. Eng. Part I J. Syst. Control Eng. **220**, 183–199 (2006)
11. Wolter, S., Franz, R., Jungblut, T., Möller, R., Bruder, T.: Aktive Anbindungsimpedanzen: Ein Anwendungsszenario echtzeitfähiger Lagermodelle. DVM-Workshop Elastormerbauteile, Weinheim (2011)
12. Facchinetti, A., Bruni, S.: Hardware-in-the-loop hybrid simulation of pantograph–catenary interaction. J. Sound Vib. **331**(12), 2783–2797 (2012)
13. Atzrodt, H., Mayer, D., Koch, T., Vrbata, J.: Development of a multiaxial test environment for highly dynamically loaded mechatronic components. In: Symposium on Structural Durability SoSDiD 2017, Deutschland: Darmstadt, p. 15 (2017)
14. Bartl, A., Karamooz Mahdiabadi, M., Rixen, D.J.: Conception of a noise and vibration hardware-in-the-loop test bed. In: Proceedings of IMAC, a Conference and Exposition on Structural Dynamics (2017)

15. Traphöner, P., Olma, S., Kohlstedt, A., Jäker, K.-P., Trächtler. A.: Universelle Entwicklungs- und Prüfumgebung für mechatronische Fahrzeugachsen. Wissenschaftsforum Intelligente Technische Systeme (WInTeSys), pp. 11–12, May 2017 (2017)
16. Olma, S., Kohlstedt, A., Traphöner, P., Jäker, K.-P., Trächtler, A.: Observer-based nonlinear control strategies for Hardware-in-the-Loop simulations of multiaxial suspension test rigs. Mechatronics **50**, 212–224 (2018)
17. Antoulas, A.C.: Approximation of large-scale dynamical systems. In: Advances in Design and Control. Society for Industrial & Applied Mathematics, US (2005)
18. Hansmann, J., Kaal, W., Seipel, B., Melz, T.: Tuneable spring element for an adaptable vibration absorber. In: ATZ Worldwide, no. 03, pp. 36–39 (2012)
19. Åström, K.J., Wittenmark, B.: Adaptive control. Courier Corporation (2013)
20. Millitzer, J., Ranisch, C., Tamm, C., Bartolozzi, R., Melz, T.: An approach for the design of a mechanical Hardware-in-the-Loop interface. In: Symposium für Smarte Strukturen und Systeme 4SMARTS, Deutschland, Braunschweig (2017)
21. Millitzer, J., et al.: Recent developments in hardware-in-the-loop testing. In: Barthorpe, R. (ed.) Model Validation and Uncertainty Quantification, Volume 3. Conference Proceedings of the Society for Experimental Mechanics Series. Springer, Cham (2019). https://doi.org/10.1007/978-3-319-74793-4_10
22. Elliott, S.J., Rafaely, B.: Frequency-domain adaptation of causal digital filters. IEEE Trans. Sig. Process. **48**(5), 1354–1364 (2000). https://doi.org/10.1109/78.839982
23. Hybrid test solutions. https://www.hytestsolutions.com/. Accessed 30 Dec 2020

Identifying and Mastering Legal Uncertainty Concerning Autonomous Systems

Laura Joggerst[(✉)] and Janine Wendt

Chair of Civil Law and Business Law, TU Darmstadt, Darmstadt, Germany
{joggerst,wendt}@jus.tu-darmstadt.de

Abstract. The level of uncertainty concerning the use of autonomous systems is still very high. This also poses a liability risk for manufactures, which can impede the pace of innovation. Legal uncertainty also contributes to this factor. This paper will discuss existing legal uncertainty. The identified uncertainty can stem from different sources. Categorizing these sources will be our first step when trying to master legal uncertainty. On the basis of these categories, we will be able to evaluate where the focus for mastering legal uncertainty should lie. This approach promises to identify true legal uncertainty, which can only be mastered by new legislation, and separate it from other forms of legal uncertainty which can stem from unclear legal guidelines or uncertainty regarding the application and scope of existing rules and guidelines. Mastering the latter could be possible by specifying said existing rules and guidelines or even by clarifying the scope of their application, a much less drastic solution. Establishing how to deal with different categories of legal uncertainty will then contribute to minimizing liability risks for manufacturers.

Keywords: Legal uncertainty · Autonomous systems · Liability

1 Product Compliance

The term product compliance is derived from the general term of compliance and refers to the specific laws and regulations that have to be met when designing and marketing products [1]. The obligations, which are combined under the umbrella term of product compliance vary with product, customer and country in which the product is marketed [2]. For our discussion, we will focus on markets and therefore regulations in Europe and specifically Germany. The underlying premise for any attempt at successful product compliance is to be aware of the applicable regulations and the scope of these rules. In general, these rules will be derived from public law, civil law or criminal law.

The preventative character of public law, forbids the marketing of any unsafe products. Therefore, compliance with these regulations has to be assured in order to even market products. The Product Safety Act is the central legal framework in this context. In severe cases, market surveillance authorities can get involved and order a company to stop marketing a product, or even order a recall of dangerous products.

© The Author(s) 2021
P. F. Pelz and P. Groche (Eds.): ICUME 2021, LNME, pp. 145–153, 2021.
https://doi.org/10.1007/978-3-030-77256-7_13

The rules of civil law are concerned with the compensation for damages. The central regulations are the Product Liability Act and product liability derived from general tort law, paragraph 823 section 1 of the German Civil Code. Liability claims due to defective products cost money and can potentially damage the reputation of the producer, and should therefore be avoided.

The obligations resulting from civil law can be divided into three categories for liability on the bases of the Product Liability Act, and four for product liability derived from general tort law. Producers need to ensure a safe product throughout the stages of design and manufacturing, as well as instruct users, if necessary, on the limitations of the product and possible dangers [3, marg. 12]. For the producer's liability derived from tort law, the producer also needs to ensure effective product monitoring after the product is marketed.

In some extreme cases, criminal law can also become relevant. If, for example, people are severely injured, human failure can be punished with criminal penalties.

Following this brief overview of the relevant regulation, we will now discuss different categories of legal uncertainty. This chosen approach aims at developing more differentiated solutions on how to overcome legal uncertainty. As part of the solution, we will discuss the role an effective product compliance management system can play in mastering this uncertainty and where it has its limitations.

2   Categories of Legal Uncertainty

Legal uncertainty is not a set expression with a specific meaning attached to it. It is rather open to interpretation. Depending on the level of legal and technical expertise, the term will have slightly different meaning to the practitioner.

Starting from a legal point of view, the sources of legal liability can be manifold. We will also discuss what we call perceived and true legal liability in this article.

What can often result in uncertainty for developers and producers when applying laws and regulation, is the abstract nature of their wording. Since every case needs to be subsumed under the regulation individually, uncertainty can arise concerning the applicability of the regulation. In the case of the Product Safety Act, this can lead to uncertainty concerning the level of safety a product needs to meet, in order to be deemed as a product which is safe for marketing. In the case of civil liability, this can lead to uncertainty concerning the obligations the developer and producer have to meet in order to comply with what is legally expected of them.

In order to make legal regulation more accessible to practitioners, technical standards and norms have been developed. These norms and standards are based on experience and expertise in the field of their application. They are set by private standardization committees and represent the current state of the art. They are an important guideline for practitioners, but mostly not legally binding [4, 5]. Nevertheless, the state of the art displayed in these standards serves as a minimum safety level a product needs to provide. Compliance with this minimum safety level indicates that the producer or developer has done everything that can be expected of him in order to achieve a safe product that will not cause harm to any users or third parties [6, marg. 16]. There are some exceptions, as compliance with norms and standards concerning harmonized products results in the assumption that the product is safe [7].

Still, even if technical standards are more detailed and concrete, their development in the standardization committees takes time. Newer and innovative technical possibilities will often not be represented by these standards until they are widespread in the industry. This delay can then again result in legal uncertainty. All of the aforementioned cases of legal uncertainty will be referred to in this article as perceived legal uncertainty. However, this does not constitute any judgment, as this case of uncertainty also has the potential to hinder innovation of developers and producers.

The most problematic source of legal uncertainty occurs when the legal framework is actually not applicable to a certain type of product. This is what we would call true legal uncertainty. In such cases, the current legal regulation is not applicable anymore or is still part of an ongoing scientific discussion [8]. In this article, we will not go into the depth of this debate, but rather try to focus on the solutions, that are currently being discussed.

2.1 Legal Framework Is Applicable

In the case that the current legal framework is applicable, we can differentiate between three cases of uncertainty, depending on the combination of the sources of uncertainty. We assume that the generally applicable legal regulations are known.

Uncertainty in Application Concerning Specifics of Legal Obligation
Since the abstract wording of legal regulation can be difficult to apply to individual cases, as already discussed, this can result in uncertainty about the extent of the legal obligations producers and developers need to comply with. As far as technical standards exist, they offer guidance as to the current state of the art on the market. Producers should therefore meet at least the minimum safety level implied by this state of the art technology. This implies that the product is safe to be marketed and therefore interference by market surveillance authorities becomes unlikely. The same goes for possible liability claims, at least if the product proves to be safe during the use phase. Therefore, product monitoring is always an essential obligation that should not be neglected.

This would be the simplest case of perceived legal uncertainty and the standard solution to obtaining certainty in the development process.

Uncertainty in Application Concerning Specifics of Legal Obligation and Uncertainty About What Technical Standards Apply
In some cases, existing standards might not be fitting, or an alternative design choice promises a better and safer product. Developers and producers are not limited by technical standards, but, as they choose to deviate from the suggested procedure, a detailed documentation of the design process and choices becomes more important. It is also worth mentioning, that producers are compelled to choose the design alternative that promises the safest possible product. If solutions proposed by technical standards do not represent, what is feasible with state of the art technology, a different product design alternative has to be chosen [9, 6, marg. 16]. Developers cannot only choose to deviate from technical standards, but might also be obliged to do so.

Although technical standards, for the most part, are not legally binding, non-compliance can come with a certain liability risk. In order to minimize this risk, producers should be able to provide reasoning for the deviation from technical standards. It is recommended, that the design process and especially the selection procedure of the final product design is well documented as part of a product compliance management system. Special attention should also be given to the product monitoring aspect of the compliance management system.

Even if the product then shows to be defective during the use phase, the developer can prove that he has done everything that could be reasonably expected of him to ensure a safe product.

Uncertainty in Application Concerning Specifics of Legal Obligation and no Technical Standards Apply

Especially for the design of innovative and new products, technical standards might not be available because they have not been developed yet. It is worth mentioning that it is not necessary or practical to apply technical standards which were developed for a specific product to a similar appearing product [10]. In these cases, perceived legal uncertainty will naturally be high, as producers cannot rely on technical standards in order to assure, that the marketed product will at least be assumed to be safe.

Although there are no technical standards to offer guidance, producers do not operate in a legal vacuum, since the legal framework is still applicable. The producer needs to prove in ways other than by complying with technical standards that the product is safe. It is advised, that producers therefore implement a risk evaluation as described in the RAPEX-Guideline [11]. Although the RAPEX-Guideline is not legally binding, it at least provides useful information on how to carry out a risk assessment for newly developed products [12].

This risk assessment should also be documented in detail as part of a product compliance management system.

2.2   Legal Framework Is not Applicable: True Legal Uncertainty

True legal uncertainty arises when the legal framework is no longer applicable.

The current developments in robotics and AI show, that the legal framework will have to adapt in order to apply to the changing technology [13]. With products becoming more and more complex, connected and "intelligent", the current legal framework will reach its limits.

The limitations mostly derive from the scope of application of the current product liability regime. Legal definitions and concepts that were developed during a time, when software for example still played no significant role. Since then, technology has evolved, whereas the legal framework has not. Some legal concepts, such as the definition of a defective product, which plays a central role in the Product Liability Act, now have to be revised. Currently, a product is deemed defective if it does not provide the level of safety, that can reasonably be expected of the product. The relevant time for this consideration is limited to the marketing of the product. Any responsibilities to monitor the product after this point can only be derived from general tort law, not from the Product Liability Act. With placing the product onto the market, the producer needs to ensure the product is safe.

Truly autonomous systems for example, adapt and change during their use phase. This change can hardly be predicted by the manufacturer at the time the product is marketed. It is not clear, how the definition of a defective product should be applied to, or even be applicable to autonomous systems [12]. For truly autonomous products that operate and adapt without human interaction, the current regulation therefore needs to be adjusted, or entirely new regulation may be necessary [14, p. 11; 15]. Even if legislation is adapted, which can be expected, important research and development is done at a much earlier stage. In the interest of having safe products on the market, as well as not hindering innovation done by developers, legal obligations should be clear at the time when innovative products are developed. Addressing the issue of regulation for AI driven systems is an important step towards safe autonomous products in the future.

3 Solutions

The aforementioned cases of perceived and true legal uncertainty have to be met with different approaches in order to manage said uncertainty.

In the three cases described as perceived uncertainty, the main tool should be implementing an effective product compliance management system. What this can and should entail will be part of the following discussion.

As for the case of true legal uncertainty, a product compliance management system has a limited use. An adaption of regulation or a new regulation is needed to truly master legal uncertainty.

The whitepaper of the European Commission [14] has sparked the discussion about how AI can be regulated in the future. The most prominent feature of the proposal would be the risk-based approach to regulation. Joining the discussion on how this risk-based approach could actually be implemented, the German Data Ethics Commission has published an opinion on the subject [16]. In the following, we will also give an overview as to what this approach implies.

3.1 Product Compliance Management System

An effective product compliance management system should be organized in such a way that legal obligations are integrated into the company's procedures and processes [17]. The legal obligations developed by German jurisdiction, as mentioned before, concern product design, manufacturing, instruction of the user and product monitoring. These obligations can be translated into the three categories prevention, detection and response.

Prevention
Producers are faced with the task of preventing harmful products from being marketed and as a result preventing damage to life, body and property of users and third parties [18, 19]. Consequently, products need to be safe by design [3, marg. 15]. Tools to aid this task can be approval and release processes, quality control processes for in-house production chains as well as quality control for incoming parts from suppliers and provisional samples of supplier-parts [20]. In the context of supplier quality control, auditing procedures should be taken seriously. Traceability of products and product

parts also plays a role in the task of preventing harm from users. A legal obligation for one aspect of traceability can be found in paragraph 6 section 1 of the Product Safety Act. Consumer products must be labeled with contact information of the producer and information for the identification of the product. In the case of a product recall, the latter helps consumers to identify if their product is affected.

Detection

An effective product compliance management system should be able to detect harmful products on the market, before damage occurs. The obligation to monitor a product during the use phase should also be used for the purpose of gaining useful insights for the potential improvement of the product design. An important requirement for product monitoring is a system for complaint management [21]. Incoming complaints need to be assessed, evaluated and documented. In addition to this passive form of product monitoring, producers also need to actively assess the safety of the product on the market. This can be done by inspecting sample products, checking for insights provided by newer state of the art technology [22] and also by accident statistics, product reviews or user forums on the internet [23].

If the product monitoring detects a potential danger for users or third parties, a response-system should step in.

Response

The response-system should entail a set procedure defining which actions have to be taken in which event. The actions necessary will vary depending on the severity of the potential danger. In some cases, it can be enough to simply inform users of potential risks and how to avoid them [24]. In cases of high risk for life or body of users or third parties, a product recall can be the only acceptable response. It is worth mentioning that market surveillance authorities can intervene if the measures taken by the producer are not efficient enough to prevent potential danger for users or third parties. These measures can go from prohibiting further sales of a product to a product recall.

3.2   Risk-Based New Regulation for Artificial Intelligence

The solution for true legal uncertainty can only be an adjustment of the current legal framework, new guidelines for their application or a new regulation. The latter is proposed by the European Commission. The new regulation for artificial intelligence revolves around the idea of a risk-based approach. In the following, systems based on artificial intelligence will also be referred to as autonomous systems.

The General Idea

The idea behind the risk-based approach is to regulate according to the potential dangers that can come with the use of artificial intelligence. This is simply the application of the general principle of proportionality [25].

How to categorize the many possible autonomous systems according to their potential danger for users or others is the main question on which the success of the new regulatory framework depends.

Proposal of the European Commission

In the white paper on artificial intelligence, the European Commission mentions a risk-based approach for a new regulatory framework for artificial intelligence. The risk assessment should be made depending on the application area and purpose of the autonomous system [14, p. 20]. In the white paper, the only differentiation made is between systems and products with "high risk" and "others". In order to achieve a practical solution, the framework should entail a more nuanced inspection of the systems and products to be regulated. The white paper also lacks a proposal on how the risk-based approach should be put into practice.

Opinion of the Data Ethics Commission

The German Data Ethics Commission proposes a much more differentiated risk-based approach for the regulation of artificial intelligence, or algorithmic systems, being the term used in the proposal. A criticality pyramid should be the base for the new regulatory framework. The pyramid is comprised of five categories. Each category represents a level of risk and is assigned different regulatory obligations [16, p. 177].

On the first level, applications with zero or negligible potential for harm are assigned no special measures.

The second level is made up of applications with some potential for harm. The regulatory measures for this risk-level are i.e., transparency obligations, publication of risk assessments or monitoring procedures such as audit procedures, or disclosure obligations towards supervisory bodies.

Level 3 applications are such with regular or significant potential for harm. The proposed measures to regulate these types of applications are, in addition to the measures, which apply to level 2 applications, ex-ante approval procedures. This approval could be implemented by having to license products. As also stated by the Data Ethics Commission, the approval might have to be reviewed and renewed throughout the further development of the product and its life-cycle.

Level 4 applications pose serious potential for harm. The already mentioned obligations for Level 2 and 3 applications could be supplemented by requirements to enable simultaneous oversight of the application by supervisory institutions.

Products and applications which pose an intolerable potential for harm should be totally or partially banned, according to the Data Ethics Commission [16, p. 180].

The opinion of the German Data Ethics Commission proposes a much more concrete and differentiated approach to the idea of risk-based regulation of algorithmic systems. The challenge for the new legal framework will be to define the properties of products and systems that fall into the different categories and are then assigned specific risks and therefore regulatory measures.

4   Conclusions

In conclusion, we have found, that most legal uncertainty can be minimized with the tools provided by laws or by technical standards. Where uncertainty in the application of law or technical standards arises, an effective product compliance system will help master the remaining uncertainty and therefore minimize liability risk for producers.

In the case of true legal uncertainty, the solution has to come from legislators. Initial proposals for the idea of a risk-based regulation have been presented. The discussion and work on new regulation for artificial intelligence has only just begun. Especially for the definition of product properties and risk categories, more interdisciplinary work should be done. We will be able to see the results of that work in the proposal for the regulation of artificial intelligence of the European Commission, which is awaited with great anticipation.

References

1. See for compliance in general: LG München I, decision of 10.12.2013 - 5HK O 1387/10
2. Wagner, E., Ruttloff, M.: Product Compliance - Dos and Don'ts für Praktiker. In: Betriebs-Berater (BB), 1288–1294 (1289) (2018)
3. BGH, decision of 16.06.2009 - VI ZR 107/08
4. Deutscher Bundestag, Drucksache 17/6276, Entwurf eines Gesetzes über die Neuordnung des Geräte- und Produktsicherheitsrechts, 42
5. Giesberts, L., Gayger, M.: Sichere Produkte ohne technische Normen. In: Neue Zeitschrift für Verwaltungsrecht (NVwZ), 1491–1495 (1492) (2019)
6. BGH, decision of 09.09.2008 - VI ZR 279/06
7. Menz, S., Klindt, T. (eds.) Produktsicherheitsgesetz ProdSG, § 4, marg. 3, 2. Edition, Munich (2015)
8. See for further discussion: Joggerst, L., Wendt, J.: Die Weiterentwicklung der Produkthaftungsrichtline. In: Zeitschrift für Innovations- und Technikrecht (InTeR), 13–17 (2021)
9. BGH, decision of 18.09.1984 - VI ZR 223/82 = Neue Juristische Wochenschrift (NJW) 1985, 49
10. OLG Frankfurt, decision of 05.07.2018 - 6 U 28/18, marg. 5
11. Commission implementing Decision (EU) 2019/ 417 of 8 November 2018 laying down guidelines for the management of the European Union Rapid Information System 'RAPEX' established under Article 12 of Directive 2001/95/EC on general product safety and its notification system, notified under document C(2018) 7334
12. For a further discussion, see: Joggerst, L., Wendt, J.: Product safety requirements for innovative products. In: Pelz, P., Groche, P. (eds.) Uncertainty in Mechanical Engineering, pp. 225–231. Springer, Darmstadt (2021)
13. von Westphalen: Produkthaftungsrechtliche Erwägungen beim Versagen Künstlicher Intelligenz (KI) unter Beachtung der Mitteilung der Kommission COM(2020) 64 final, in: Verbraucher und Recht (VuR), 248–254 (249) (2020)
14. European Commission: White Paper on Artificial Intelligence, 19.2.2020 – COM(2020) 65 final
15. Seehafer, A., Kohler, J.: Künstliche Intelligenz: Updates für das Produkthaftungsrecht? In: Europäische Zeitschrift für Wirtschaftsrecht (EuZW), 213–218 (218) (2020)
16. Data Ethics Commission: Opinion of the Data Ethics Commission (2019)
17. For Compliance Management: Sonnenberg, T.: Compliance-Systeme in Unternehmen, Einrichtung, Ausgestaltung und praktische Herausforderungen. In: Juristische Schulung (JuS), 917–922 (917) (2007)
18. BGH, decision of 28.04.1952 - III ZR 118/51 = NJW 1952, 1050 (1051)
19. BGH decision of 17.06.1997 - VI ZR 156/96 = NJW 1997, 2517 (2529)
20. Wagner, E., Ruttloff, M., Miederhoff, R.: Product Compliance – Ein wesentlicher Baustein für Compliance-Organisationen. In: Corporate Compliance Zeitschrift (CCZ), 1–7 (3) (2020)

21. Foerste, U. (ed.) Produkthaftungshandbuch, part 2, § 24, marg. 379 (2012)
22. BGH, decision of 17.03.1981 - VI ZR 286/78, marg. 34
23. Hauschka, C., Klindt, T.: Eine Rechtspflicht zur Compliance im Reklamationsmanagement? In: NJW, 2726–2729 (2729) (2007)
24. Wagner, G.: Münchner Kommentar zum BGB, 8. Edition, § 823, marg. 1004
25. Unger, O.: Grundfragen eines neuen europäischen Rechtsrahmens für KI. In: Zeitschrift für Rechtspolitik (ZRP), 234–237 (236) (2020)

Uncertainty Quantification

Identification of Imprecision in Data Using ϵ-Contamination Advanced Uncertainty Model

Keivan Shariatmadar[1]([✉]), Hans Hallez[1], and David Moens[2]

[1] M-Group, Campus Bruges, KU Leuven, Bruges, Belgium
{keivan.shariatmadar,hans.hallez}@kuleuven.be
https://iiw.kuleuven.be/brugge/m-group
[2] LMSD, Campus De Nayer Sint-Katelijne-Waver, KU Leuven, Leuven, Belgium
david.moens@kuleuven.be
https://www.mech.kuleuven.be/en/mod

Abstract. One of the importance of the contamination uncertainty model is to consider in-determinism in the uncertainty. We consider this advanced property and develop two methods. These methods identify if there is imprecision in a given model or data. In the first approach, we build two different—a probability distribution and an interval—models for a test function f via given data/model. Then, we identify the level of imprecision by assessing, so-called model trust, $\epsilon \in (0, 1)$ in the contamination model whether the weight is higher for the probabilistic/interval model or not. In the second approach, we calculate the lowest and highest previsions for the test function and identify the imprecision interval out of them. We further discuss and show the idea via two simple production and clutch design problems to illustrate our novel results.

Keywords: Imprecision · ϵ-Contamination · Uncertainty · Indecision

1 Introduction

Dealing with uncertainty is one of the problems which is needed for the problems under uncertainty. The uncertainty is present because of lack of information or data. One of the uncertainty models is probabilistic (data-driven or analytical) model. These models' intentions are to represent e.g., agents' beliefs (agent like human, machines, or robots) about the domain/area they are operating in, which describe and even determine the actions they will take or decide in a diversity of situations or realisations [38]. Probability theory provides a normative system for *reasoning* and decision making in the face of uncertainty. Bayesian or precise probability models have the property that they are purely *decisive* i.e., a Bayesian agent always has an optimal choice when faced with several alternatives, whatever his state of information is, see e.g., [19,38]. While many may view this as an advantage, it is not always realistic. There are two problems, Gilboa [17] offers historical surveys with (precise) probabilities as a

© The Author(s) 2021
P. F. Pelz and P. Groche (Eds.): ICUME 2021, LNME, pp. 157–172, 2021.
https://doi.org/10.1007/978-3-030-77256-7_14

model to describe uncertainty: (i) the interpretation is not clear or at least, the consequences in the real world are not clear. Therefore, we want an operational and behavioural model (ii) the model is unique and static while the real model behaviour is dynamic. In any precise decision problem, there is always an optimal solution. You can—beholding some degenerate cases—decide between two actions. The idea whether there is a fair price or is not (either to accept/buy or reject/sell a gamble) is not vital, the possibility of *indecision* is rather important [19,38]. Imprecise probability (data-driven, grey/white box) models deal with said issues by explicitly allowing for indecision while retaining the normative, coherent stance of the Bayesian approach, see for more details, [5,19,38,42,44].

In this paper, our main goal is to answer a question about the existence of the imprecision in a data or model i.e., how to know that there is imprecision in the uncertainty made via the given data or model? In this section, we describe the advanced uncertainty modelling in depth via some simple examples to understand the concepts and especially the generic theory of lower and upper previsions. In our recent works [36–40], we have focused on the novel approach to make decisions under different types of imprecise uncertainties in linear optimisation problems (as one of the applications). We proposed two different solutions under two decision criteria—Maximinity and Maximality i.e., the worst-case solutions (the least risky solutions) and less conservative solutions (more optimal solutions). With these approaches, we can always decide based on the applications and preferences (from the final decision maker) to choose whether the more optimal (more risky) solutions or less risky (less optimal) solutions[1]. In the next Sect. 1.1. first, we give an overview of the state-of-the-art and history about the uncertainty. Second, in Sect. 2.1. we explain the uncertainty briefly under Walley's integration [42].

1.1   Literature Status and History

There is a long history about using imprecise probability models starting from the middle of the 19th century [38]. For instance, in *probabilistic logic*: it was already known to George Boole [4] that the result of probabilistic inferences may be a set of probabilities (an imprecise probability model), rather than a single probability. In 1920, Keynes [22] worked on an explicit interval estimate method to probabilities. Work on imprecise probability models proceeded in the 20th century, by A. Kolmogorov [23] in 1933, B. Koopman [24] in 1940, C. A. B. Smith [41] in 1961, I. J. Good [18] in 1965, A. Dempster [13] in 1967, H. Kyburg [21] in 1969, B. de Finetti [16] in 1975, G. Shafer [34] in 1976, P. M. Williams [48] in 1978, I. Levi [26] in 1980, P. Walley [42] in 1991, T. Seidenfeld [33], and G. de Cooman [5,44] in 1999. In 1990, P. Walley's published the reference book: *Statistical Reasoning with Imprecise Probabilities* [42] representing the theory of *imprecise probability*. He also interpreted the subjective probabilities as accepting/buying and rejecting/selling prices in gambling. In 1990 some

---

[1] The risk is the distance between the worst-case solution and the less conservative solutions e.g., in the linear optimisation problem, the risk is the distance between the objective function at maximin point and the maximal solutions.

important works published by Kuznetsov [25] and Weichselberger [45,46] about
the interval probabilities. Also Weichselberger generalizes the Kolmogorov's work
[23] in 1933. In 2000, R. Fabrizio [30] presented the robust statistics. In 2004,
T. Augustin [1] provided non-parametric statistics. In 2008, the important concept about Choquet integration is proposed by G. de Cooman [9]. This work
together with the work of P. Huber [20] about two-monotone and totally mono-
tone capacities have been the foundation of artificial intelligence. Moreover, in
2008, G. de Cooman and F. Hermans [8] proposed imprecise game theory (as the
extension of the work of Safer and Vovk [35]). Dealing with missing or incomplete
data, leading to so-called partial identification of probabilities, is proposed by G.
de Cooman and C. F. Manski [10,27]. Another application in network domain
so-called *credal nets* were proposed by F. Cozman [6,7] which are essentially
Bayesian nets with imprecise conditional probabilities.

The paper is organised as follows. In the next Sect. 2 we explain the theory
of imprecise probability and show the differences between precise and imprecise uncertainties via several simple examples. An advanced—ϵ-contamination—
model as well as two novel methods to identify imprecision are discussed in
Sect. 3. In Sect. 4 we propose a numerical production problem to illustrate the
results. We conclude and discuss the future works in Sect. 5.

2 Uncertainty

Generally, uncertainty is the consequence of lack of data, information, or knowledge. Conventional methods of introducing uncertainty into a problem, ignore
the following cases: (a) imprecision, (b) mixed or combined precise and imprecise
models, or (c) choosing best imprecise models for the available amount of data.
In this paper, we consider (a) and (b) to propose two methods to identify if
there is imprecision in a given uncertainty model or not. In this section, we first
explain the difference between precise and imprecise uncertainty. To understand
this better, we illustrate these concepts via several simple examples. Second, we
use define a prevision operator to measure the uncertainty. We interpret lower
and upper prevision operators to quantify the imprecise uncertainty. Finally, we
define an advanced mixed/combined model to identify imprecision in a given
uncertainty model (analytical or data-driven model) in two ways.

2.1 Interpretation of Lower and Upper Previsions

Most of the above mentioned works on imprecise probability theory was introduced by Walley [42]. In this paper, we follow the terminology and school of
thought of Walley [42,43] who follows the tradition of Frank Ramsey [29], Bruno
de Finetti [12] and Peter Williams [50] in trying to establish a rational model
for a subject's beliefs and reasoning. In the subjective interpretation of Walley,
the upper and lower previsions/expectations for gamblers are seen as prices. A
gambler's highest desirable buying price and the lowest desirable selling price,
respectively. In gambling, which is about exchanging of gambles, assume that a

gambler (decision maker) wants to make a profit whether (s)he wants/accepts to buy or sell a gamble. By knowing the highest desirable price to buy the gamble and the lowest desirable price to sell the gamble, (s)he can make any desirable decision to not to lose money.

Generally, a decision maker's lower prevision/expectation $\underline{P}(\cdot)$ is the highest acceptable price α to buy a gamble/utility function f. In other words, $\underline{P}(f)$ is supremum price to buy the gamble f. Mathematically $\underline{P}(f)$ is defined as:

$$\underline{P}(f) := \sup_{\alpha \in \mathbb{R}} \{\alpha : f - \alpha \geq 0\}, \tag{1}$$

and the upper prevision/expectation $\overline{P}(\cdot)$ is the lowest acceptable price β to sell the gamble f. In other words, $\overline{P}(f)$ is infimum price to sell the gamble f. Mathematically $\overline{P}(f)$ is defined as:

$$\overline{P}(f) := \inf_{\beta \in \mathbb{R}} \{\beta : \beta - f \geq 0\}. \tag{2}$$

In classical probability theory, the upper and lower previsions are coincided: $\underline{P}(f) = \overline{P}(f) := P(f)$. Then $P(f)$ is interpreted as the gambler's *fair price* for the gamble f. The price that the decision-maker accepts to but f for any lower price and sell it for any higher price than $P(f)$. The gap between $\underline{P}(f)$ and $\overline{P}(f)$ is called *imprecision* or *indecision*. This is the main difference between precise and imprecise probability theories—as shown in Fig. 1, imprecise models allow for *indecision*/imprecision. Such gaps arise naturally e.g., in betting markets which happen to be financially illiquid due to asymmetric information, for more information see [21, 26]. As an interpretation, for instance in gambling (which is about exchanging of a gamble f), $\overline{P}(f)$ is the lowest desirable price to sell the gamble f. In other words, if a gambler knows the lowest acceptable price of a gamble then (s)he can accept any higher price than $\overline{P}(f)$. To explain the importance and deeper view about the in-deterministic uncertainty, in the next Sect. 2.2., different types of uncertainty, as well as some simple examples, will be talked to clarify the distinction between precise and imprecise uncertainty. In Sect. 3 a general overview about modelling uncertainty via one of the advanced models called ϵ-contamination as well as two methods (imprecision identification methods) will be discussed. Next, a simple example will be given to illustrate the results in Sect. 4. Conclusions and further discussions will be in Sect. 5.

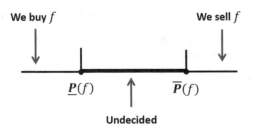

Fig. 1. Highest buying and lowest selling prices for a gamble f

2.2   Classification of Uncertainty

There are four levels of certainty (or uncertainty) about knowledge or data. In Fig. 2, four levels of these certainties or uncertainty are illustrated from known knowns (knowledge) towards unknown unknowns (imprecise uncertainty). Our main focus here is on the *Unknown unknowns* where the unknown data is not precise. In other words, the probability of an event or a phenomenon is vague. In real-life problems, the nature of the uncertainties are usually imprecise uncertainties and one of the sources of the imprecisions, in which we have researched about, is human, also weather, traffic, and so on, [39]. One of the interesting purposes in almost all of those real-life problems is to find the best choice under some conditions dealing with the uncertainties. In other words, one of the major problems is to make the best (optimal) decision based on the restrictions (uncertainty, constraints, and so on) within some criteria. Mathematically, the idea can be formulated as optimising a goal function under an uncertain domain given by constraints. But the important point is how to deal with the uncertainty? Even more importantly, how to know that the uncertainty is not deterministic? To understand deeper the idea of the existence of indeterminism in the uncertainty, let's point out three real-life examples.

Fig. 2. Precision vs. imprecision

2.3   Probability Under Different Conditions–Travelling to Work

Assume the problem of driving a car each day from home to work and back over a (long) distance. Consider there are two possible routes. Typically, one would measure the duration of travel for both routes over some period, let's say a year and compute probability or cumulative distribution function (CDF) from the data. The goal is–using the computed CDF results in some tool–to decide, based on the probability, which route is beneficial. As we have seen from a real database [36] sent by a factory here in Belgium[2], the CDF functions differ and are not unique. Consequently, a single CDF function cannot capture the true distribution because of the indeterministic parameters influencing this duration, e.g., weather conditions, human (driver) mood, or traffic status that might change during travel which one path might be highly influential by these weather, driver, or traffic conditions in contrast to the other path. The variation on the CDF of one path might be much higher than for the other which is not possible to model via one single CDF. It is, therefore, better to capture this uncertainty by use of an advanced model (considering the indeterminism).

One of the best models to describe the imprecise uncertainty for this problem is sets of distributions functions which is called *probability box (p-box)* [15]. This model is developed and discussed briefly under an optimisation problem in [37], which is the most informative model. Another alternative is to use the contamination model, which is simpler and doesn't require lots of data, we will discuss it in the next section. In this case, every CDFs is collected in a set which is bounded from below and above, called *upper* and *lower* bounds, where can model *variations* in the probabilities (imprecision). The variation on the probability of the duration for both routes is illustrated in Fig. 3 where the full and dashed lines represent the probability under different conditions. In the case of the driving example, subset division can be made based on whether or traffic conditions (obtained from a weather or traffic database), as well.

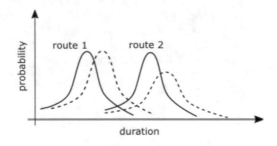

Fig. 3. Probability density of the duration for both routes 1 and 2

This, in turn, allows for more robust decision making in the future (which path to take) based on the imprecision in the data, captured by the advanced models[3]. Again, the main question is how to measure this imprecision and find out there is imprecision in the data (or model), generically?

---

[2] Because of the confidentiality about the agreement, we can not make the names and details of the database public unless under an official confidentiality agreement.

[3] Generally, p-box uncertainty model, described above, belong to *coherent upper* and *lower previsions* family, see e.g., for details and terminology [28,42].

2.4   Probability Under Different Conditions–Diagnosis and Treatment

Logical decision making is a major part of all sciences, engineering and decision-based professions, where scientists, engineers or specialists apply their knowledge or beliefs in a given area to make optimal decisions. However, the decision under uncertainty is one of more advance topics compared to the deterministic decisions. Even more challenging where the uncertainty is not precise. For example, in the medical science area, decision making often involves a diagnosis and the selection (decision) of appropriate (optimal) treatment under a vague data—meaning, the data is not large enough or incomplete because of several restrictions such as expensive tests, test-case limitation, missing data, or unknown unmeasurable parameters, to gather enough data—where we call *life involved* (high-risk) problems. The nature of uncertainty is not unique. In other words, for instance, the uncertainty is not the same from one patient to another. In these kinds of areas, when the uncertainty is imprecise, we do not have a single (optimal) decision to make, however, it is very important to know at least the extreme cases e.g., the *worst/best cases*[4].

2.5   Probability Under Different Conditions–Clutch Design

Another example, in the mechanical engineering area, is decision making (usually) about a design of a component under some conditions such as selections of right parameters for a design–for instance, a clutch design–e.g., diameters, friction disks, friction coefficient (uncertain parameter), torque capacity, speed, gear parameters, cooling system parameters (uncertain parameter), and so on. To design a safe clutch pack, in one hand, the engineer needs to make a safe decision i.e., tries to find the worst-case solution to avoid the risks, on the other hand, concerning the total cost of ownership, he/she needs to decide to have minimum cost i.e., less conservative solutions. In both examples, the nature of uncertainty is not unique. In other words, for instance, the uncertainty is not the same from one clutch to another e.g., the friction coefficient is not known and is changing in different temperature ranges (coming from energy loss by friction or oil condition) and different geometry see Fig. 4.

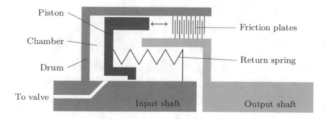

Fig. 4. Wet-plate clutch of an automatic transmission

---

[4] It is also interesting to know what are possible *less conservative* cases/decisions.

Modelling those unknowns is not possible via classical uncertainty because of the imprecision like the uncertainties. Knowing in advance, the presence of the imprecision might help to choose the right uncertainty model and have more robust and stable (optimal) decisions. Then the important question is that is there an imprecision in a problem or data? How to find out that there are random fluctuations in a problem/data? We discuss this in detail in the next section.

3 $\epsilon-$Contamination Model

To a Bayesian[5] analyst, the distinction between fixed, random, and mixed models boils down to a specification of the number of stages in a given hierarchical model. One of these mixed models is called ϵ-contamination model. This model is more advanced than the interval model [40] i.e., it links the precise model to the imprecise model. This model is recommended to be used to also analyse if there is imprecision in a given uncertainty model or not[6]. Furthermore, the ϵ-contamination model is easier to build as well as implement compared to the p-box (or other imprecise models e.g., possibility distribution model [40]). In literature [2], several classes of prior distribution have been proposed but the most commonly used one is the contamination class e.g., works of Good [18], Huber [20], Dempster [14], Rubin [31], and Berger [3] to mention a few. In particular, it is concerned with what they call the posterior robustness[7]. The idea is to acknowledge the prior uncertainty by specifying a class/set M of possible prior distributions and then investigating the robustness of the posterior distribution as the prior varies over M. It had been mentioned by Berger [3] and Huber [20], to work with the contamination class of priors when investigating posterior robustness. They proposed the contamination class of combining elicited prior—termed the base prior—with a contamination class of arbitrary priors. These approaches are popular with Bayesian sensitivity analysis—first, to elicit an additive probability measure P, and then consider possible inaccuracies in the assessment[8] of P, [42]. Those contamination models, achieve statistical efficiency and robustness simultaneously, however, not much attention has been paid to this framework in non-deterministic advanced uncertainty cases (pure non-probabilistic such as intervals or high-dimensional cases like ϵ-contamination or probability box). In the next section, we explain the ϵ-contamination model for a given probability measure E and an imprecise interval model $\underline{E}$.

3.1   Definition

ϵ-contamination model $\underline{P}(\cdot)$ is described as a convex combination of two uncertainty models: (i) linear prevision model—Probabilistic model, e.g., Normal

---

[5] In this paper, without any intention, we call a researcher who works on the deterministic uncertainty framework, a Bayesian analyst.

[6] We can decide if a pure precise model could be suitable or not.

[7] Which was different from the robustness defined by White [47].

[8] The second step is called constructing a neighbourhood set of P.

distributed model E, and (ii) lower prevision (imprecise) model e.g., interval vacuous model $\underline{E}$, which is described as follows:

$$\underline{P}(f) = (1 - \epsilon)E(f) + \epsilon\underline{E}(f) \tag{3}$$

where E is the set of dominating linear previsions by $\underline{E}$ i.e., $E \in \mathcal{M}\{\underline{E}\} = \{E : \forall f \in \mathcal{L}(\mathcal{Y}), E(f) \geq \underline{E}(f)\}$, for a given interval $[a, b] \subset \mathbb{R}$, the $\underline{E}(f) := \min_{y \in [a,b]} f(y)$, and $0 < \epsilon < 1$ is called (here) level of *model-trust/importance*[9]. One question is, how to build or get the ϵ-contamination model? Let's consider a simple example. We need to build two models (out of given data or model), one probabilistic and one imprecise model. Assume, there is a %60 chance (precise model) of having heavy traffic in a road A around time t, where t varies between 1:00 and 2:00 o'clock, i.e., $t \in [1, 2]$ hours (imprecise model). We are not sure about the time t: sometimes $t = 13 : 00$ and sometimes $t = 14 : 00$. Suppose we have an equal belief to the precise and the imprecise models, i.e., $\epsilon = 0.5$. Therefore, the uncertainty model for a given test-function f in this problem becomes the average of both models,

$$\underline{P}(f) = 0.5E(f) + 0.5\underline{E}(f) = 0.5\big(E(f) + \underline{E}(f)\big).$$

3.2   Rationale

One of the important properties for this model is that this model considers both probabilistic (a probability measure E) and non-probabilistic (an interval $[a, b]$) models where we can tune it by choosing the right trust value (ϵ). This needs some expert knowledge or historical information about selecting the right level. However, since the problem is convex, we can always generate all possible outcomes for all $\epsilon \in (0, 1)$, mathematically[10]. In many real-life problems e.g., said traffic problem, we have both, a variation, and a guess or chance in the real-life problems. The variation can be found via a robustness test or experiment. By the time, with enough information, via e.g., sensitivity analysis, or reliability tests/experiments we can also obtain the percentage of beliefs about the unknown parameter, event, realisation, or phenomenon. Mathematically, to find an interval model we need the lower and upper values of the realisation which is varying between them i.e., the two boundary values are enough to build the interval. For the probabilistic model, normally, we need more data to get those percentages and guesses. But to consider both models, current classical (precise) uncertainty models are not able to handle and deal with both models, simultaneously. We believe that to start moving towards advanced uncertainty (after interval case) the ϵ-contamination model is one of the best models to use in many real problems and applications [40].

---

[9] ϵ is also called *tuning* parameter or weight factor.

[10] For instance, we can easily calculate the outcome of the convex combination of two points which is a line between the two points.

3.3 Imprecision Identification–Method I

Another importance of the ϵ-contamination model is to distinguish between imprecision and precision i.e., the question is how to identify the imprecision in a given problem or data? How to find out that there are random fluctuations in a problem/data? The answer is given via this ϵ-contamination model as follows. From a given model or available data (database), we first assume a known outcome $\underline{P}(f)$ for a given real-valued test function f[11]. Then we build a probability distribution as well as the variation interval for the test function f via given data. By calculation the expected values for f in both cases–Probability and Interval–we know $E(f)$ and $\underline{E}(f)$. Finally, we solve the Eq. (3) to find the ϵ. If $0 \leq \epsilon \leq 0.5$ then there is less chance (less than %50) of having imprecision in the data or the model otherwise, there is imprecision with the probability of higher than %50.

3.4 Imprecision Identification–Method II

Assume an interval $[a, b] \in \mathbb{R}$ and a probability distribution are given via a data (a database) or model. Another method to identify whether there is imprecision in the given data (via a database) or model[12], is to calculate the lower as well as the upper previsions for a chosen test function f as follows via (3):

$$\text{the lower prevision } \underline{P}(f) \text{ is } : (1 - \underline{\epsilon})E(f) + \underline{\epsilon}\,\underline{E}(f), \text{and} \qquad (4)$$

$$\text{the upper prevision } \overline{P}(f) \text{ is } : (1 - \overline{\epsilon})E(f) + \overline{\epsilon}\,\overline{E}(f) \qquad (5)$$

where $\underline{\epsilon}, \overline{\epsilon} \in (0, 1)$, the upper prevision in the given interval $[a, b]$ is $\overline{E}(f) := \max_{y \in [a,b]} f(y)$. If $\exists \epsilon^* = \max\{\underline{\epsilon} \in \underline{\mathcal{E}}, \overline{\epsilon} \in \overline{\mathcal{E}}\}$ where $\underline{\epsilon}$ satisfies in (4),

$$\underline{\epsilon} \in \underline{\mathcal{E}} := \{\underline{\epsilon}_i : \underline{\epsilon}_i \in (0, 1)\} \text{ and } \overline{\epsilon} \text{ satisfies in (5)}$$

$$\overline{\epsilon} \in \overline{\mathcal{E}} := \{\overline{\epsilon}_i : \overline{\epsilon}_i \in (0, 1)\}$$

such that $\underline{P}(f) < \overline{P}(f)$ then there is imprecision in the uncertainty model with probability of ϵ^*, and the imprecision interval is $[\underline{P}(f), \overline{P}(f)]$.

4 Numerical Example

4.1 Chocolate Production Problem

Consider a chocolate manufacturer which produces two types of chocolate $\mathfrak{A}$ and $\mathfrak{B}$. Both chocolates require Milk and Cacao only (for simplicity). Each unit of

---

[11] This can be done by an expert or historical data, to have the simplest case that we know the outcome of the realisation which is given via both uncertainty models.

[12] These models as discussed, could be estimated from the existing data or the available model under uncertainty. These estimations are not the aim of this paper but for instance the interval can be estimated via a sensitivity analysis and the probability distribution can be calculated via a normal distribution fitted to the data/model.

$\mathfrak{A}$ requires 1 unit of Milk and Y_1 units of Cacao. Each unit of $\mathfrak{B}$ requires 1 unit of Milk and Y_2 units of Cacao. The company capacity has a total of 12 units of Cacao (no limit for milk). On each sale, the company makes profit of €1 per unit $\mathfrak{A}$ and €1 per unit $\mathfrak{B}$. The goal is to maximise profit (how many units of $\mathfrak{A}$ and $\mathfrak{B}$ should be produced respectively). Mathematically, the problem is modelled as a linear programming problem:

$$
\max \quad x_1 + x_2
$$
$$
\text{s.t} \quad \begin{cases} \mathbf{Y_1} x_1 + \mathbf{Y_2} x_2 \le 12 \\ x_1, x_2 \ge 0 \end{cases} \tag{6}
$$

Assume that there are two sources of uncertainties: (i) a priory probabilistic information (about dealing with *experiments* that numerically describes the number of desired outcomes)–obtained by a historical data, expert knowledge, or sensitivity analysis–and (ii) a set of realisations obtained via e.g., reliability analysis (about robustness/variation). Suppose (i) the probabilistic models are given with distribution functions $N_1 := N(\mu_1 = 7.5, \sigma_1 = 1)$ and $N_2 := N(\mu_2 = 9.5, \sigma_2 = 1)$ about how likelihood we need the amount of cacao for both chocolates $\mathfrak{A}$ and $\mathfrak{B}$ in one year and (ii) also we know that the amounts of cacao for both chocolates $\mathfrak{A} \in [7, 8]$ and $\mathfrak{B} \in [9, 10]$ are varying. In other words, the problem is to maximise profit under the ϵ-contamination uncertain constraint for Cacao, which has likelihood amounts given by the normal probability distributions and varying in the assumed lower and upper values, shown in Table 1.

Table 1. Uncertain Chocolate production problem—ϵ-contamination uncertainty

	Milk	Cacao	Profit per unit
$\mathfrak{A} := x_1$	1	$\mathbf{N(7.5, 1)}$, $[\mathbf{7, 8}]$	1
$\mathfrak{B} := x_2$	1	$\mathbf{N(9.5, 1)}$, $[\mathbf{9, 10}]$	1
Capacity	No limit	12	

Since, in this problem we do not have the lower and upper expected values then we use method II to identify if there is imprecision in this example or not. The lower and upper previsions are defined as follows,

$$
(1 - \underline{\epsilon})\big((x_1 + x_2)|_{7.5x_1 + 9.5x_2 \le 12}\big) + \underline{\epsilon}\big((x_1 + x_2)|_{8x_1 + 10x_2 \le 12}\big), \text{ and} \tag{7}
$$
$$
(1 - \overline{\epsilon})\big((x_1 + x_2)|_{7.5x_1 + 9.5x_2 \le 12}\big) + \overline{\epsilon}\big((x_1 + x_2)|_{7x_1 + 9x_2 \le 12}\big) \tag{8}
$$

For instance, to maximise the profit $(x_1 + x_2)$, from (7) we have: $(1 - \underline{\epsilon})\frac{5}{3} + \underline{\epsilon}\frac{3}{2}$ and from (8) we have: $(1 - \overline{\epsilon})\frac{5}{3} + \overline{\epsilon}\frac{12}{7}$, where for all $\overline{\epsilon}$, $\underline{\epsilon} \in (0, 1)$ the profit for upper prevision (8) is higher than (7) and the $\epsilon^* = \max(0, 1) \approx 0.9999$, meaning with the high probability 99.99 percent there is imprecision in the given model (6). There are many conditions such as traffic, weather, or human behaviour/mood could affect e.g., transportations delays and consequently the exact amount of stock (Milk) or the exact availability of warehouse capacity. Furthermore, this inexact amount of stock or warehouse capacity is dynamically changing from day-to-day. Therefore, using only probabilistic (truncated) distributions for this problem will result in the suboptimal solution and the ϵ-contamination model is the suitable model for (6).

4.2   Clutch Design Problem

Back to the Clutch Design example discussed in Sect. 2.5, one of the main ideas is to have a maximum torque transfer from one side of the clutch to the other side. We simplify the problem as follows. Assume we want to design the clutch to have a maximum friction torque τ_f defined as:

$$\tau_f := \mu R N A P_k \Delta \omega \tag{9}$$

where R, A, N are the radius, area, and the number of friction disks, respectively. P_k is the internal oil clutch pressure (pushing the friction plates towards each other to close by increasing the pressure and open by decreasing it), $\Delta \omega$ is the slip speed, and μ is the uncertain friction coefficient. If the spring force f_s is higher than the friction force $P_k A$, then the clutch is open otherwise it is closed. This is controlled via the pressure P_k to create a smooth closing (opening) with less torque loss. As defined in (9), this pressure depends on the friction coefficient μ. Currently, the friction coefficient is estimating, and it is a fixed value however there are many disturbances e.g., oil temperature, the air in the oil, centrifugal force, oil leakage, and so on, changing the friction coefficient. We use the data provided by the work of Schneider [32]. In the given test data (durability tests), we have seen that μ is varying between 0.09 and 0.18 (interval model). Also a normal distribution function for μ can be estimated as $N(\mu = 0.11, \sigma = 1)$. We calculate the lower and upper prevision for the following linear optimisation problem for a closing clutch.

$$\begin{array}{ll} \max_{\mu} & \mu R N A P_k \Delta \omega \\ \text{such that} & f_s < P_k A \\ & \mu, R, N, A, P_k, \Delta \omega \geq 0 \end{array} \quad \rightarrow \quad \begin{array}{ll} \max_{t} & t \\ \text{such that} & f_s < P_k A, \\ & \mu R N A P_k \Delta \omega \geq t \end{array} \tag{10}$$

The lower and upper previsions are defined as follows:

$$(1 - \bar{\epsilon}) t_{|0.11 R N A P_k \Delta \omega \geq t} + \bar{\epsilon}\, t_{|0.18 R N A P_k \Delta \omega \geq t}$$

$$(1 - \underline{\epsilon}) t_{|0.11 R N A P_k \Delta \omega \geq t} + \underline{\epsilon}\, t_{|0.09 R N A P_k \Delta \omega \geq t}$$

Assume, $R = 0.1\,\mathrm{m}, N = 2, A = 0.001132\,\mathrm{m}^2, P_k = 5\,\mathrm{bar}, \Delta \omega = 7.5\,\mathrm{m/s}$, then

$$(1 - \bar{\epsilon}) t_{|93.39 \geq t} + \bar{\epsilon}\, t_{|152.82 \geq t} \tag{11}$$

$$(1 - \underline{\epsilon}) t_{|93.39 \geq t} + \underline{\epsilon}\, t_{|76.41 \geq t} \tag{12}$$

For instance, to maximise the objective t, from (11) we have: $(1-\bar{\epsilon})93.39+\bar{\epsilon}152.82$ and from (12) we have: $(1-\underline{\epsilon})93.39+\underline{\epsilon}76.41$, where for all $\bar{\epsilon}$, $\underline{\epsilon} \in (0,1)$ the profit for upper prevision (11) is higher than (12) and the $\epsilon^* = \max(0,1) \approx 0.9999$, meaning with the high probability 99.99 percent there is imprecision in the given model (10). So, we need to consider an imprecise model for the friction coefficient μ rather that a fixed estimated value.

5   Conclusion

In this paper we consider two methods to identify if there is imprecision in a given problem under uncertainty with some degrees. The problem either is given via a database (black-box) or analytically (white-box) where there is uncertainty in either case, e.g., an unknown parameter where we know about distribution or variation in the parameter (in the model or the measured data). We use one of the advanced uncertainty—ϵ-contamination—models to identify the imprecision in the given data or model under uncertainty via two methods. If the lowest and the highest expected values on the problem are given (e.g., by a decision-maker) then we use method (I) proposed in Sect. 3.3 to search for $\epsilon \in (0,1)$. Otherwise, if the expected values are not available, then we proposed method (II) discussed in Sect. 3.4 to search for the $\epsilon^* \not\subseteq \emptyset$. In both methods, the chance (degree) of having the imprecision is determined by the ϵ. That is up to the final decision maker to decide whether using the imprecise uncertainty model is more optimal when the chance is low e.g., lower than %50, or not. The approach here to analyse and identify the existence of imprecision is a fundamental decision before modelling the uncertainty. By knowing that, we can decide to choose the best uncertainty model for the problem under uncertainty. This will avoid having further issues such as instability, inaccuracy, or wrong results from the model with wrong uncertainty model, and will help to have a more stable and accurate model for any decision (or design) problem. In both methods I and II, the problem is linear and convex i.e., the proposed methods are not NP-hard.

References

1. Augustin, T., Coolen, F.: Nonparametric predictive inference and interval probability. J. Stat. Plan. Infer. **124**(2), 251–272 (2004). https://doi.org/10.1016/j.jspi.2003.07.003
2. Baltagi, B.H., Bresson, G., Chaturvedi, A., Lacroix, G.: Robust linear static panel data models using ϵ-contamination. Center Policy Res. **239** (2017). https://surface.syr.edu/cpr/239
3. Berger, J.O.: Statistical decision theory and Bayesian analysis, 2nd edn. Springer, New York (1985)
4. Boole, G.: The Laws of Thought. Dover Publications, New York (1847, reprint 1961)
5. de Cooman, G., Aeyels, D.: Supremum preserving upper probabilities. Inf. Sci. **118**(1–4), 173–212 (1999)

6. Cozman, F.G.: Credal networks. Artif. Intell. **120**, 199–233 (2000). https://doi.org/10.1016/S0004-3702(00)00029-1

7. Cozman, F.G.: Graphical models for imprecise probabilities. Int. J. Approx. Reason. **39**(2–3), 167–184 (2005). https://doi.org/10.1016/j.ijar.2004.10.003

8. de Cooman, G., Hermans, F.: Imprecise probability trees: bridging two theories of imprecise probability. Artif. Intell. **172**, 1400–1427 (2008). https://doi.org/10.1016/j.artint.2008.03.001

9. de Cooman, G., Troffaes, M.C., Miranda, E.: n-Monotone exact functionals. J. Math. Anal. Appl. **347**(1), 143–156 (2008). https://doi.org/10.1016/j.jmaa.2008.05.071

10. de Cooman, G., Zaffalon, M.: Updating beliefs with incomplete observations. Artif. Intell. **159**(1–2), 75–125 (2004). https://doi.org/10.1016/j.artint.2004.05.006

11. de Finetti, B.: Teoria delle Probabilità. Einaudi, Turin (1970)

12. de Finetti, B.: Theory of Probability: A Critical Introductory Treatment. Wiley, Chichester (1974). English translation of [11], two volumes

13. Dempster, A.P.: Upper and lower probabilities induced by a multivalued mapping. Ann. Math. Stat. **38**(2), 325–339 (1967). https://doi.org/10.1214/aoms/1177698950

14. Dempster, A.: Examples relevant to the robustness of applied inferences. In: Gupta, S.S., Moore, D.S. (eds.) Statistical Decision Theory and Related Topics, pp. 121–138. Academic Press (1977). https://doi.org/10.1016/B978-0-12-307560-4.50010-7. Supported in part by National Science Foundation Grant MCS75-01493

15. Destercke, S., Dubois, D., Chojnacki, E.: Unifying practical uncertainty representations: I. Generalized p-boxes. Int. J. Approx. Reason. **49**, 649–663 (2008)

16. de Finetti, B.: Theory of Probability, vol. 2. Wiley, Hoboken (1975)

17. Gilboa, I., Postlewaite, A.W., Schmeidler, D.: Probability and uncertainty in economic modeling. J. Econ. Perspect. **22**(3), 173–88 (2008). https://doi.org/10.1257/jep.22.3.173

18. Good, I.J.: The Estimation of Probabilities: An Essay on Modern Bayesian Methods. M.I.T. Press Research Monographs. M.I.T. Press, Cambridge (1965). https://books.google.be/books?id=wxLvAAAAMAAJ

19. Hermans, F.: An operational approach to graphical uncertainty modelling. Ph.D. thesis, Ghent University (2012)

20. Huber, P.J., Strassen, V.: Minimax tests and the Neyman-Pearson lemma for capacities. Ann. Stat. **1**(2), 251–263 (1973). https://doi.org/10.1214/aos/1176342363

21. Kyburg, Jr., H.E.: Probability Theory. Prentice-Hall, Englewood Cliffs (1969)

22. Keynes, J.M.: A Treatise on Probability. Macmillan, London (1921)

23. Kolmogorov, A.N.: Grundbegriffe der Wahrscheinlichkeitsrechnung. Springer, Heidelberg (1933). English translation: Foundations of the Theory of Probability. Chelsea, New York (1950)

24. Koopman, B.: The bases of probability. Bull. Am. Math. Soc. **46**, 763–774 (1940)

25. Kuznetsov, V.P.: Interval Statistical Models. Radio i Svyaz Publication, Moscow (1991). (in Russian)

26. Levi, I.: The Enterprise of Knowledge. An Essay on Knowledge, Credal Probability, and Chance. MIT Press, Cambridge (1980). https://doi.org/10.2307/2184951

27. Manski, C.F.: Partial Identification of Probability Distributions. Springer, New York (2003)

28. Miranda, E.: A survey of the theory of coherent lower previsions. Int. J. Approx. Reason. **48**, 628–658 (2008). https://doi.org/10.1016/j.ijar.2007.12.001

29. Ramsey, F.P.: Truth and probability (1926). In: Braithwaite, R.B. (ed.) The Foundations of Mathematics and other Logical Essays, chap. VII, pp. 156–198. Taylor & Francis Group, London (1931)
30. Ríos, D., Ruggeri, F.: Robust Bayesian Analysis. Springer, Heidelberg (2000). https://doi.org/10.1007/978-1-4612-1306-2
31. Rubin, H.: Robust Bayesian estimation. In: Gupta, S.S., Moore, D.S. (eds.) Statistical Decision Theory and Related Topics, pp. 351–356. Academic Press, Cambridge (1977). https://doi.org/10.1016/B978-0-12-307560-4.50023-5
32. Schneider, T., Voelkel, K., Pflaum, H., Stahl, K.: Friction behavior of pre-damaged wet-running multi-plate clutches in an endurance test. Lubricants **8**(7) (2020). https://doi.org/10.3390/lubricants8070068
33. Seidenfeld, T.I.: The fiducial argument. Ph.D. thesis, Columbia University (1976)
34. Shafer, G.: A Mathematical Theory of Evidence. Princeton University Press, Princeton (1976)
35. Shafer, G., Vovk, V.: Probability and Finance: It's Only a Game!. Wiley, New York (2001)
36. Shariatmadar, K., Debrouwere, F., Misra, A., Versteyhe, M.: Improved uncertainty modelling of process variations in a smart industry manufacturing facility by use of probability box uncertainty models. In: International Conference on Stochastic Processes and Algebraic Structures – From Theory Towards Applications. Mälardalen University, Vasteras, Sweden (2019)
37. Shariatmadar, K., Versteyhe, M.: Linear programming under p-box uncertainty model. In: Machines, pp. 84–89. MDPI AG, TU Delft, Netherlands, August 2019. https://doi.org/10.1109/ICCMA46720.2019.8988632
38. Shariatmadar, K., Arrigo, A., Vallée, F., Hallez, H., Vandevelde, L., Moens, D.: Day-ahead energy and reserve dispatch problem under non-probabilistic uncertainty. Energies **14**(4) (2021). https://doi.org/10.3390/en14041016
39. Shariatmadar, K., De Ryck, M., Driesen, K., Debrouwere, F., Versteyhe, M.: Linear programming under ϵ-contamination uncertainty. Comput. Math. Methods **2**(2) (2020). https://doi.org/10.1002/cmm4.1077
40. Shariatmadar, K., Versteyhe, M.: Numerical linear programming under non-probabilistic uncertainty models—interval and fuzzy sets. Int. J. Uncertainty Fuzziness Knowl. Based Syst. **28**(03), 469–495 (2020). https://doi.org/10.1142/S0218488520500191
41. Smith, C.A.B.: Consistency in statistical inference and decision. J. Roy. Stat. Soc. Ser. B Methodol. **23**(1), 1–37 (1961). http://www.jstor.org/stable/2983842
42. Walley, P.: Statistical Reasoning with Imprecise Probabilities, Chapman and Hall, Monographs on Statistics and Applied Probability, vol. 42. Taylor & Francis (1991). https://books.google.be/books?id=Nk9Qons1kHsC
43. Walley, P.: Measures of uncertainty in expert systems. Artif. Intell. **83**(1), 1–58 (1996). https://doi.org/10.1016/0004-3702(95)00009-7
44. Walley, P., de Cooman, G.: Coherence of rules for defining conditional possibility. Int. J. Approx. Reason. **21**, 63–107 (1999). https://doi.org/10.1016/S0888-613X(99)00007-9
45. Weichselberger, K.: The theory of interval-probability as a unifying model for uncertainty. Int. J. Approx. Reason. **24**, 149–170 (2000). https://doi.org/10.1016/S0888-613X(00)00032-3
46. Weichselberger, K.: Elementare Grundbegriffe einer allgemeineren Wahrscheinlichkeitsrechnung I: Intervallwahrscheinlichkeit als umfassendes Konzept. Physica, Heidelberg (2001). https://doi.org/10.1007/978-3-642-57583-9_3. In cooperation with Augustin, T. and Wallner, A

47. White, H.: A heteroskedasticity-consistent covariance matrix estimator and a direct test for heteroskedasticity. Econometrica **48**, 817–838 (1980). https://doi.org/10.2307/1912934
48. Williams, P.M.: On a new theory of epistemic probability. Br. J. Philos. Sci. **29**(4), 375–387 (1978). htttps://doi.org/10.1093/bjps/29.4.375
49. Williams, P.M.: Notes on conditional previsions. Technical report, School of Math and Physics Science, University of Sussex, February 1975. Published as [50]
50. Williams, P.M.: Notes on conditional previsions. Int. J. Approx. Reason. **44**, 366–383 (2007). https://doi.org/10.1016/j.ijar.2006.07.019. Original: [49]

Forward vs. Bayesian Inference Parameter Calibration: Two Approaches for Non-deterministic Parameter Calibration of a Beam-Column Model

Maximilian Schaeffner[1]([✉]), Christopher M. Gehb[1], Robert Feldmann[1], and Tobias Melz[1,2]

[1] Mechanical Engineering Department, System Reliability,
Adaptive Structures and Machine Acoustics SAM,
Technical University of Darmstadt, Darmstadt, Germany
`maximilian.schaeffner@sam.tu-darmstadt.de`
[2] Fraunhofer Institute for Structural Durability and System Reliability LBF,
Darmstadt, Germany

Abstract. Mathematical models are commonly used to predict the dynamic behavior of mechanical structures or to synthesize controllers for active systems. Calibrating the model parameters to experimental data is crucial to achieve reliable and adequate model predictions. However, the experimental dynamic behavior is uncertain due to variations in component properties, assembly and mounting. Therefore, uncertainty in the model parameters can be considered in a non-deterministic calibration. In this paper, we compare two approaches for a non-deterministic parameter calibration, which both consider uncertainty in the parameters of a beam-column model. The goal is to improve the model prediction of the axial load-dependent lateral dynamic behavior. The investigation is based on a beam-column system subjected to compressive axial loads used for active buckling control. A representative sample of 30 nominally identical beam-column systems characterizes the variations in the experimental lateral axial load-dependent dynamic behavior. First, in a forward parameter calibration approach, the parameters of the beam-column model are calibrated separately for all 30 investigated beam-column systems using a least squares optimization. The uncertainty in the parameters is obtained by assuming normal distributions of the separately calibrated parameters. Second, in a Bayesian inference parameter calibration approach, the parameters are calibrated using the complete sample of experimental data. Posterior distributions of the parameters characterize the uncertain dynamic behavior of the beam-column model. For both non-deterministic parameter calibration approaches, the predicted uncertainty ranges of the axial load-dependent lateral dynamic behavior are compared to the uncertain experimental behavior and the most accurate results are identified.

Keywords: Bayesian inference · Uncertainty · Parameter calibration · Frequency response · Active buckling control

© The Author(s) 2021
P. F. Pelz and P. Groche (Eds.): ICUME 2021, LNME, pp. 173–190, 2021.
https://doi.org/10.1007/978-3-030-77256-7_15

1 Introduction

In engineering science, mathematical models are of utmost importance to predict the dynamic behavior of structures or to improve the structural design. The necessity for accurate mathematical models arises from the ever-increasing virtualization of the product development process and the need to assess dynamic performance under all possible environmental and operating conditions for stability evaluation, designing robust structures or for appropriate controller design [1–3]. During the modeling process, assumptions and simplifications have to be made that determine the form of a model and its parameters. Thereby, model uncertainty and parameter uncertainty are almost unavoidable and omnipresent [4]. Consequently, the numerical predictions of the mathematical model are also uncertain. It follows that the quantification and reduction of model and parameter uncertainty is necessary for reliable and adequate numerical predictions. In this paper and as a gambit for reliable and adequate numerical predictions, we focus on reducing parameter uncertainty while shelving the consideration of model uncertainty. Reducing the parameter uncertainty is achieved i.e. via model parameter calibration [5–7]. Thus, the mathematical model and its numerical predictions are adjusted to the experimentally observed dynamic behavior.

Parameter calibration is commonly achieved by solving an optimization problem to find deterministic values for each model parameter to be calibrated that best fit the chosen calibration criteria. For example in [8], parameters of a friction model were identified for a direct-drive rotary torque motor using the Novel Evolutionary Algorithm optimization. Experimental data from a test rig was used within the optimization process and two objective functions were minimized for different parameter sets. As a result, the calibrated parameters are stated as deterministic values. Similar approaches but using for example genetic and particle swarm optimization algorithms for parameter calibration can be found in [6] and [9] for models of mechanical servo systems. In these studies, the remaining parameter uncertainty after calibration is not taken into account. Hence, deterministic optimization approaches are searching for the best fitting parameter values and then treating the parameters as known and fixed.

In contrast, non-deterministic calibration approaches aim to achieve statistical consistency between model prediction and experimental data [10,11]. The calibrated parameters are stated as distributions representing the remaining parameter uncertainty. In [12], the model parameters of a historic masonry monument FE model were calibrated using a non-deterministic calibration approach. BAYESIAN inference parameter calibration with MARKOV CHAIN MONTE CARLO leads to calibrated parameters resulting in reduced uncertainty in the model prediction. Another example for non-deterministic parameter calibration can be found in [5]. BAYESIAN inference was successfully used to calibrate parameters for several friction models, but inconclusive for the parameters of a LuGRE-friction model. Successful use of BAYESIAN inference to calibrate parameters of a LuGRE-friction model was conducted in [13]. Although only 3 out of 7 model parameters could be calibrated, the uncertainty of the model prediction was reduced considerably.

In this paper, the parameters of a beam-column system subjected to axial loading and prone to buckling, [14,15], are calibrated using two different approaches for non-deterministic parameter calibration, namely a forward parameter calibration approach [14] and a BAYESIAN inference parameter calibration approach [10,16]. The calibrated beam-column model is used to design an active buckling control, which is intended to increase the maximum bearable load of the beam-column [14,15]. The accuracy and credibility of the mathematical beam-column model is essential for the successful application of the active buckling control.

The experimental data for the parameter calibration of the beam-column model is obtained by measuring the axial load-dependent lateral dynamic behavior of 30 nominally identical beam-column systems subjected to varying axial loading. In the forward parameter calibration approach, the parameters of the beam-column model are calibrated separately for all 30 investigated beam-column systems using a least squares optimization. The uncertainty in the parameters is obtained by assuming normal distributions of the separately calibrated parameters. The non-deterministic BAYESIAN inference parameter calibration approach statistically correlates the beam-column model predictions with the experimental data. Additionally, it enables to quantify and reduce the parameter uncertainty concurrently. The uncertainty in the parameters is obtained by sampling from the posterior distributions via MARKOV CHAIN MONTE CARLO (MCMC) sampling. Both approaches increase the beam-column model prediction credibility since the parameters are not assumed as deterministic and, hence, better represent the typically non-deterministic reality.

This paper is organized as follows: The beam-column system and its corresponding mathematical model are introduced in Sect. 2 and Sect. 3. The model calibration approaches are performed and compared in Sect. 4. Additionally, the model predictions with non-calibrated and calibrated parameters are presented. Finally, conclusions and proposed future work are given in Sect. 5.

2   System Description

This section introduces the investigated beam-column system used for active buckling control in [14,15]. In the concept for active buckling control, a slender beam-column subjected to a compressive axial load F_x is stabilized by two piezo-elastic supports at both beam-column ends, as depicted in the schematic sketch in Fig. 1.

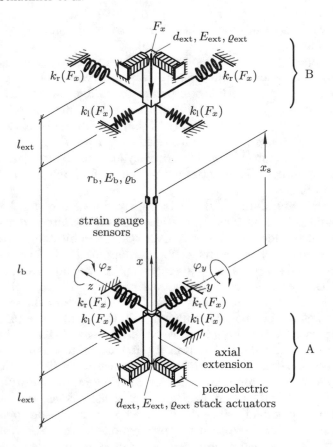

Fig. 1. Schematic sketch of the axially loaded beam-column with elastic supports, piezoelectric stack actuators and strain gauge sensors [14]

The slender beam-column has length $l_b = 400\,mm$, circular solid cross-section with constant radius $r_b = 4\,mm$, Young's modulus $E_b = 75.8 \cdot 10^3\,N/mm^2$ and density $\varrho_b = 2.79 \cdot 10^{-3}\,g/mm^3$. The lower beam-column end is fixed at support A. The upper beam-column end at support B is free to move in longitudinal x-direction and is used to apply the axial load F_x. The piezoelectric stack actuators are integrated in the lateral load path via axial extensions (subscript ext) with length $l_{ext} = 8.1\,mm$, which are connected to the beam-column ends. The axial extensions have quadratic cross-sections with edge length $d_{ext} = 12\,mm$ and relatively high bending stiffness with Young's modulus $E_{ext} = 210.0 \cdot 10^3\,N/mm^2$ and density $\varrho_{ext} = 7.81 \cdot 10^{-3}\,g/mm^3$. Piezoelectric stack actuators exert active lateral forces in positive and negative y- and z-direction to the beam-column's axial extensions by applying the actuator voltages $V_{pz,y/z}(t)$. The resulting active bending moments may act in arbitrary directions at the lower and upper beam-column ends. This is realized by piezo-elastic supports A at location $x = 0$ and B at $x = l_b$ [17]. Finally, four strain gauge sensors at the sensor position $x_s = l_b/2$

are used to measure the surface strains $\varepsilon_{s,y/z}(t)$ due to bending to calculate the lateral deflection of the beam-column in y- and z-direction.

Figure 2a) shows the experimental beam-column with circular solid cross-section. Its lower and upper ends are connected to the piezo-elastic supports A and B, shown in detail in Fig. 2b), which are fixed to a baseplate and a parallel guidance, respectively.

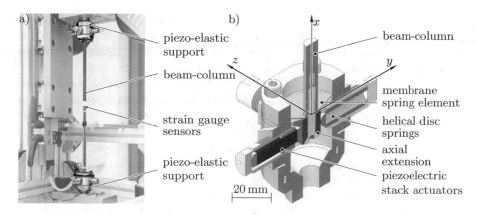

Fig. 2. a) Beam-column system for active buckling control, b) sectional view of piezo-elastic support with x-, y-and z-directions [14]

The piezo-elastic supports are designed to provide the elastic boundary conditions of the beam-column and to include the piezoelectric stack actuators to influence the lateral deflections [17]. The elastic boundary conditions are represented by the axial-load-dependent lateral stiffness $k_l(F_x)$ and rotational stiffness $k_r(F_x)$ in Fig. 1. Figure 2b) shows a sectional view of the piezo-elastic support A. Two differently shaped membrane spring elements bear the axial and lateral loads and allow rotations in any plane perpendicular to the beam-column's longitudinal x-axis. The two piezoelectric stack actuators exert lateral forces in y- and z-direction to the beam-column's axial extensions at a distance l_{ext} from the beam-column ends, as shown in Fig. 1. The piezoelectric stack actuators are mechanically prestressed by allocated helical disk springs.

The lateral dynamic behavior of the beam-column system is strongly dependent on the axial load $F_x(t)$. For the experimental characterization of the axial load-dependent lateral dynamic behavior without active buckling control, the beam-column system is loaded by static axial loads $F_x(t) = $ const. and separately excited by broadband white noise via the piezoelectric stack actuator voltages $V_{pz,y/z}(t)$ in y- and z-direction, which results in the measured beam-column surface strains $\varepsilon_{s,y/z}(t)$. The experimental beam-column transfer functions

$$G_y^{\mathcal{D}}(F_x, \Omega) = \frac{\varepsilon_{s,y}(F_x, \Omega)}{V_{pz,y}(\Omega)} \quad \text{and} \quad G_z^{\mathcal{D}}(F_x, \Omega) = \frac{\varepsilon_{s,z}(F_x, \Omega)}{V_{pz,z}(\Omega)} \quad (1)$$

in y- and z-direction are obtained from the experimental data (superscript $\mathcal{D}$) for a harmonic excitation with angular frequency Ω [18]. The experimental beam-column transfer functions $G^{\mathcal{D}}_{y/z}(F_x, \Omega)$ (1) are estimated by the *tfestimate* algorithm in the MATLAB® *System Identification Toolbox* [19].

3   Mathematical Model of the Active Beam-Column

As in many applications relying on model predictions, the performance of the active buckling control is primarily determined by the quality of the underlying mathematical model used for the model-based controller synthesis, as e.g. in [14,15]. In particular, the model of the beam-column system shown in Figs. 1 and 2 has to properly describe the complex boundary conditions created by the piezo-elastic supports and has to include the piezoelectric stack actuators and strain gauge sensors. Furthermore, the axial load-dependency of the beam-column that is prone to buckle has to be considered. This section introduces the finite element (FE) model of the beam-column with piezo-elastic supports, actuators and sensors as well as the beam-column transfer function that is later used for parameter calibration in Sect. 4.

The mathematical FE model of the slender beam-column with piezo-elastic supports used for active buckling control in an experimental test setup is derived in detail in [14,15]. The resulting state space beam-column model in the LAPLACE domain according to [18] is given by

$$
\begin{aligned}
s\,\boldsymbol{x}(s) &= \boldsymbol{A}(F_x)\boldsymbol{x}(s) + \boldsymbol{B}\boldsymbol{u}(s) \\
\boldsymbol{y}(s) &= \quad \boldsymbol{C}_y\ \boldsymbol{x}(s).
\end{aligned}
\tag{2}
$$

The state vector $\boldsymbol{x}(t)$ of the FE beam-column model is composed of the FE displacement vector and its derivative and describes the lateral deflection of the slender beam-column along the x-axis.

The axial load-dependent system matrix $\boldsymbol{A}(F_x)$ contains the FE mass matrix, FE damping matrix and the FE stiffness matrix. Damping is modeled by RAYLEIGH proportional damping [20], and the boundary conditions resulting from the piezo-elastic supports and the surrounding structure, see Fig. 1, are included in the FE stiffness matrix, which is linearly dependent on the axial load F_x [14]. Potential non-linearities and resulting model uncertainty, especially concerning the membrane spring elements, are purposefully neglected since the linear dependency of the system matrix $\boldsymbol{A}(F_x)$ is required for the concept of the gain-scheduled H_∞ controller used for active buckling control.

The external forces of the piezoelectric stack actuators acting on the beam-column are modeled by the term $\boldsymbol{B}\,\boldsymbol{u}(s)$ in (2), where the voltage input matrix $\boldsymbol{B}$ allocates the piezoelectric stack actuator forces to the lateral degrees of freedom of the first and last nodes of the FE model. The beam-column input vector

$$
\boldsymbol{u}(t) = \begin{bmatrix} V_{\mathrm{pz},y}(t) \\ V_{\mathrm{pz},z}(t) \end{bmatrix}.
\tag{3}
$$

includes the actuator voltages $V_{\mathrm{pz},y}(t)$ and $V_{\mathrm{pz},z}(t)$ that are simultaneously applied to both piezoelectric stack actuators in supports A and B in y- and z-direction. According to Fig. 1, the strain gauge sensors at sensor position $x_\mathrm{s} = l_\mathrm{b}/2$ measure the surface strains $\varepsilon_{\mathrm{s},y}(t)$ and $\varepsilon_{\mathrm{s},z}(t)$ in the beam-column center due to bending in y- and z-direction. Consequently, the surface strains are used as the beam-column model output

$$\boldsymbol{y}(t) = \begin{bmatrix} \varepsilon_{\mathrm{s},y}(t) \\ \varepsilon_{\mathrm{s},z}(t) \end{bmatrix} \tag{4}$$

calculated from the state vector $\boldsymbol{x}(t)$ and the output matrix $\boldsymbol{C}_y$ according to (2), as derived in [14].

Equation (2) is the state space representation of the beam-column model that is used to derive the beam-column transfer function and to analyze the lateral dynamic behavior of the beam-column system. The $[2 \times 2]$ matrix of beam-column model (superscript $\mathcal{M}$) transfer functions

$$\boldsymbol{G}^{\mathcal{M}}(F_x, s) = \begin{bmatrix} G_y^{\mathcal{M}}(F_x, s) & 0 \\ 0 & G_z^{\mathcal{M}}(F_x, s) \end{bmatrix} = \frac{\boldsymbol{y}(F_x, s)}{\boldsymbol{u}(s)} = \boldsymbol{C}\left(\mathbf{I}s - \boldsymbol{A}(F_x)\right)^{-1} \boldsymbol{B} \tag{5}$$

describes the transfer behavior from the actuator voltages $\boldsymbol{u}(s)$ (3) to the beam-column strains (4) $\boldsymbol{y}(s)$ [21].

4   Model Parameter Calibration Approaches

This section presents two non-deterministic calibration approaches for calibrating the parameters of the beam-column model introduced in Sect. 3. Calibrating parameters with experimental data is necessary to adjust the beam-column model to the experimental data, i.e. the measured lateral dynamic behavior, and to achieve a reliable and adequate model prediction.

With both presented approaches, not only one optimally fitted model parameter set is achieved. Instead, parameter uncertainty is taken into account and quantified. The parameters of the beam-column model are calibrated by comparison of the experimental transfer functions (1) with the numerical beam-column model transfer functions (5). With zero initial conditions and the conversion $s = j\,\Omega$ for the beam-column model (5), [18], the experimental and the beam-column model transfer functions are

$$G^{\mathcal{D}}(F_x, \Omega) = \frac{\varepsilon_\mathrm{s}(F_x, \Omega)}{V_{\mathrm{pz}}(\Omega)} \quad \text{and} \quad G^{\mathcal{M}}(F_x, \Omega) = \frac{y(F_x, \Omega, \boldsymbol{\theta})}{u(\Omega, \boldsymbol{\theta})}, \tag{6}$$

which are obtained for the six measured axial loads $F_{x,1} = 337\,\mathrm{N}$, $F_{x,2} = 500\,\mathrm{N}$, $F_{x,3} = 1000\,\mathrm{N}$, $F_{x,4} = 1500\,\mathrm{N}$, $F_{x,5} = 2000\,\mathrm{N}$ and $F_{x,6} = 2500\,\mathrm{N}$. The lowest axial load that is realizable due to the dead weight of the test setup in Fig. 2a) is $F_{x,1} = 337\,\mathrm{N}$. As was shown in [14,22], the lateral dynamic behavior of the beam-column system in y- and z-direction is very similar, so that the parameters in both directions and both piezo-elastic supports are assumed to be identical and are calibrated as one, see Fig. 1. Hence, no distinction between y- and z-direction is made for the calibration approaches.

The parameters that were shown to have a strong influence on the beam-column lateral dynamic behavior, [14,23], are calibrated to fit the numerical beam-column model transfer functions $G^{\mathcal{M}}(F_x, \Omega)$ with the experimental beam-column transfer functions $G^{\mathcal{D}}(F_x, \Omega)$, see (6). In addition to [23], a variance-based sensitivity analysis is applied to further reduce the parameter space in order to avoid identifiability issues with parameters [12,13]. The calibrated parameters are

$$\boldsymbol{\theta} = [k_{\mathrm{r,e}},\ k_{\mathrm{r,g}},\ \beta,\ \zeta_1,\ m_{\mathrm{s}},\ k_{\mathrm{s}}] \tag{7}$$

with the elastic and geometric rotational support stiffness $k_{\mathrm{r,e}}$ and $k_{\mathrm{r,g}}$ that describe the axial-load dependent rotational support stiffness shown in Fig. 1 via $k_{\mathrm{r}}(F_x) = k_{\mathrm{r,e}} + F_x \cdot k_{\mathrm{g,r}}$, the piezoelectric force constant β, the modal damping ratio of the first mode of vibration ζ_1, and the sensor mass and stiffness m_{s} and k_{s}. The parameters of the beam-column model in (2) that are assumed to be fixed for all investigated beam-column systems are given in [14].

4.1 Forward Parameter Calibration

Parameter calibration is commonly achieved by solving an optimization problem to find deterministic values for each parameter to be calibrated that best fit the chosen calibration criteria. The experimental data for calibration mostly stems from one investigated system. In contrast, solving multiple optimization problems for varying but nominally identical beam-column systems complements statistical information and provides the opportunity to quantify the parameter uncertainty, as was shown in [14]. In this investigation, the forward parameter calibration is performed for 30 nominally identical beam-column systems that originate from the combination of five identically instrumented beam-columns, three combinations of lower and upper piezo-elastic supports and two different sets of piezoelectric stack actuators. Thus, the applied component variation combines the effects of uncertainty in manufacturing, assembly and mounting of the beam-column systems. For each of the 30 beam-column systems, the parameters $\boldsymbol{\theta}$ (7) in y-and z-direction are varied to solve the least squares curve fitting problem

$$\min_{\boldsymbol{\theta}} \left(\sum_{i=1}^{6} \left|\left| G^{\mathcal{D}}(F_{x,i}, \Omega) - G^{\mathcal{M}}(F_{x,i}, \Omega, \boldsymbol{\theta}) \right|\right|_2^2 \right) \tag{8}$$

with the experimental and numerical beam-column transfer functions $G^{\mathcal{D}}(F_{x,i}, \Omega)$ and $G^{\mathcal{M}}(F_{x,i}, \Omega, \boldsymbol{\theta})$ from (6) for the six measured axial loads $F_{x,i}$. The model calibration is performed for all six measured axial loads F_x at once so that the axial load-dependency is well captured by the model. The least squares curve fitting problem (8) is solved by the *lsqnonlin* algorithm in the MATLAB® *Optimization Toolbox* [24] by varying the parameters to be calibrated $\boldsymbol{\theta}$ within a specified range, i.e. the parameters' prior bounds. The resulting 30 separately calibrated parameter sets for the y- and z-directions are combined in one sample of $N = 60$ calibrated parameter sets shown as histograms in Fig. 3.

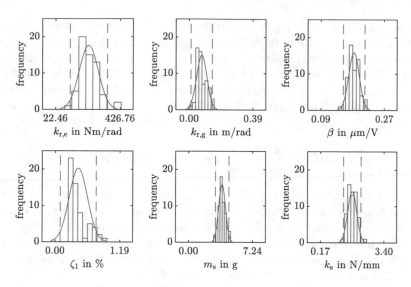

Fig. 3. Histograms of the $N = 60$ calibrated parameter sets θ and normal distribution fits $p_{\mathcal{N}}(\theta)$ with 95% interpercentiles $(--)$

The sample of calibrated parameter sets are approximated by normal distributions (subscript $\mathcal{N}$) that are used to describe the parameter uncertainty associated with each parameter of the six calibrated parameters θ. Hence, the forward parameter calibration results in the parameter distributions The fitted normal probability density functions (pdf) $p_{\mathcal{N}}(\theta)$ as well as the 95% interpercentiles given by $\mu_\theta \pm 1.96\,\sigma_\theta$ are also shown in Fig. 3. The corresponding mean values μ_θ and standard deviations σ_θ are summarized in Table 1.

Table 1. Normal distribution fits for the $N = 60$ calibrated parameter sets θ

Parameter	μ_θ	σ_θ
$k_{\mathrm{r,e}}$ in Nm/rad	230.52	59.55
$k_{\mathrm{r,g}}$ in m/rad	0.08	0.03
β in μm/V	0.18	0.01
ζ_1 in %	0.43	0.17
m_{s} in g	3.75	0.39
k_{s} in N/mm	1.75	0.22

4.2   Bayesian Inference Parameter Calibration

BAYESIAN inference can be used as non-deterministic calibration approach to calibrate uncertain parameters by correlating the model predictions with experimental data, i.e. a number of $n = 1, \cdots, N$ measurements, by solving an inverse problem [25, 26]. In this paper, the BAYESIAN inference approach statistically connects the beam-column model and the experimental data (6) to adequately calibrate the varied parameters $\boldsymbol{\theta}$ (7). Current knowledge of the system and its parameters is updated with new information obtained from experimental tests. Thus, the parameter uncertainty is reduced and quantified by inference from the posterior distribution [10, 25]. Using BAYES' Theorem [25, 27], the posterior parameter distribution given the experimental results can be stated as

$$P(\boldsymbol{\theta}, \mathcal{M}|\mathcal{D}) = \frac{L(\mathcal{D}|\boldsymbol{\theta}, \mathcal{M}) \times P(\boldsymbol{\theta})}{P(\mathcal{D})} \propto L(\mathcal{D}|\boldsymbol{\theta}, \mathcal{M}) \times P(\boldsymbol{\theta}). \tag{9}$$

Since the denominator $P(\mathcal{D})$ in (9), the total probability or evidence, is typically not computable with reasonable effort and is only normalizing the result anyway [5], it is more practical to sample from a proportional relationship of the posterior parameter distribution $P(\boldsymbol{\theta}, \mathcal{M}|\mathcal{D})$. If no further information regarding the prior distributions $P(\boldsymbol{\theta})$ in (9) is available, uninformative priors between certain upper and lower bounds are assumed [5, 12]. The likelihood function

$$L(\mathcal{D}|\boldsymbol{\theta}, \mathcal{M}) = c^{-1} \prod_{n=1}^{N} \exp\left(E_{\text{stat},n}(\boldsymbol{\theta}) + E_{\max,n}(\boldsymbol{\theta}) + E_{\text{ef},n}(\boldsymbol{\theta})\right) \tag{10}$$

represents the probability of observing experimental data $\mathcal{D}$ given a set of parameters $\boldsymbol{\theta}$ for the mathematical model $\mathcal{M}$ [11, 25]. Here, the beam-column transfer functions obtained from the experimental data are defined as $\mathcal{D} := G^{\mathcal{D}}(F_x, \Omega)$ and the beam-column model transfer functions as $\mathcal{M} := G^{\mathcal{M}}(F_x, \Omega, \boldsymbol{\theta})$ according to (6). The scaling factor c^{-1} is constant and crosses out when calculating the acceptance ratio, see Algorithm 1. In this paper, the error function as an essential element of the likelihood function $L(\mathcal{D}|\boldsymbol{\theta}, \mathcal{M})$ in (10) consists of the three parts

$$E_{\text{stat},n}(\boldsymbol{\theta}) = \sum_{i=1}^{6} \left(\frac{\left(G_n^{\mathcal{D}}(F_{x,i}, \Omega \to 0) - G_n^{\mathcal{M}}(F_{x,i}, \Omega = 0, \boldsymbol{\theta})\right)^2}{\sigma_{\text{stat}}^2} \right) \tag{11a}$$

$$E_{\max,n}(\boldsymbol{\theta}) = \sum_{i=1}^{6} \left(\frac{\left(G_n^{\mathcal{D}}(F_{x,i}, \Omega = \omega_1) - G_n^{\mathcal{M}}(F_{x,i}, \Omega = \omega_1, \boldsymbol{\theta})\right)^2}{\sigma_{\max}^2} \right) \tag{11b}$$

$$E_{\text{ef},n}(\boldsymbol{\theta}) = \sum_{i=1}^{6} \left(\frac{\left(\omega_{1,n}^{\mathcal{D}}(F_{x,i}) - \omega_{1,n}^{\mathcal{M}}(F_{x,i}, \boldsymbol{\theta})\right)^2}{\sigma_{\text{ef}}^2} \right) \tag{11c}$$

that are chosen to be the squared errors between experimental data and model predictions, cf. [25], for the static deflection E_{stat}, the maximum amplitude at resonance $E_{\max}$ and the first eigenfrequency E_{ef} summed up over the six measured axial loads $F_{x,i}$. The variances σ_{stat}^2, $\sigma_{\max}^2$ and σ_{ef}^2 are related to the measurement errors of the corresponding measured values, that are assumed to be independent and normally distributed with zero mean [25]. In order to achieve reasonable results, the variances σ_{stat}^2, $\sigma_{\max}^2$ and σ_{ef}^2, which have a strong effect on the parameter calibration results, were chosen to be double the measurement errors of the corresponding measured values.

In this paper, the parameter space is explored using MARKOV CHAIN MONTE CARLO (MCMC) sampling to approximate the posterior parameter distributions $P(\boldsymbol{\theta}, \mathcal{M}|\mathcal{D})$ by drawing multiple samples from these posterior parameter distributions. That is, the histograms of the parameters $\boldsymbol{\theta}$ of all random samples produce the approximated posterior parameter distributions $P(\boldsymbol{\theta}, \mathcal{M}|\mathcal{D})$ [25, 28]. By application of MCMC sampling, it is possible to circumvent the calculation of the denominator $P(\mathcal{D})$ in (9) and, thus, to enable a mathematically more efficient application of the BAYESIAN inference parameter calibration approach using the METROPOLIS-Algorithm [25], as summarized in Algorithm 1.

Algorithm 1: METROPOLIS-Algorithm

Initialize parameters $\boldsymbol{\theta}_0$ to set chain start point

for $m = 1, \ldots, M$ **do**

    Generate proposed parameters: $\boldsymbol{\theta}^* \sim \mathcal{N}(\boldsymbol{\theta}_{m-1}, \boldsymbol{C})$

    Compute acceptance probability:

$$\alpha = \min\left(1, \frac{L(\mathcal{D}|\boldsymbol{\theta}^*, \mathcal{M}) \times P(\boldsymbol{\theta}^*)}{L(\mathcal{D}|\boldsymbol{\theta}_{m-1}, \mathcal{M}) \times P(\boldsymbol{\theta}_{m-1})}\right)$$

    Sample from uniform distribution: $u_\alpha \sim \mathcal{U}(0, 1)$

    Accept/Reject decision:

    if $u_\alpha < \alpha$ **then**

        $\theta_m = \theta^*$

    else

        $\theta_m = \theta_{m-1}$

    end

end for

return $\boldsymbol{\theta}_0, \ldots, \boldsymbol{\theta}_M$

The number of M chain elements is generated iteratively. The proposal distribution $\mathcal{N}(\boldsymbol{\theta}_{m-1}, \boldsymbol{C})$ is the probability of moving to a point in the parameter space and is chosen to be GAUSSIAN centered at the current parameter set $\boldsymbol{\theta}_{m-1}$ with covariance matrix $\boldsymbol{C}$. The covariance matrix $\boldsymbol{C}$ governs the dispersion in terms of how far the proposal parameter set $\boldsymbol{\theta}^*$ moves from the current parameter set $\boldsymbol{\theta}_{m-1}$. Subsequently, the acceptance probability α is calculated by dividing the posterior probability given the proposed parameter set $\boldsymbol{\theta}^*$ by the posterior probability given the current parameter set $\boldsymbol{\theta}_{m-1}$. Each iteration ends with the decision, if the proposed parameter set $\boldsymbol{\theta}^*$ is accepted or the current parameter set $\boldsymbol{\theta}_{m-1}$ is kept. Each result is a sample of the posterior distributions and the histograms of the calibration candidate parameters for all samples represent the approximated posterior distributions.

Figure 4 depicts the parameter calibration results obtained from $M = 50.000$ MCMC runs, where an acceptance rate of 27% was achieved. On the diagonal, the parameter distributions are shown as histograms representing approximations of the marginal posterior distributions for the calibrated parameters $\boldsymbol{\theta}$ (7). The lower off-diagonal depicts distribution contour plots for the joint probability distribution pairs of the parameters. The upper off-diagonal depicts posterior sample pairs of the parameters. Furthermore, the narrow histograms graphically depict the knowledge gain and the uncertainty reduction of the posterior parameter ranges compared to the prior parameter bounds.

4.3   Comparison of Calibration Results

For a comparison and assessment of the uncertainty inherent to the calibration parameters, it is first advisable to examine the posterior distributions obtained by the forward parameter calibration and the BAYESIAN inference parameter calibration, as depicted in Fig. 3 and Fig. 4. In general, the posterior distributions obtained by the forward parameter calibration approach are more dense than those obtained by the BAYESIAN inference parameter calibration approach. This exception might have been provoked by the bad fitting of the normal distribution for ζ_1, as can be seen in Fig. 3.

Table 2 summarizes the prior and posterior uncertainty in form of parameter bounds and 95% interpercentiles after parameter calibration as well as the reduction of parameter uncertainty in $\boldsymbol{\theta}$ (7). The parameter ranges covering the 95% interpercentiles are reduced by 42.3% to 79.0% for the forward parameter calibration approach and by 8.3% to 79.3% for the BAYESIAN inference parameter calibration approach compared to the prior bounds. Only for the modal damping ratio of the first mode of vibration ζ_1, a larger reduction is achieved by the BAYESIAN inference approach.

While both approaches can be compared by the posterior distributions shown in Fig. 3 and Fig. 4, the posterior distribution obtained by BAYESIAN inference provides additional information about covariances displayed on the off-diagonals. These covariance plots can reveal statistical dependencies between the parameters. Inadequate modeling can cause these dependencies and help detect model uncertainty.

Aside from parameter uncertainty, the assessment and comparison of the model predictions for the axial load-dependent lateral dynamic behavior of the beam-column system provide valuable insights into the two calibration approaches. Figure 5 depicts the envelopes of amplitude $|G^{\mathcal{M}}(F_{x,i}, \Omega, \boldsymbol{\theta})|$ and phase responses $\arg G^{\mathcal{M}}(F_{x,i}, \Omega, \boldsymbol{\theta})$ for the prior distributions as well as both calibration approaches that were generated using 50.000 samples of the resulting calibrated parameters $\boldsymbol{\theta}$ within their 95% interpercentiles in Monte-Carlo simulations. Additionally, the envelopes for the 60 measured amplitude $|G^{\mathcal{D}}(F_{x,i}, \Omega)|$ and phase responses $\arg G^{\mathcal{D}}(F_{x,i}, \Omega)$ are shown. For the purpose of simplicity and clarity, the simulated and measured envelopes are only shown for the selected axial loads $F_{x,1} = 337\,\text{N}$, $F_{x,4} = 1500\,\text{N}$ and $F_{x,6} = 2500\,\text{N}$.

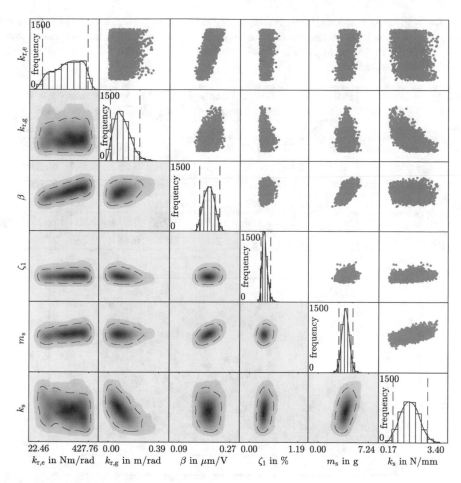

Fig. 4. Posterior distributions with 95% interpercentiles ($--$) for the parameters $\boldsymbol{\theta}$ on the diagonal, bivariate joint distributions on the lower off-diagonal and posterior samples on the upper off-diagonal

Here, the area covered by the envelopes can be seen as a measure of prediction uncertainty. Compared to the envelopes generated using the prior bounds, the area covered by the posterior envelopes associated with the forward parameter calibration and the BAYESIAN inference parameter calibration approach is reduced by 43.0% and 86.7%, respectively. The experimental data envelopes are almost entirely contained in the posterior envelopes and the axial load-dependency is well captured by both posterior envelopes. It is evident, that the BAYESIAN inference approach comes closest to the experimental data envelopes and, therefore, outperforms the forward parameter calibration approach. This assessment stands in contrast to the observations made for the parameter distributions that initially suggested a greater reduction of uncertainty for most

Table 2. Prior and posterior uncertainty of the calibration parameters θ (7) in form of bounds and 95% interpercentiles for the forward parameter calibration approach and the BAYESIAN inference approach as well as uncertainty reduction from prior bounds to 95% interpercentiles

Parameter	Prior bounds		Forward calibration 95% interpercentiles			BAYESIAN inference 95% interpercentiles		
	min	max	min	max	red	min	max	red
$k_{\mathrm{r,e}}$ in Nm/rad	22.46	426.76	113.81	347.23	42.3%	45.84	416.40	8.3%
$k_{\mathrm{r,g}}$ in m/rad	0.00	0.39	0.01	0.14	68.3%	0.01	0.24	42.5%
β in μm/V	0.09	0.27	0.15	0.21	67.8%	0.16	0.23	62.0%
ζ_1 in %	0.00	1.19	0.09	0.76	43.9%	0.27	0.51	79.3%
m_{s} in g	0.00	7.24	2.99	4.51	79.0%	3.01	5.03	72.1%
k_{s} in N/mm	0.17	3.40	1.31	2.19	72.8%	0.45	2.78	27.8%

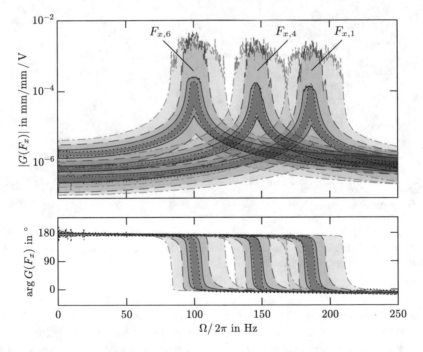

Fig. 5. Envelopes of amplitude and phase response obtained by sampling from prior distribution ( ), posterior distribution resulting from the forward parameter calibration approach ( ) and the BAYESIAN inference approach ( ) as well as envelope of the 60 measured amplitude and phase responses ( )

parameters caused by the forward parameter calibration approach. This illustrates, that a reduction of parameter uncertainty alone is not able to indicate a reduction of prediction uncertainty.

5   Conclusion

This paper presents two non-deterministic parameter calibration approaches for the model of a beam-column system to predict the lateral dynamic behavior. The beam-column system is intended to investigate active buckling control via piezo-elastic supports. The parameters are calibrated to achieve model predictions that are statistically consistent with the experimental data. Therefore, instead of having a potentially over-fitted deterministic value for each calibrated parameter, the inherent uncertainty of the beam-column system is considered. The parameters to be calibrated are selected based on their uncertainty and sensitivity on the beam-column model predictions. Parameters with negligible uncertainty or sensitivity are not calibrated to reduce the computational effort and avoid identifiability issues. With the reduced number of parameters to be calibrated, the forward parameter calibration and the BAYESIAN inference parameter calibration are performed as non-deterministic parameter calibration approaches. The parameter ranges covering the 95% interpercentiles, indicating the parameter uncertainty, are reduced by 42.3% to 79.0% for the forward parameter calibration approach and by 8.3% to 79.3% for the BAYESIAN inference parameter calibration approach compared to the non-calibrated prior bounds. The parameter uncertainty of one parameter, the modal damping ratio of the first mode of vibration ζ_1, is reduced significantly more by the BAYESIAN inference parameter calibration approach. This leads not only to a reduction of the uncertainty in the beam-column model predictions by up to 86.7% comparing to predictions obtained with non-calibrated prior bounds, but also to an almost identical representation of the experimental data. The forward parameter calibration approach leads to a reduction of the uncertainty in the beam-column model predictions by up to 43.0% compared to the prediction obtained with non-calibrated prior bounds with significantly wider envelopes around the experimental data. The BAYESIAN inference approach comes closest to the experimental data envelopes and, therefore, outperforms the forward parameter calibration approach. Nevertheless, by reducing the parameter uncertainty, the model prediction accuracy and credibility can be increased with both non-deterministic parameter calibration approaches. The consideration of model uncertainty in future investigations may further increase the accuracy and credibility of the beam-column model predictions. By that, biased results for the parameter calibration may be avoided by simultaneously keeping the physical meaning of the parameters.

Acknowledgement. The authors like to thank the Deutsche Forschungsgemeinschaft (DFG, German Research Foundation) for funding this project within the Sonderforschungsbereich (SFB, Collaborative Research Center) 805 "Control of Uncertainties in Load-Carrying Structures in Mechanical Engineering" – project number: 57157498. In addition, the authors acknowledge Lei Xu for his assistance during this project.

References

1. Herold, S., Jungblut, T., Kurch, M.: A systematic approach to simulate active mechanical structures. In: Multi-Disciplinary Simulations - The Future of Virtual Product Development (2009)
2. Tamm, C., Thiel, J., Bartel, T., Atzrodt, H., Herold, S.: Methodisches vorgehen zur auslegung des vibro-akustischen verhaltens eines fahrzeugs. In: Wiedemann, M., Misol, M., Melz, T. (eds.) Smarte Strukturen und Systeme, pp. 95–106 (2016)
3. Platz, R., Götz, B.: Non-probabilistic uncertainty evaluation in the concept phase for airplane landing gear design. In: Barthorpe R.J., Platz, R., Lopez, I., Moaveni, B., Papadimitriou, C. (eds.) Model Validation and Uncertainty Quantification, Conference Proceedings of the Society for Experimental Mechanics Series, vol. 3. Springer International Publishing, Cham (2017)
4. Pelz, F.P., Pfetsch, M.E., Kersting, S., Kohler, M., Matei, A., Melz, T., Platz, R., Schaeffner, M., Ulbrich, S.: Types of Uncertainty, pp. 27–45. Springer International Publishing, Cham (2021). (in Print)
5. Green, P.L., Worden, K.: Modelling friction in a nonlinear dynamic system via Bayesian inference. In: Allemang, R., de Clerck, J., Niezrecki, C., Wicks, A. (eds.) Special Topics in Structural Dynamics, vol. 6, pp. 543–553. Springer, New York (2013)
6. Liu, D.-P.: Parameter identification for lugre friction model using genetic algorithms. In: Proceedings of 2006 International Conference on Machine Learning and Cybernetics, Piscataway NJ. IEEE (2006)
7. Mollineaux, M.G., van Buren, K.L., Hemez, F.M., Atamturktur, S.: Simulating the dynamics of wind turbine blades: Part i, model development and verification. Wind Energy 16(5), 694–710 (2013)
8. Wang, X., Lin, S., Wang, S.: Dynamic friction parameter identification method with lugre model for direct-drive rotary torque motor. In: Mathematical Problems in Engineering, pp. 1–8 (2016)
9. Wenjing, Z.: Parameter identification of lugre friction model in servo system based on improved particle swarm optimization algorithm, pp. 135–139 (2007)
10. Kennedy, M.C., O'Hagan, A.: Bayesian calibration of computer models. J. R. Stat. Soc. Ser. B (Stat. Methodol.) 63(3), 425–464 (2001)
11. Gehb, C.M.: Uncertainty evaluation of semi-active load redistribution in a mechanical load-bearing structure. Dissertation, Technische Universität Darmstadt, Darmstadt (2019)
12. Atamturktur, S., Hemez, F.M., Laman, J.A.: Uncertainty quantification in model verification and validation as applied to large scale historic masonry monuments. Eng. Struct. 43, 221–234 (2012)
13. Gehb, C., Atamturktur, S., Platz, R., et al.: Bayesian Inference Based Parameter Calibration of the LuGre-Friction Model. Exp Tech 44, 369–382 (2020). https://doi.org/10.1007/s40799-019-00355-7
14. Schaeffner, M.: Quantification and evaluation of uncertainty in active buckling control of a beam-column subject to dynamic axial loads. Dissertation, Technische Universität Darmstadt, Darmstadt (2019)
15. Schaeffner, M., Platz, R.: Gain-scheduled H_∞ buckling control of a circular beam-column subject to time-varying axial loads. Smart Mater. Struct. 27(6), 065009 (2018)
16. van Buren, K.L., Mollineaux, M.G., Hemez, F.M., Atamturktur, S.: Simulating the dynamics of wind turbine blades: Part ii, model validation and uncertainty quantification. Wind Energy 16(5), 741–758 (2013)

17. Enss, G.C., et al.: Device for bearing design elements in lightweight structures (Festkörperlager). DE 10 2015 101 084 A1
18. Fuller, C.R., Elliott, S.J., Nelson, P.A.: Active Control of Vibration. Academic Press Inc. (1997)
19. Ljung, L.: System Identification Toolbox^{TM}. User's Guide. Natick, MA, Math-Works (2017)
20. Preumont, A.: Vibration control of active structures, An introduction. Springer, Berlin (2011)
21. Skogestad, S., Postlethwaite, I.: Multivariable Feedback Control. John Wiley & Sons (2001)
22. Schaeffner, M., Platz, R., Melz, T.: Adequate mathematical beam-column model for active buckling control in a tetrahedron truss structure. In: Model Validation and Uncertainty Quantification, Conference proceedings of the Society for Experimental Mechanics series, vol. 3, pp. 323–332. Springer International Publishing (2020)
23. Li, S., Goetz, B., Schaeffner, M., Platz, R.: Approach to prove the efficiency of the Monte Carlo method combined with the elementary effect method to quantify uncertainty of a beam structure with piezo-elastic supports. In: Papadrakakis, M., Papadopoulos, V., Stefanou, G. (eds.) Proceedings of the 2nd International Conference on Uncertainty Quantification in Computational Sciences and Engineering (UNCECOMP 2017), pp. 441–455. Institute of Structural Analysis and Antiseismic Research School of Civil Engineering National Technical University of Athens (NTUA) (2017)
24. Mathworks Inc., ed. Optimization toolbox^{TM}, User's Guide. Natick, MA, Math-Works (2018)
25. Smith, R.C.: Uncertainty quantification: theory, implementation, and applications. Computational Science Engineering. Society for Industrial and Applied Mathematics, Philadelphia (2014). ISBN 978-1-611973-21-1
26. Nagel, J.B.: Bayesian techniques for inverse uncertainty quantification. Dissertation, ETH Zürich (2017)
27. Bayes, T.: An essay towards solving a problem in the doctrine of chances. By the late Rev. Mr. Bayes, F.R.S communicated by Mr. Price, in a letter to John Canton, A.M.F.R.S Philosophical Transactions of the Royal Society of London, vol. 53, pp. 370–418 (1763)
28. Riddle, M., Muehleisen, R.T.: A guide to Bayesian calibration of building energy models. In: ASHRAE/IBPSA-USA (2014)

Surrogate Model-Based Uncertainty Quantification for a Helical Gear Pair

Thomas Diestmann[1](✉), Nils Broedling[1], Benedict Götz[2], and Tobias Melz[3]

[1] Robert Bosch GmbH, Robert-Bosch-Campus 1, 71272 Renningen, Germany
thomas.diestmann@de.bosch.com
[2] Fraunhofer Institute for Structural Durability and System Reliability LBF,
Bartningstr. 47, 64289 Darmstadt, Germany
[3] Fachgebiet Systemzuverlässigkeit, Adaptronik und Maschinenakustik SAM,
Technische Universität Darmstadt, Magdalenenstr. 4, 64289 Darmstadt, Germany

Abstract. Competitive industrial transmission systems must perform most efficiently with reference to complex requirements and conflicting key performance indicators. This design challenge translates into a high-dimensional multi-objective optimization problem that requires complex algorithms and evaluation of computationally expensive simulations to predict physical system behavior and design robustness. Crucial for the design decision-making process is the characterization, ranking, and quantification of relevant sources of uncertainties. However, due to the strict time limits of product development loops, the overall computational burden of uncertainty quantification (UQ) may even drive state-of-the-art parallel computing resources to their limits. Efficient machine learning (ML) tools and techniques emphasizing high-fidelity simulation data-driven training will play a fundamental role in enabling UQ in the early-stage development phase.

This investigation surveys UQ methods with a focus on noise, vibration, and harshness (NVH) characteristics of transmission systems. Quasi-static 3D contact dynamic simulations are performed to evaluate the static transmission error (TE) of meshing gear pairs under different loading and boundary conditions. TE indicates NVH excitation and is typically used as an objective function in the early-stage design process. The limited system size allows large-scale design of experiments (DoE) and enables numerical studies of various UQ sampling and modeling techniques where the design parameters are treated as random variables associated with tolerances from manufacturing and assembly processes. The model accuracy of generalized polynomial chaos expansion (gPC) and Gaussian process regression (GPR) is evaluated and compared. The results of the methods are discussed to conclude efficient and scalable solution procedures for robust design optimization.

Keywords: Transmission design · Uncertainty quantification ·
Generalized polynomial chaos expansion · Gaussian process regression

© The Author(s) 2021
P. F. Pelz and P. Groche (Eds.): ICUME 2021, LNME, pp. 191–207, 2021.
https://doi.org/10.1007/978-3-030-77256-7_16

1   Introduction

The design of industrial transmission systems is characterized by conflicting key performance indicators (KPIs), which need to be optimized simultaneously. This multi-objective optimization problem of the main objectives efficiency, NVH, and weight is constrained by the requirement of minimum tooth flank and tooth root load carrying capacity [18]. The challenge is, moreover, the expansive evaluation of the KPIs and the impact of uncertainty due to manufacturing tolerances. The definition of tolerances [1] and gear accuracy grade [3] is especially crucial in this design process. On the one hand, an upgrade in the gear accuracy grade drives the cost up to +80% [21]; on the other, it has a significant influence on the design objectives. However, the influences of manufacturing tolerances often rely on experience or are not fully considered [5].

The influence of tolerances on objective functions has been investigated in previous works, ranging from engineering tolerances to a prior assumptions on approximation model form and its parametrization. The impact of geometrical deviation on the objective unloaded kinematic transmission error is analyzed in [6]. The tooth contact analysis algorithm determines the kinematic relationships of each tooth meshing, neglecting elastic deformations under load. Due to the simplicity of the underlying model, a Monte-Carlo simulation is used for statistical analysis. In [5], Brecher considers multiple design objectives, which are evaluated by a finite element-based simulation. The uncertain parameters are discretized, and a full factorial approach is pursued to calculate the mean and the variance of the objectives.

In this study, the application of surrogate models for uncertainty quantification in transmission design is proposed. A surrogate or meta model approximates the expansive objective function. To enable robust optimization of real transmission systems in an early design stage, it is essential to reduce the number of expansive objective function evaluations to a minimum. Therefore, the focus of this study is the data-efficiency of surrogates, i.e. a high model accuracy based on small data sets of high-fidelity simulations. Gaussian process regression (GPR) and generalized polynomial chaos expansion (gPC) have received much attention in the field of uncertainty quantification [9] and are analyzed in combination with different sampling methods employing two exemplary applications (see Fig. 1). The surrogates are used for global sensitivity analysis and forward uncertainty propagation. Global sensitivity analysis helps to determine the importance of each model input in predicting the response. It can help to reduce the complexity of surrogates by considering only key inputs. In forward uncertainty propagation, the effect of the uncertain inputs on the response is evaluated. In the exemplary analyzed applications, the inherently variable tolerances are aleatoric uncertainties of the model's inputs with a probability distribution defined by the production process. For reasons of clarity, the study is limited to a single KPI, the NVH indicator transmission error of a helical gear pair. The peak-to-peak transmission error calculation via loaded tooth contact analysis (LTCA) is described in Sect. 2.

Section 3 introduces numerical methods for uncertainty quantification covering the above mentioned surrogate models GPR (Sect. 3.1) and gPC (Sect. 3.2), sampling methods (Sect. 3.3) and Sobol indices for global sensitivity analysis (Sect. 3.4). In Sect. 4 these methods are applied and evaluated regarding data-efficiency in example applications with two (Sect. 4.1) and five (Sect. 4.2) uncertain input parameters followed by the conclusion in Sect. 5.

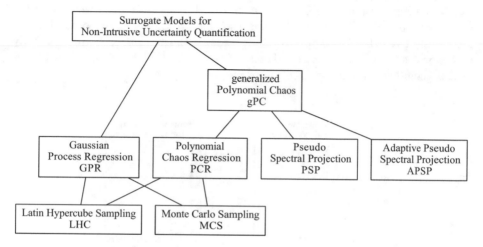

Fig. 1. Selected surrogate models and sampling methods for numerical forward uncertainty quantification and global sensitivity analysis

2   Evaluation of NVH Performance

The static transmission error is a measure of gear excitation. It is a typical objective function for the NVH performance of a gear pair in an early design stage. The TE is defined as the gear pairs load-dependent deviation of the ideal kinematic angular displacement. The deviations are mainly caused by elastic deformation, flank modifications, tooth manufacturing variances, misalignment of the gears, and pitch deviations. Typically the angular transmission error is projected onto the line of action by multiplying it with the base radius r_{b0} (1):

$$TE = (\theta_0 - \theta_1 \frac{z_1}{z_0})r_{b0} \qquad (1)$$

in which the subscripts 0 and 1 respectively refer to the pinion and wheel gear, θ is the angular displacement, and z represents the number of teeth. The absolute transmission error value depends on the respective position during a gearing cycle. Relevant for the excitation of a transmission system is the varying part. Therefore, either the truncated spectral representation of the cycle or the difference between maximum and minimum, the peak-to-peak transmission error, are commonly used as NVH indicators (see Fig. 2) [11].

To derive the TE, the relative angular position of pinion and wheel is evaluated under load for a discrete number of positions θ_0 during a gearing cycle. The exact tooth contact and the deformation are computed via loaded tooth contact analysis. This calculation is not standardized and a variety of methods of different fidelities exists.

Essentially a full FE approach can be taken. Well-established commercial FE solvers are equipped with universal contact algorithms. The contact zone requires a high-resolution FE mesh to model the interaction between meshing gears precisely. Flank and profile modifications are implemented onto the surface mesh. Even though computationally expansive, this approach automatically incorporates the effects of extended contact zones due to deformation, the impact of rim geometry, and the interaction of neighboring teeth.

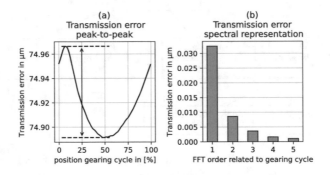

Fig. 2. NVII indicator transmission error: evaluation of a gearing cycle

In transmission design, the utilization of tailored methods for specific gear types is state of the art. In the first step, the contact lines are discretized. An explicit description of the ideal helical gear pairs contact line is available exploiting the kinematic equations. The contact point's stiffnesses can be analytically approximated. The stiffness components are composed of tooth bending, gear body deformation, and Hertzian deformation. The gear flank is separated into slices in this approach. Empirical constants characterize the interconnection of these slices and the gear body deformation. Hence, the analytical approximation of the stiffness has limited validity.

Higher fidelity methods use FE models to calculate the stiffness [8]. In a preprocessing step, a unit force is applied to every point of the FE mesh involute surface respectively to obtain influence coefficients. The stiffness matrix of the discrete contact points can be deduced from these influence coefficients. Therefore, a linear system of equations is given. The relative displacements of the contact points are corrected according to the flank modifications. The influence coefficients of a macro geometry design can be reused for different gear mesh positions and the evaluation of flank modifications. This system must be solved iteratively to find the loaded points of contact.

In this study, the multibody simulation (MBS) solver Adams with the plug-in GearAT is used, an LTCA implementation based on FE influence coefficients. Contact simulations are performed under quasi-static loading conditions. The choice of an MBS solver enables the evaluation of dynamic KPIs in further investigations.

3   Numerical Methods for Uncertainty Quantification

Numerical methods for uncertainty quantification determine the impact of uncertain input variables on the objective function. A primary result of forward uncertainty propagation is determining the output distribution's central moments, such as the statistical measures mean and standard deviation. Generally, this problem cannot be solved in closed form, particularly when the objective function's explicit representation is unknown. Yet, the dominant approach is to treat the inputs as random variables and thus convert the original deterministic system into a stochastic system [27]. This work aims to survey suitable UQ approaches for transmission systems. This class of high-dimensional engineering problems requires computationally expensive simulations to predict physical system behavior with high-fidelity models. In the present work, GPR and gPC as surrogate models for uncertainty quantification are investigated. This approach is classified as a non-intrusive method since the objective function is treated as a black box model [23].

3.1   Gaussian Process Regression

Gaussian Process regression (GPR) or Kriging is a non-parametric model that is not committed to a specific functional form. It was first introduced in [15]. A Gaussian Process (GP) is a distribution over functions, sharing a joint Gaussian distribution defined by a mean and covariance function (2) [19].

$$f(x) \sim \mathcal{GP}(m(x), k(x, x')) \tag{2}$$

A common assumption is a zero mean and squared exponential (SE) covariance function. The covariance function defines the entries of the covariance matrix K. For the d-dimensional variables x_p and x_q the anisotropic squared exponential function (3) has a separate length scale l_i in each input dimension i. Adjacent points have a covariance close to unity and are therefore strongly correlated. In Fig. 3a four realizations of a GP are displayed. The grey shaded area represents the 95% confidence interval of the function values.

$$\text{cov}(f(\mathbf{x_p}), f(\mathbf{x_q})) = k(\mathbf{x_p}, \mathbf{x_q}) = \exp\left(-\sum_{i=1}^{d} \frac{1}{2l_i} |x_{p,i} - x_{q,i}|^2\right) \tag{3}$$

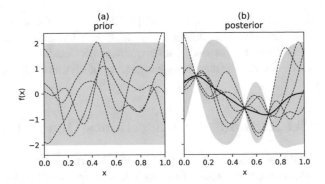

Fig. 3. Gaussian Process

A given training data set is represented by X sample points with $\mathbf{y} = \mathbf{f}(X) + \mathcal{N}(0, \sigma_n^2)$ being the observations of the underlying function with additional Gaussian noise. The hyperparameters of the GP are determined by maximizing the marginal log likelihood function (4). In this case, the observations are computer simulations. Therefore, the noise σ_n in (4) can be set to zero [20]. For the anisotropic SE kernel, the hyperparameters are the length scales l_i. The inverse of the length scales can be interpreted as activity parameter with lager l_i relating to less relevant input parameters [7].

$$\log p(\mathbf{y}|X) = -\tfrac{1}{2}\mathbf{y}^T(K + \sigma_n^2 I)^{-1}\mathbf{y} - \tfrac{1}{2}\log|K + \sigma_n^2 I| - \tfrac{n}{2}\log 2\pi \qquad (4)$$

The prediction $\mathbf{f}_*$ of the underlying function at X_* requires the extension of the joint distribution, see (5).

$$\begin{bmatrix} \mathbf{y} \\ \mathbf{f}_* \end{bmatrix} \sim \mathcal{N}\left(\mathbf{0}, \begin{bmatrix} K(X,X) + \sigma_n^2 I & K(X,X_*) \\ K(X_*,X) & K(X_*,X_*) \end{bmatrix}\right) \qquad (5)$$

The conditional distribution $\mathbf{f}_*$ given the data $\mathbf{y}, X$, and the sample points X_* in case of noiseless observations is:

$$\begin{aligned} \mathbf{f}_*|X_*, X, \mathbf{f} \sim \mathcal{N}(&K(X_*,X)K(X,X)^{-1}\mathbf{f}, \\ &K(X_*,X_*) - K(X_*,X)K(X,X)^{-1}K(X,X_*)) \end{aligned} \qquad (6)$$

The mean of this distribution are the predicted values $\mathbf{f}_*$. A strength of a GPR is the capability of additionally supplying a measure of uncertainty about the predictions. In Fig. 3b, the expected function values are represented by the solid line. The 95% confidence interval is again shaded in grey. The uncertainty in the prediction increases with the distance to the three given noise-free data samples.

3.2   Generalized Polynomial Chaos Expansion

In contrast to GPR, a generalized polynomial chaos expansion (gPC) model is a parametric surrogate. Given a model with random independent input parameters $\mathbf{X} = \{X_1, X_2, ...X_d\}$ with the probability density function $f_{\mathbf{X}}(\mathbf{x})$, the output $Y = F(\mathbf{X})$ is also a random variable. Assuming a finite variance of Y, the function can be expressed by an infinite series of polynomials (7). To make this problem computationally tractable, the function is approximated by a truncated series.

$$F(\mathbf{X}) = \sum_{j \in \mathbb{N}}^{\infty} y_j \mathbf{\Psi_j}(\mathbf{X}) \approx \tilde{M}(\mathbf{X}) = \sum_{\alpha \in \mathbb{N}^d}^{P} y_\alpha \mathbf{\Psi_\alpha}(\mathbf{X}) \qquad (7)$$

The multivariate polynomials $\mathbf{\Psi_\alpha}(\mathbf{X})$ are defined by the tensor product of univariate polynomials $\psi_{\alpha_i}^{(i)}(x_i)$ of the Askey scheme (8). In the initial PC formulation Ghanem [10] used Hermite polynomials as an orthogonal basis. Xui extended this approach to gPC by using polynomials from the Askey scheme, depending on the underlying probability distribution (see Table 1) in order to achieve optimal convergence [28].

$$\mathbf{\Psi_\alpha}(\mathbf{x}) = \prod_{i=1}^{d} \psi_{\alpha_i}^{(i)}(x_i) \qquad (8)$$

Table 1. Types of gPC and their underlying random variables [27]

Distribution	gPC basis polynomials	Support
Gaussian $\mathcal{N}(0,1)$	Hermite	$(-\infty, \infty)$
Gamma $\Gamma(\alpha, \lambda = 1)$	Laguerre	$[0, \infty)$
Beta $\mathcal{B}(a, b)]$	Jacobi	$[a, b]$
Uniform $\mathcal{U}(-1, 1)$	Legendre	$[a, b]$

The orthogonality is transmitted from the univariate polynomials to the multivariate polynomials:

$$\mathbb{E}[\mathbf{\Psi_\alpha}(\mathbf{X})\mathbf{\Psi_\beta}(\mathbf{X})] = \int_{\mathcal{D}_{\mathbf{X}}} \mathbf{\Psi_\alpha}(\mathbf{x})\mathbf{\Psi_\beta}(\mathbf{x})f_{\mathbf{X}}(\mathbf{x})d\mathbf{x} = \delta_{\alpha\beta} \quad \forall \alpha, \beta \in \mathbb{N}^d \qquad (9)$$

in which $\delta_{\alpha\beta}$ represents the Kronecker delta.

The total number of y_α-coefficients is determined by input dimension d and the maximum polynomial degree p of the truncated series:

$$P = \frac{(D+p)!}{D!p!} - 1 \qquad (10)$$

The computation of these coefficients can be carried out by ordinary least-squares analysis. As a rule of thumb, the number of samples should be about $n \approx 2 - 3 * P$ to avoid overfitting [16]. These models will be referred to as PCR. The coefficients can also be computed by the spectral projection making use of the orthogonality of the polynomials. The integral in Eq. (11) can be approximated via a sparse grid quadrature using the Smolyak algorithm [26]. An increased level of sparse grid integration leads to a higher number of simulation model evaluations. Models based on this approach will be referred to as pseudo-spectral projection (PSP). In contrast, adaptive pseudo-spectral projections models (APSP) use an anisotropic sparse grid in which the importance of each input dimensionality will be evaluated before expanding the sparse grid with further samples per epoch.

$$\mathbf{y}_\alpha = \mathbb{E}[Y(\mathbf{X}), \mathbf{\Psi}_\alpha(\mathbf{X})] = \int_{\mathcal{D}_\mathbf{X}} Y(\mathbf{X})\mathbf{\Psi}_\alpha(\mathbf{X})f_\mathbf{X}(\mathbf{x})d\mathbf{x} \tag{11}$$

Once a gPC model is trained, it can be used as a computationally efficient surrogate to propagate the uncertainty via Monte-Carlo simulation. The mean, variance and Sobol indices can be directly computed from the coefficients $\mathbf{y}_\alpha$ [16].

3.3 Sampling Methods

In numerical sampling the n input sample points $\mathbf{x_i} = [x_i^{(1)}, x_i^{(2)}, ...x_i^{(d)}]$ for a d-dimensional design of experiment $\mathbf{X} = [\mathbf{x_1}, \mathbf{x_2}, ...\mathbf{x_n}]^T$ are set. There are several sampling methods to set up computer experiments [24]. Two of the most common ones are compared in this study: Monte-Carlo sampling (MCS) and Latin hypercube sampling (LHC).

In MCS, the samples are randomly drawn according to their probability distribution. A pseudo-random number generator outputs a number in the interval $[0, 1]$, projected into the physical domain by the input variables underlying cumulative distribution function. The samples are independent, potentially leading to clustering of sample points, whereas part of the design space is poorly sampled. LHC, in contrast, is a stratified random sampling strategy that was introduced by McKay [17], yielding samples that are better distributed in the design space. Each input variable is partitioned into n intervals of equal probability. A random sample is drawn from every interval before shuffling the samples separately in each dimension.

3.4 Sensitivity Analysis

As an initial step to any UQ study or optimization task, sensitivity analysis helps determine how important each model input is in predicting the response and may reduce the number of uncertain parameters. In this study, Sobol indices as a global variance-based method are used [22]. The first order indices describe the share of variance of Y due to a given input parameter X_d.

$$S_d = \frac{\mathrm{Var}(\mathbb{E}(Y|X_\mathrm{d}))}{\mathrm{Var}(Y)} \tag{12}$$

Higher-order indices quantify the amount of total variance associated with the interaction of parameters. The calculation of Sobol indices is based on surrogates in this study (cf. [16]). The Sobol indices of a PC model can be computed analytically from the coefficients $\mathbf{y}_\alpha$. For GP models, in contrast, a Monto-Carlo integration approach is applied using the open-source python library ASLib [12].

4   Application of UQ Methods in Transmission Design

This investigation's main objective is to develop a data-efficient surrogate model to determine the impact of tolerances on the NVH indicator TE. A data-efficient surrogate provides a high model accuracy based on a small data set. The tolerances are treated as a random input variable. The TE is calculated via LTCA.

This investigation includes the analysis of a helical gear pair of transmission ration $i = 3.2$. The input torque is applied to the smaller pinion gear. The output speed defines the angular displacement of the wheel. The peak-to-peak TE is calculated during a single gear mesh cycle (see Fig. 2a). The bearing system for pinion and wheel is modeled as a rotation joint with only one degree of freedom. Therefore, the possible tolerances reduce to the relative position of the axis and deviations of the gear flank geometry. The simplicity of this system enables large-scale DoEs.

The considered parameter ranges may be applicable for transmission systems of traction eAxles. Recommendations for admissible values for shaft deviations can be found in [1]. A distinction is made for in-plane deviation $f_{\Sigma\delta}$ and out-of-plane deviations $f_{\Sigma\beta}$ (see Fig. 4a). Assuming gear quality grade, $Q = 6$ leads to $f_{\Sigma\delta} = 29.47\,\mu\text{m}$ and $f_{\Sigma\beta} = 14.74\,\mu\text{m}$. For practical applications of real transmission systems, the assumptions for axis parallelism would rather be determined by a statistical tolerance analysis. The flank deviations are modeled as variation in the gears microgeometry in terms of profile barreling C_α (see Fig. 4b) and lead crowning C_β (see Fig. 4c).

In the following, two example applications are presented comparing the different surrogates.

4.1   Analysis with Two Uncertain Parameters

In this analysis, the influence of axis parallelism variations on the TE is evaluated. The position of the bearing seats induces axis parallelism. Tolerances of injection molded parts tend to be uniformly distributed due to the wear of the tool over time [13]. Therefore, both uncertain parameters are assumed to be uniformly distributed within the tolerances $f_{\Sigma\delta}$ and $f_{\Sigma\beta}$. The input torque is set to 100 Nm. Further fixed parameters are listed in Table 2 - 4.1.

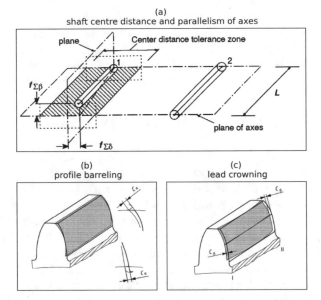

Fig. 4. Considered uncertain parameters according to ISO definition [1,2]. a) $f_{\Sigma\delta}$, $f_{\Sigma\beta}$ and A_{ae}/A_{ai}, b) C_α c) C_β

Table 2. Uncertain parameters in unit μm. All varying parameters are assumed to be uniformly distributed.

name	Symbol	4.1		4.2a		4.2b	
		min	max	min	max	min	max
In-plane deviation	$f_{\Sigma\delta}$	−29.47	29.47	−29.47	29.47	−29.47	29.47
Out-of-plane deviation	$f_{\Sigma\beta}$	−14.74	14.74	−14.74	14.74	−14.74	14.74
Center displacement	A_{ae}/A_{ai}	0		−17.5	17.5	−17.5	17.5
Profile barreling	C_α	3		3	8	5.25	5.75
Lead crowning	C_β	5		5	10	7.25	7.75

Training and Validation of a Reference Model. PCR and GPR are based on a randomized sampled data set (see Sect. 3.3). It is thus not sufficient to compare the models based on a single random training set. It is desired to train models several times with an identical sample size to evaluate the consistency of the method. The computation of the high number of required LTCA simulations is very time-consuming. To avoid this computational burden, a precise reference model based on a large DoE is developed to substitute the actual LTAC simulation. A PCR and a GPR model are trained with a training set size of $n_{\text{tr}} = 5000$ and validated by a test set consisting of $n_{\text{te}} = 500$ samples. The training and the test data sets are sampled independently via LHC. The model's accuracy is commonly assessed by the coefficient of determination r^2, with (13) [4].

$$r^2 = 1 - \frac{RMSE}{\sigma_{\text{te}}} \tag{13}$$

The root mean square error ($RMSE$) between surrogate prediction and test data is set into relation with test data objective functions standard deviation σ_{te}. Therefore, the coefficient of determination is independent of the output scale and converges towards one with increased model accuracy. Additionally, the distribution of the relative error ϵ_{te} (see (14)) is displayed in form of a histogram, in which Y_{te} represents the actual TE of the test data and Y_{m} the surrogate model's predictions.

$$\epsilon_{te} = 1 - \frac{Y_{\text{m}}}{Y_{\text{te}}} \tag{14}$$

The Q-Q plot in Fig. 5a indicates a good correlation of test data and surrogate model prediction. The GPR model is slightly more accurate than the PCR at $r^2 = 0.99935$. The relative error is smaller than 3% for every test sample and for the GPR less than 2% in 99.6% of the test cases (see Fig. 5b). Mean and standard deviation are in accordance within 1×10^{-4} µm. These values and the probability distribution, which is approximated by a histogram, serve as a reference solution for the surrogate models based on smaller sample sizes (see Fig. 5c). The GPR is sufficiently accurate and will substitute LTCA. Hereafter, it is referred to as the reference model.

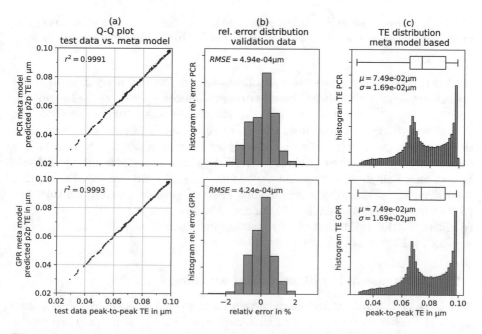

Fig. 5. Properties of reference surrogate models based on 5000 training and 500 test data samples. The top row shows results of the polynomial chaos regression model, the bottom row of the Gaussian process regression model.

Data-Efficiency Analysis of the Surrogate Models. First, PCR and GPR models are compared based on Monte-Carlo sampling. The sample size ranges from 15 to 100. For each sample size, the DoE generation and model training is repeated 100 times. The coefficient of determination r^2 is calculated with regards to the test set consisting of $n_{te} = 500$ samples. The results are presented in box plots, see Fig. 6. With an increasing number of training samples, both models increase in accuracy, whereas the spread between models of the same sample size decreases. Overall, GPR is superior to PCR. Neglecting outliers GPR models require a sample size of at least $n_{tr} = 70$ to achieve $r^2 > 0.95$, PCR models require $n_{tr} = 90$.

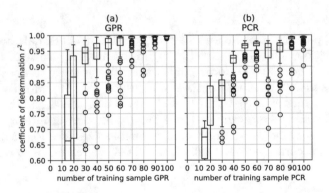

Fig. 6. Accuracy of surrogate model based on Monte-Carlo sampling considering two uncertain parameters. a) Gaussian process regression b) polynomial chaos regression

Next, the DoEs are generated via Latin hypercube sampling. The observed trends are analogous to MSC-based surrogate models (see Fig. 7). In general, accuracy is improved through LHC sampling. Assuming that the complexity of the underlying functions does not differ significantly between different gear designs, these plots help to choose the correct number of samples to achieve the desired accuracy. For PSP and APSP models, the training sets are predefined. Hence, only a single result per sample size can be evaluated, see Fig. 7c. For this problem with two uncertain parameters, spectral projection methods turn out to be very data-efficient. With a training sample size of $n_{tr} = 31$, APSP has a coefficient of determination of $r^2 = 0.969$, predicting $\mu = 7.50 \times 10^{-2}\,\mu m$ and $\sigma = 1.73 \times 10^{-1}\,\mu m$.

4.2   Analysis with Five Uncertain Parameters

The analysis is extended by the uncertain parameters center distance allowance A_{ae}/A_{ei}, profile barreling C_α and lead crowning C_β. The identical microgeometry is applied to both pinion and wheel gear. The assumptions for tolerance width and parameter distribution are summarized in Table 2 - 4.2a. For clarity, all parameters are assumed to be uniformly distributed. Therefore Legendre

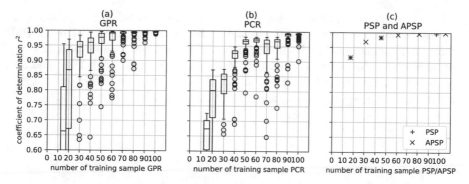

Fig. 7. Accuracy of surrogate model based on Latin hypercube sampling in a) and b) considering two uncertain parameters. a) Gaussian process regression b) polynomial chaos regression c) generalized polynomial chaos expansion based on pseudo-spectral projection and adaptive pseudo-spectral projection

polynomials are used as a basis for the PC surrogates, see Table 1. If information about the actual input parameters distribution is available, this assumption needs to be adapted, affecting the output distribution. The resulting flank topologies are in compliance within the tolerance definition of gear accuracy grade $Q = 6$, see [3]. The simulation model setup is identical to Sect. 4.1, the input torque is set to 100 Nm.

Training and Validation of a Reference Model. Analogous to the previous section, a reference surrogate model is created. A GPR is trained with a Latin hypercube DoE composed of $n_{tr} = 5000$, resulting in a coefficient of determination of $r^2 = 0.9998$. Even though the absolute error ($RMSE = 1.49 \times 10^{-3}\,\mu m$) of this model is higher compared to the case with two uncertain parameters (cf. Sect. 4.1), r^2 is closer to one since the error is set into relation with the standard deviation of the data, which is also increased due to the additional parameter, see (13).

Data-Efficiency Analysis of the Surrogate Models. Figure 8 shows comparisons of the surrogate types regarding model accuracy. For each sample size, the regression models are trained again with data sets from 100 different Latin hypercube DoEs. In this case, accuracy converges in significantly fewer iterations. Additionally, the range of achieved model accuracy with identical sample size is remarkably narrower. A possible reason for this unexpected result could be the dominant effect of the additional uncertain parameters on the objective function. The coefficient of determination r^2 is higher if the regression models fit the effect well, while the standard deviation σ_{te} is increased, even though more uncertain parameters are considered, see (13).

However and regarding data-efficiency, the model ranking is equivalent to the case of two uncertain parameters. Since the sample grid points for the PSP model are extended uniformly in every uncertain parameter dimension, the step size between models of higher accuracy is reasonably large. With five uncertain parameter dimensions, a level $k = 1$ of sparse grid integration results in a grid consisting of 11 sample points, a level $k = 2$ in 71, and level $k = 3$ in 341 sample points.

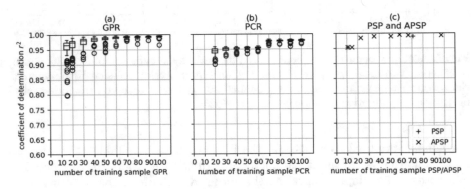

Fig. 8. Accuracy of surrogate model based on Latin hypercube sampling in a) and b) considering five uncertain parameters. a) Gaussian process regression b) polynomial chaos regression c) generalized polynomial chaos expansion based on pseudo-spectral projection and adaptive pseudo-spectral projection

The APSP model with a training set size of $n_{tr} = 23$ is compared to the reference model. Figure 9a shows a comparison of the first-order Sobol indices (see Sect. 3.4) of the uncertain parameters. For the APSP model, the Sobol indices can be directly derived by the polynomial coefficients, whereas for the reference model first, a DoE needs to be set up as described in Sect. 3.4. A good concordance between the reference model and the data-efficient APSP is achieved despite the small number of training samples of the APSP model. Given the assumptions made in Table 2 - 4.2a, the influence of the parameters defining the flank topology is dominant. If the tolerance range of C_α and C_β is reduced to $0.5\,\mu m$ (see Table 2 - 4.2b), the axis parallelism, defined by $f_{\Sigma\delta}$ and $f_{\Sigma\beta}$, has a similar influence on the peak-to-peak TE, see Fig. 9b. As expected, the center distance allowance has the most negligible impact since an involute gear still fulfills the law of gearing as the center distance is changed.

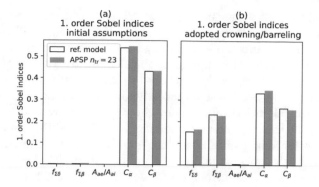

Fig. 9. First order Sobol indices: reference model vs. data-efficient surrogate model

Next, a Latin hypercube DoE with 100.000 samples is used to propagate the uncertainty of input parameters through the models. Figure 10 shows the approximated peak-to-peak TE distribution of the reference model and the data-efficient APSP via a histogram. The APSP model predicts the mean value and the standard deviation very precisely. In case tolerances are neglected, the nominal peak-to-peak TE value of 2.157×10^{-1} µm is close to the mean value because the distribution is only slightly skewed.

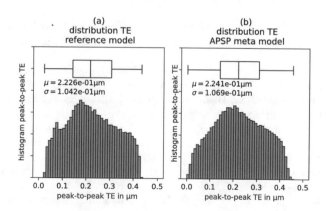

Fig. 10. TE distribution: reference model vs. data-efficient APSP surrogate model with $n_{\mathrm{tr}} = 23$

5   Conclusion

In this study, the potential of high-fidelity data-efficient surrogate modeling for uncertainty quantification in transmission design is investigated. Selected surrogate types are developed for a simple helical gear transmission system to enable forward uncertainty propagation and global sensitivity analysis for the NVH indicator peak-to-peak transmission error. Based on identical training sample set sizes, Gaussian process regression is advantageous compared to polynomial chaos regression regarding model accuracy for the presumed manufacturing tolerances. Our evaluation of sampling methods indicates that Latin hypercube sampling is particularly beneficial compared to Monte-Carlo sampling for smaller sample sizes. Notably, it was found that polynomial chaos models based on adaptive pseudo-spectral projection generally show faster convergence rates of model accuracy than the investigated regression models GPR and PCR in both example applications. Additionally, mean, variance, and Sobol indices can be determined efficiently due to the polynomial formulation of this model. Therefore, it is recommended to pursue this surrogate type for future design robustness analysis and explore its scalability for additional uncertain parameters and further KPIs. The use of surrogate models significantly reduces the required number of high-fidelity simulations for forward uncertainty quantification enabling robustness evaluation of KPIs in an early design state for optimization.

References

1. ISO TR 10064-3: Recommendations relative to gear blanks, shaft centre distance and parallelism of axes (1996)
2. ISO 21771 2007-09 Gears - Cylindrical involute gears and gear pairs - Concepts and geometry (2007)
3. ISO 1328-1:2013-09 Cylindrical gears - ISO system of flank tolerance classification (2013)
4. Blatman, G., Sudret, B.: An adaptive algorithm to build up sparse polynomial chaos expansions for stochastic finite element analysis. Prob. Eng. Mech. **25**(2), 183–197 (2010)
5. Brecher, C., Löpenhaus, C., Brimmers, J.: Function-oriented tolerancing of tooth flank modifications of beveloid gears. Procedia CIRP **43**, 124–129 (2016)
6. Bruyère, J., Dantan, J.Y., Bigot, R., Martin, P.: Statistical tolerance analysis of bevel gear by tooth contact analysis and Monte Carlo simulation. Mech. Mach. Theory **42**(10), 1326–1351 (2007)
7. Forrester, A.I., Keane, A.J.: Recent advances in surrogate-based optimization. Progress Aerosp. Sci. **45**(1–3), 50–79 (2009)
8. Früh, P.: Dynamik von Zahnradgetrieben: Modellbildung, Simulation und experimentelle Analyse, Shaker (2008)
9. Ghanem, R., Higdon, D., Owhadi, H.: Handbook of uncertainty quantification, vol. 6. Springer, Heidelberg (2017)
10. Ghanem, R.G., Spanos, P.D.: Stochastic finite elements: a spectral approach. Springer-Verlag, Heidelberg (1991)
11. Heider, M.K.: Schwingungsverhalten von Zahnradgetrieben: Beurteilung und Optimierung des Schwingungsverhaltens von Stirnrad- und Planetengetrieben. Ph.D. thesis, Verl. Dr. Hut, München (2012). OCLC: 848060576
12. Herman, J., Usher, W.: SALib: an open-source Python library for sensitivity analysis. J. Open Source Softw. **2**(9), 97 (2017)
13. Klein, B.: Statistische Tolerierung. Vieweg+Teubner Verlag (2002)
14. Korff, M., Hellenbroich, S., Terlinde, S., Vogt, S.: Statistical Methods in Gear Design. Examples from daily work and advanced studies. In: VDI-Berichte, Conference Proceedings, vol. 2255.1, pp. 445–456. VDI-Verlag (2015)

15. Krige, D.G.: A statistical approach to some basic mine valuation problems on the Witwatersrand. J. South. Afr. Inst. Min. Metall. **52**(6), 119–139 (1951)
16. Le Gratiet, L., Marelli, S., Sudret, B.: Metamodel-based sensitivity analysis: Polynomial chaos expansions and gaussian processes (2017)
17. McKay, M., Beckman, R., Conover, W.: A Comparison of the three methods for selecting values of input variable in the analysis of output from a computer code. Technometrics (United States) **21**(2) (1979)
18. Parlow, J.C.: Entwicklung einer Methode zum anforderungsgerechten Entwurf von Stirnradgetrieben. Ph.D. thesis (2015)
19. Rasmussen, C.E., Williams, C.K.I.: Gaussian processes for machine learning. In: Adaptive Computation and Machine Learning. MIT Press, Cambridge (2006)
20. Sacks, J., Welch, W.J., Mitchell, T.J., Wynn, H.P.: Design and analysis of computer experiments. Statistical science, pp. 409–423 (1989)
21. Schlecht, B.: Maschinenelemente 2, vol. 2. Pearson Deutschland GmbH (2010)
22. Sobol, I.M.: Sensitivity analysis for non-linear mathematical models. Math. Modell. Comput. Experiment **1**, 407–414 (1993)
23. Son, J., Du, Y.: Comparison of intrusive and nonintrusive polynomial chaos expansion-based approaches for high dimensional parametric uncertainty quantification and propagation. Comput. Chem. Eng. **134** (2020)
24. Tong, C.: Refinement strategies for stratified sampling methods. Reliab. Eng. Syst. Safety **91**(10–11), 1257–1265 (2006)
25. Weber, C., Banaschek, K.: Formänderung und Profilrücknahme bei gerade-und schrägverzahnten Stirnrädern. Schriftenreihe Antriebstechnik **11** (1953)
26. Xiu, D.: Efficient collocational approach for parametric uncertainty analysis. Commun. Comput. Phys. **2**(2), 293–309 (2007)
27. Xiu, D.: Numerical Methods for Stochastic Computations: A Spectral Method Approach. Princeton University Press (2010)
28. Xiu, D., Karniadakis, G.E.: The Wiener-Askey polynomial chaos for stochastic differential equations. SIAM J. Sci. Comput. **24**(2), 619–644 (2002)

Approach to Assess Basic Deterministic Data and Model Form Uncertainty in Passive and Active Vibration Isolation

Roland Platz[✉]

College of Engineering, Architectural Engineering, Penn State University,
University Park, PA 16802, USA
rxp5110@psu.edu
https://www.ae.psu.edu/index.aspx

Abstract. This contribution continues ongoing own research on uncertainty quantification in structural vibration isolation in early design stage by various deterministic and non-deterministic approaches. It takes into account one simple structural dynamic system example throughout the investigation: a one mass oscillator subject to passive and active vibration isolation. In this context, passive means that the vibration isolation only depends on preset inertia, damping, and stiffness properties. Active means that additional controlled forces enhance vibration isolation. The simple system allows a holistic, consistent and transparent look into mathematical modeling, numerical simulation, experimental test and uncertainty quantification for verification and validation. The oscillator represents fundamental structural dynamic behavior of machines, trusses, suspension legs etc. under variable mechanical loading. This contribution assesses basic experimental data and mathematical model form uncertainty in predicting the passive and enhanced vibration isolation after model calibration as the basis for further deterministic and non-deterministic uncertainty quantification measures. The prediction covers six different damping cases, three for passive and three for active configuration. A least squares minimization (LSM) enables calibrating multiple model parameters using different outcomes in time and in frequency domain from experimental observations. Its adequacy strongly depends on varied damping properties, especially in passive configuration.

Keywords: Vibration isolation · Passive and active damping
Data uncertainty · Model form uncertainty

1 Introduction to Data and Model Form Uncertainty

Awareness and quantifying uncertainty in mathematical modeling, experimental test, and model verification and validation in early design stage are essential in structural dynamic application. The author recognizes the fact that, not rarely, uncertainty quantification approaches and documentation tend to lack transparency and comprehensibility, together with reluctance in consistently consolidating mathematic, stochastic and engineering terminology. This makes it often

© The Author(s) 2021
P. F. Pelz and P. Groche (Eds.): ICUME 2021, LNME, pp. 208–223, 2021.
https://doi.org/10.1007/978-3-030-77256-7_17

impractical, or too difficult and time consuming to transfer and apply uncertainty quantification measures to common and real engineering problems.

Generally, literature subdivides uncertainty in *aleatoric* and *epistemic* uncertainty, [19,20,24]. Aleatoric uncertainty is irreducible and mostly characterized by probabilistic distribution functions. It is presumed intrinsic randomness of an outcome. Epistemic uncertainty is reducible and occurs due to lack of knowledge, or insufficient of incomplete data or models, [2]. Data includes model parameters and state variables, a model determines the functional relation of data. The German Collaborative Research Center SFB 805 "Control of Uncertainty in Load-Carrying Structures in Mechanical Engineering", which funds this work, distinguishes between data and model form uncertainty. Currently, the SFB 805 discusses a third characteristic, structure uncertainty. Data uncertainty may appear as *stochastic uncertainty* or *incertitude*, [11]. In case of stochastic uncertainty, probabilistic measures like BAYES-inferred MONTE CARLO simulations process known or assumed distribution functions of data. In case of incertitude, non-probabilistic measures like FUZZY and interval analysis process membership functions and intervals. *Ignorance* of uncertainty prevails if neither stochastic uncertainty nor incertitude are taken into account.

Model form uncertainty expresses unknown, incomplete, inadequate or unreasonable functional relations between the model input and output, model parameters and state variables when compared to observations from real experimental test. The scope and complexity of the model also have an impact on the severity of the uncertainty. The dilemma the designer encounters in early stage design, before calibration, verification and validation processes start, is the extent of uncertainty. The works [5,7] introduce a general relation between a real observation from experiments and a mathematical model to identify model form uncertainty. An observation reflects the measured physical outcome, mostly as states like forces, displacement, accelerations etc. A model must reflect the same outcome, which depends on data and the chosen functional relations. The difference between the outcome of the observation and the model is a combination of quantified model deviation and measurement uncertainty, including noise. Deterministic or non-deterministic approaches estimate the deviation, for example as a discrepancy function.

As a first step, prior to develop discrepancy functions, this paper quantifies basic deviations between experimental and numerical simulated outcomes of a one mass oscillator's passive and active vibration isolation capability. The investigated one mass oscillator is equipped with a velocity feedback controller that realizes passive and active damping. Considering this particular structural dynamic example, Platz et al. so far investigated the influence of data uncertainty on the vibrational behavior in frequency domain by numerical simulations in [16–18] in. Lenz et al. [8] conducted experimental investigations with regard to data uncertainty of the same system, introduced in [15]. The investigations covered data uncertainty in frequency domain.

The current paper looks upon experimental data and model form uncertainty in frequency <u>and</u> in time domain. First, it presents the derivation of the analytical excitation and response models. Second, the author explains the test setup, followed by discussing measured data, resp. measuring uncertainty. Third, remaining deviations between experimental observation and predictions by mathematical models after calibrating selected model parameters disclose model form uncertainty.

2 Analytical Model

A one mass oscillator is the most simple representation of a vibrating rigid body system to describe linear passive and active vibration isolation for many structural dynamic systems. For example, it is often used for first numerical estimation of a driving car's vertical dynamic behavior, [23], Fig. 1.

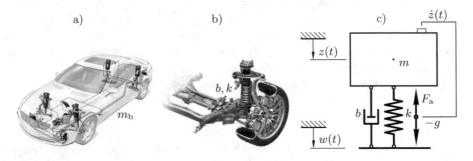

Fig. 1. Derivation of a one mass oscillator, a) automobile (©Auto Reporter/Mercedes Benz) with total mass m_b, b) front suspension leg (Mercedes Benz) with damping b and stiffness k, mass of suspension leg neglected, c) one mass oscillator model with position excitation $w(t)$ of a massless base point, and active vibration isolation by active velocity feedback control force $F_a = -g\,\dot{z}$

The analytical mathematical model represents only one fourth of the car's chassis and one suspension leg, with the chassis mass m and the suspension leg's damping and stiffness properties: damping coefficient b and the stiffness k in passive configuration. For active configuration, a velocity feedback with gain g is added to provide an active force, Fig. 1c. The absolute vertical displacement $z(t)$ of the mass and the base point displacement excitation function $w(t)$ depend on time t. The base point is assumed without inertia, represented only by a horizontal line. For example, $w(t)$ represents driving on an uneven bumpy road.

The inhomogeneous differential equation of motion for mass m

$$\ddot{z}(t) + \left[2\,D_p\,\omega_0 + \frac{g}{m}\right]\dot{z}(t) + \omega_0^2\,z(t) = 2\,D_p\,\omega_0\,\dot{w}(t) + \omega_0^2\,w(t) = \omega_0^2\,r(t) \quad (1)$$

of the one mass oscillator includes the damping ratio D_p from passive damping, $0 < D_p < 1$, as well as the angular eigenfrequency ω_0

$$2\,D_p\,\omega_0 = \frac{b}{m}, \text{ and } \omega_0^2 = \frac{k}{m}. \quad (2)$$

The term $r(t)$ in (1) is the general expression of the excitation function multiplied by ω_0^2. In this particular case it is the linear combination of a damper-spring base point excitation, [6].

2.1 Vibration Excitation and Responses

With $w(t) \neq 0$ and $\dot{w}(t) \neq 0$ in (1), the solution $z(t)$ is a linear combination of free and forced vibration responses in time domain. In frequency domain, the forced vibrational response depends on the magnitude and frequency content of the excitation source. The base point excitation $w(t)$ triggers $z(t)$. It is assumed that $z(t)$ is the response of a step function resulting from an initial impulse applied on the rigid frame, Sect. 3.

2.2 Time Domain

The unit step function

$$
\sigma(t - t_0) = \begin{cases} 0 & \text{for } t < t_0 \\ 1/2 & \text{for } t = t_0 \\ 1 & \text{for } t > t_0 \end{cases} \tag{3}
$$

is an ideal excitation model of the sudden change from the state 0 for $t < t_0$ to state 1 for $t > t_0$ before or after a certain point of time t_0. The dynamic system's vibration response

$$
z(t) = r_0 \left\{ 1 - e^{-D\omega_0 t} \left[\cos \omega_D t - D\frac{\omega_0}{\omega_D} \sin \omega_D t \right] \right\} \tag{4}
$$

for $t > t_0$ is the sum of the system's free vibration response solution $z_h(t)$ of the homogeneous equation of motion (1), with initial conditions $w(t_0) = 0$ and $\dot{w}(t_0) = 0$, and the forced vibration response

$$
r_0 = \frac{1}{\omega_0} 2 D_p \dot{w}_0 + w_0. \tag{5}
$$

(5) is the assumptive particular solution $z_{ih}(t)$ of the inhomogeneous equation of motion (1) for $w_0 = w(T_i) \neq 0$ and $\dot{w}_0 = \dot{w}(T_i) \neq 0$, when the impulse ends at time T_i, [6], Sect. 4.1.

2.3 Frequency Domain

The frequency content of the excitation source determines the amplitude frequency and phase frequency response. (1) is transferred into frequency domain

$$
\left\{ -\Omega^2 + i\Omega \left(2 D_p \omega_0 + \frac{g}{m} \right) + \omega_0^2 \right\} \hat{z}_{ih}\, e^{i\Omega t} = \left\{ i\Omega\, 2 D_p \omega_0 + \omega_0^2 \right\} \hat{w}\, e^{i\Omega t} \tag{6}
$$

by adding the sine term $i\sin(\Omega t + \varphi)$ as a complex extension, and by using exponential form with the constant and complex excitation amplitude magnitude $\widehat{w}$, and the complex response amplitude magnitude $\widehat{z}_{\mathrm{ih}}$. The complex vibrational displacement response of the mass m

$$\widehat{z}_{\mathrm{ih}} = \frac{i\Omega\, 2\, D_{\mathrm{p}}\,\omega_0 + \omega_0^2}{-\Omega^2 + i\Omega\left(2\, D_{\mathrm{p}}\,\omega_0 + \dfrac{g}{m}\right) + \omega_0^2}\, \widehat{w}, \tag{7}$$

using

$$\zeta = \frac{\Omega}{m\,\omega_0^2} \quad \text{and} \quad \eta = \frac{\Omega}{\omega_0}, \tag{8}$$

results in the complex magnifying function

$$\underline{V}(\eta) = \frac{\widehat{z}_{\mathrm{ih}}}{\widehat{w}} = \frac{i\, 2\, D_{\mathrm{p}}\,\eta + 1}{1 - \eta^2 + i\,(2\, D_{\mathrm{p}}\,\eta + g\,\zeta)}, \tag{9}$$

leading to the amplitude response

$$|\underline{V}(\eta)| = \sqrt{\frac{(2\, D_{\mathrm{p}}\,\eta)^2 + 1}{(1 - \eta^2)^2 + (2\, D_{\mathrm{p}}\,\eta + g\,\zeta)^2}} \tag{10}$$

and phase response

$$\psi(\eta) = \arctan\frac{-2\, D_{\mathrm{p}}\,\eta^3 - g\,\zeta}{1 - \eta^2 + (2\, D_{\mathrm{p}}\,\eta)^2 + 2\, D_{\mathrm{p}}\,\eta\, g\,\zeta}. \tag{11}$$

3 Experimental Test Setup

Figure 2 explains the real test setup concept, with the one mass oscillator model embedded in a frame with a relatively heavy mass $m_{\mathrm{f}} \gg m$ as the real test setup concept. It contains the physical and real representation of the base point for experimental testing. The frame is excited by the force $F(t)$ due to an impulse using a modal hammer; it is connected to the ground via an elastic support with relatively low damping $b_{\mathrm{f}} \ll b$ and low stiffness $k_{\mathrm{f}} \ll k$. These properties lead to a quasi-static dynamic response of the frame after the impulse, with a relatively low first eigenfrequency $\omega_{0,\mathrm{f}} \approx 2\pi\, 1/\mathrm{s} \ll \omega_0$, compared to the first eigenfrequency ω_0 of the mass m. It is fair to assume that the forced vibration response $z(t)$ is the result of an assumed one mass oscillator.

The virtual rigid frame model with mass m_{f} in Fig. 2 is fixed by an idealized gliding support assumed to have no friction perpendicular to the z-direction. The support permits a frame movement only in z-direction. The frame is constrained by an idealized damper with the damping coefficient b_{f} and a spring with the stiffness k_{f} in z-direction. The frame suspends from a rigid mount via elastic straps vertical to the z-direction, allowing low frequency pendulum motion of the frame in z-direction, Fig. 3. This motion is the translational absolute excitation displacement $w(t)$ in z-direction, when the frame is excited by a hammer impulse. Figure 3 shows the real test setup.

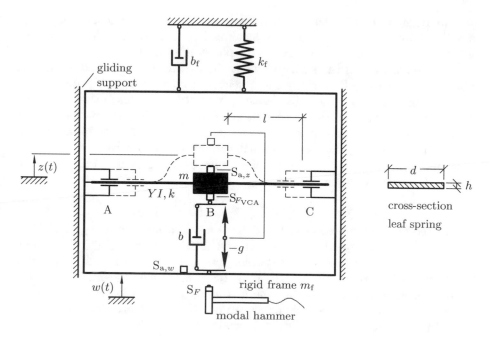

Fig. 2. One mass oscillator – schematic diagram of real test set up

Fig. 3. Physical test setup – left: assembly of leaf spring and VCA; right: hammer impulse on frame; the components are: acceleration sensor $S_{a,z}$ attached to the oscillating mass C1, two leaf springs C2 with partial stiffness $k/2$ on each side of C1, glide support C3, fixed leaf spring support C4, VCA coil support/holder C5, VCA stator, magnet outer ring C6, front/side structure of rigid frame C7a/b with total mass m_f, elastic strap C8, mount C9 to suspend the frame with elastic straps, acceleration sensor $S_{a,w}$ (hidden behind the frame) on the frame mass m_f, and force sensor S_F measuring the impulse force from the model hammer A

The frame in Figs. 2 and 3 retains two supports that fix a leaf spring at its ends at A and C, with the effective bending length l on sides A-B and B-C, and with the rigid mass m in the center position at B. The leaf spring is the practical realization of the spring elements in Fig. 1c. Its cross section area is $d\,h$, with the cross section width d and height h; its stiffness k is a function of the bending stiffness EI; E is the elastic or YOUNG'S modulus of the leaf spring made from carbon fiber reinforced polymer (CFRP), I is the area moment of inertia. The two supports at A and C are adjustable along l to tune the leaf spring's bending deflection, and eventually its effective stiffness k. A voice coil actuator (VCA) provides the passive and active damping forces F_b and F_a. The VCA's electromotive force

$$F_{\mathrm{S_{VCA}}} = F_b + F_a = b\left[\dot{z}(t) - \dot{w}(t)\right] - g\,\dot{z}(t) \tag{12}$$

is detected via the sensor $\mathrm{S_{VCA}}$, Fig. 3. Two acceleration sensors $\mathrm{S_{a,z}}$ and $\mathrm{S_{a,w}}$ measure the absolute accelerations $\ddot{z}(t)$ and $\ddot{w}(t)$ of the mass and the frame. The absolute accelerations are transformed into absolute velocities $\dot{z}(t)$ and $\dot{w}(t)$ by numerical integration in the SIMULINK-DSPACE^{TM} environment. The masses of the sensors $\mathrm{S_{a,z}}$, $\mathrm{S_{F_{VCA}}}$, and of the leaf spring, are considered parts of the oscillating mass m. Gravitational forces are neglected, the directions of $z(t)$ and $w(t)$ of the test rig are perpendicular to gravitation.

4   Experimental Models

4.1   Excitation

The frame's vibrational displacement response

$$w(t) = \frac{\check{F}(t_0)}{m_{\mathrm{f}}\,\omega_{\mathrm{D,f}}}e^{-D_{\mathrm{f}}\,\omega_{0,\mathrm{f}}\,t}\,\sin\omega_{\mathrm{D,f}}\,t \tag{13}$$

is the result from the impulse force

$$\check{F} = \int_{t_1}^{t_2} F(t)\,\mathrm{d}t \tag{14}$$

as the integral of the force function

$$F(t) = \begin{cases} \dfrac{\widehat{F}}{2}\left[\sin\left(\Omega_{\mathrm{i}}t - \dfrac{\pi}{2}\right) + 1\right] & \text{between } 0 \leq t \leq T_{\mathrm{i}} \\[4mm] 0 & \text{for } t > T_{\mathrm{i}} \end{cases} \tag{15}$$

assumed form experiments, with the peak force $\widehat{F}$ within a short time period between 0 and T_{i}. The response (13) is valid for low damping $0 < D_{\mathrm{f}} < 1$, [6, 10]. When the impulse ends at $t = T_{\mathrm{i}}$, the response (13) reaches a value that is approximated as the frame's effective displacement response step hight $w_0 = w(T_{\mathrm{i}})$, which is used to determine the step hight r_0 in (5) to calculate the step response $z(t)$ in (4).

4.2   Stiffness

The stiffness

$$k^* = \frac{12\,EI}{l^3} \tag{16}$$

derives from one leaf spring's flexural bending stiffness EI with respect to the length l between A and B, and between B and C, Fig. 2. The effective stiffness with four leaf springs becomes $k = 4 \cdot k^*$. It is the sum of four added stiffnesses, linearity assumed, with two leaf springs at each side A–B and B–C of the mass m.

4.3   Damping

The VCA provides the passive damping and active gain forces F_b and F_a (12) for passive and active vibration isolation, [15]. The LORENTZ force

$$F_{\mathrm{VCA}} = r\,l_c\,i \cdot B = \Psi\,i = b(\dot{z} - \dot{w}) + g\dot{z} \tag{17}$$

is expressed by the force constant Ψ, length of the coil l_c and the ratio r of the effective coil length, the magnetic flux density B, and the electrical current i, if B and i are perpendicular to each other. The VCA's driving electrical power is

$$P = u_{\mathrm{VCA}}\,i = F_{\mathrm{VCA}}\,v \tag{18}$$

and equivalent to the driving mechanical power $F_{\mathrm{VCA}}\,v$. The driving voltage $u_{\mathrm{VCA}} = \Psi\,v$ initiates the electromotive force F_{VCA}. The VCA's properties inductance L, Ohmic resistance R, and force constant Ψ lead to the control voltage

$$u = \Psi\,v + \frac{di}{dt}L + iR. \tag{19}$$

Eventually, the applied control voltage

$$u = \left\{ \Psi\,(\dot{z} - \dot{w}) + \frac{d}{dt}\left[\frac{1}{\Psi}\left(b(\dot{z} - \dot{w}) + g\dot{z} \right) \right] L + \frac{1}{\Psi}\left(b(\dot{z} - \dot{w}) + g\dot{z} \right)R \right\} \tag{20}$$

depends on the relative velocity $\dot{z} - \dot{w}$ between the frame and the mass to provide the passive damping force F_b, and on the absolute velocity $\dot{z}$ of the mass to provide the active force F_a. It depends directly on b and g.

4.4   Frequency Response Estimation and Coherence

The amplitude and phase estimation

$$|V(\Omega)| = |H_2(\Omega)| = \left| \frac{S_{\tilde{\tilde{z}},\,\tilde{z}}(\Omega)}{S_{\tilde{\tilde{z}},\,\tilde{w}}(\Omega)} \right|, \quad \psi(\Omega) = \angle\,H_2(\Omega) \tag{21}$$

process the auto-power and cross-power spectral densities $S_{\tilde{\tilde{z}},\,\tilde{z}}(\Omega)$ and $S_{\tilde{\tilde{z}},\,\tilde{w}}(\Omega)$ from the measured mass acceleration response $\tilde{\tilde{z}}(t)$, and from the frame acceleration excitation $\tilde{w}(t)$ from hammer excitation, averaged 5-times. The signals take into account normal perturbations from inexact and manual handling of the impulse hammer during averaging, marked by the tilde $\tilde{\ }$. The well known estimator $H_2(\Omega)$ leads to relatively small response errors in resonance compared to higher errors for anti-resonances. The coherence

$$\gamma^2(\Omega) = \frac{|S_{\tilde{z}, \tilde{w}}(\Omega)|^2}{S_{\tilde{z}, \tilde{z}}(\Omega)\, S_{\tilde{w}, \tilde{w}}(\Omega)}, \quad 0 \leq \gamma^2(\Omega) \leq 1 \tag{22}$$

determines the quality of the experimental signals, with full correlation, resp. highest estimation quality at $\gamma^2(\Omega) = 1$, and no correlation, resp. lowest estimation quality at $\gamma^2(\Omega) = 0$ between the excitation and response signals.

5   Deterministic Uncertainty Measures

5.1   Measurement and Data Uncertainty

The validity of the sensor sensitivity is checked by exciting the mass m 10 Hz with a $u = 2$ V amplitude input from the VCA, and measuring the force and acceleration outputs $F_{\mathrm{S_{VCA}}}$ and $a_{\mathrm{S_{a,z}}}$. The two outputs are converted back to voltage signals $u_{\mathrm{S_{VCA}}}$ and $u_{\mathrm{S_{a,z}}}$ via their sensitivities 0.01124 V/N and 102 V/(m/S^2), given by the manufacturers [13,14], and compared to the measured voltage outputs $u_{\mathrm{osc_{VCA}}}$ and $u_{\mathrm{osc_a}}$ from a parallel connected oscilloscope. The VCA gets its defined input signal from a signal generator.

For the validity of the measurement chain, the modal hammer hits a calibrated 1 kg mass hold in hand. The force signal excitation peak $F_{\mathrm{S_F}}$ of the hammer impulse is the input, the acceleration response peak $a_{\mathrm{S_{a,z}}}$ is the output, both are low pass filtered 500 Hz by an analogue filter. A DSPACE^{TM} and MATLAB^{TM} real time controller processes the signals. In addition, the power spectra density estimator $H_2(\Omega)$ (21) determines the measured frequency response between $0 \leq \Omega/2\pi \leq 200$ Hz. If measurement uncertainty is absent, $H_2(\Omega)$ must be 0 db over the entire frequency range.

For checking the stiffness reproducibility, the leaf-spring length $l = 0.08$ m in Fig. 2 leads to the lowest possible stiffness $k = 4k^* = 25,788$ N/m for the design according to (16), with the elastic modulus $E = 62 \cdot 10^9$ N/m^2 for CFRP, and the area moment of inertia I from $d = 4 \cdot 10^{-2}$ and $h = 0.11 \cdot 10^{-2}$ m. After measuring the frequency response (21), the screws in the fixed leaf spring support are loosened and tightened three times to again adjust the specified length l, after manually shaking the assembly for few moments. This procedure is repeated three times to see if there are any different resonance peaks according to (2) with changing stiffness from (21), after mounting and dismounting the leaf spring support.

Table 1 quantifies the measurement uncertainty after the calibration of sensors, measurement chain, and reproducibility of stiffness adjustment. With 2% and 10%, the uncertainty in force and acceleration signals remain below $\pm 15\%$ and close to $\pm 10\%$ of the uncertainty margins given by the sensor manufacturer, [12–14]. The measurement chain provides low uncertainty with $\approx 3\%$ deviation from the expected 1 kg and ≈ -0.18 dB from the expected 0 dB. The reproducibility check from assembling discloses a stiffness deviation of $\approx 6\%$.

Table 1. Measurement uncertainty from sensors sensitivity, measurement chain, and stiffness reproducibility affecting the eigenfrequency

	States				Model parameter k
	Force	Acceleration	Measurement chain		Eigenfrequency (2)
	$u_{osc_{VCA}}/u_{S_{VCA}}$	u_{osc_a}/u_{S_a}	$F_{S_F}/a_{S_{a,z}}$	$|H_2|$	$\omega_{min}/\omega_{max}$
dev.	$\approx 2\%$	$\approx 10\%$	$\approx 1.03\,\text{kg}$	$\approx -0.18\,\text{dB}$	$\approx 6\%$

5.2 Model Form Uncertainty

The relevant vibration isolation outcomes: excitation hammer force (15), frame and mass accelerations from (13) and (4), as well as phase (10) and amplitude (11) from experiments and models after calibration in time domain, are measured by experiments and calculated from mathematical models. The objective function

$$\min_{\theta \in \mathbb{R}} \frac{1}{N} \sum_n^N \sum_p^P \frac{\{y_p(X_n) - v_p(X_n, \theta)\}^2}{\max |y_p(X_n)|^2} \tag{23}$$

for model calibration, or, respectively, model updating is a least squares minimization (LSM). It uses P observation outcomes $y_p(X_n)$ and P predicted model outcomes $v_p(X_n, \theta)$ with control parameter $X_n = [t_n, f_n]$ as discrete time and frequency elements, and calibration parameters $\theta = [k, b, g]$. In this work, the outcomes may appear in time and/or in frequency domain, leading to $N = 4096$ time samples and/or $N = 2048$ frequency samples. In this example, the control parameter X_n is frequency. The number of outcomes is $P = 2$, only $|V|$ and ψ are used in (23). LSM is conducted by the particle swarm optimization (PSO) algorithm [3]; it reduces the risks of converging the LSM to a local minimum, [4]. All other relevant and remaining model parameters: oscillating mass $m = 0.9249$, frame mass $m_f = 9.33$ kg, stiffness $k_f = 722$ N/m, and damping $b_f = 10$ Ns/m, as well as state variables: excitation force amplitude $\hat{F} = 135$ N and impulse time $T_i = 5 \cdot 10^{-3}$ s, are assumed constant. However, they are potential candidates as calibration parameters in succeeding investigations.

Figure 4 displays phase and frequency responses as the outcomes for LSM for three different passive damping cases a) to c) and three different active damping cases d) to f), Table 2. The coherence $\gamma^2(f)$ (22) evaluates the quality of the experimental observations. The objective function (23) uses $y_1(f) = \psi^e(f)$ and $y_2(f) = |V^e(f)|$ from experimental observation, and $v_1(f, k, b, g) = \psi^m(f)$ and $v_2(f, k, b, g) = |V^m(f)|$ from numerical simulation using the models after parameter calibration. Figure 4 also shows the model outcome from the initially guessed parameter k after adjusting the leaf spring length l in (16), as well as b, and g for the VCA's control voltage u (20) used to specify the test rig properties. Further, it shows the deviation between the calibrated and measured outcomes. Table 2 lists the initially chosen and calibrated parameters k, b, and g, along with the deviations and remaining LSM-values after calibration. The VCA's given parameters are inductance $L = 0,003\,\text{Vs/A}$, resistance $R = 4.8\,\text{V/A}$, and force constant $\Psi = 17.5\,\text{V/A}$, [1].

Table 2. Model parameters from initial guess, after calibration, values of least square minimization (LSM), deviations between k, b, and g from initial guess and calibration

		k in N/m	b in Ns/m	g in Ns/m	LSM –
	Initial guess	32,583	41.66	0	
a)	Calibrated	32,682	41.22	0	0.0069
	Deviation in %	0.30	−1.06	–	
	Initial guess	32,583	69.44	0	
b)	Calibrated	36,639	63.87	0	0.0045
	Deviation in %	12.44	−8.01	–	
	Initial guess	32,583	97.21	0	
c)	Calibrated	42,649	87.71	0	0.0125
	Deviation in %	30.89	−9.77	–	
	Initial guess	32,583	18.30	22.00	
d)	Calibrated	32,239	23.68	11.00	0.0026
	Deviation in %	−1.06	29.43	−50.00	
	Initial guess	32,583	18.30	46.00	
e)	Calibrated	33,926	11.95	29.70	0.0013
	Deviation in %	4.12	−34.66	−35.44	
	Initial guess	32,583	18.30	68.00	
f)	Calibrated	35,219	9.14	43.62	0.0031
	Deviation in %	8.09	−50.00	−35.86	

Figure 5 shows the outcomes: excitation hammer force, frame and mass accelerations. They are not used by the objective function (23). The calibrated parameters k, b, and g do not affect the deviations between guessed vs. measured forces $F(t)$, nor do they affect the frame acceleration $\ddot{w}(t)$. The deviations are caused by multiple nonidentical experimental trials. The deviations between calibrated vs. measured oscillating mass acceleration $\ddot{z}(t)$, however, are affected by the calibration parameters.

First of all, it is worth noting that the results in Fig. 4 confirm the improved vibration isolation effect by active measures in the cases d) to f). In both experimental and numerical simulation, a strong vibration isolation effect for frequencies higher than resonance frequencies is present with increasing damping, compared to the passive cases a) to c). Second, increasing passive and active damping shows that the calibrated model parameters k, b and g deviate significantly from the initially guessed parameters used in the experiments, Table 2. Specifically, the increasing damping in the passive configuration leads to less adequate

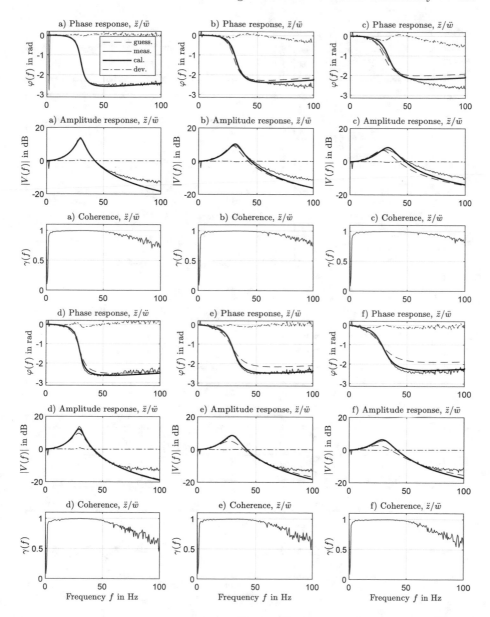

Fig. 4. Outcome, initially guessed from models, from measurement, from calibration, deviations between measurement and calibration of phase $\psi(f)$ and amplitude $|V(f)|$ frequency responses, and choherence $\gamma^2(f)$ in frequency domain for six damping cases a) to d), Table 2

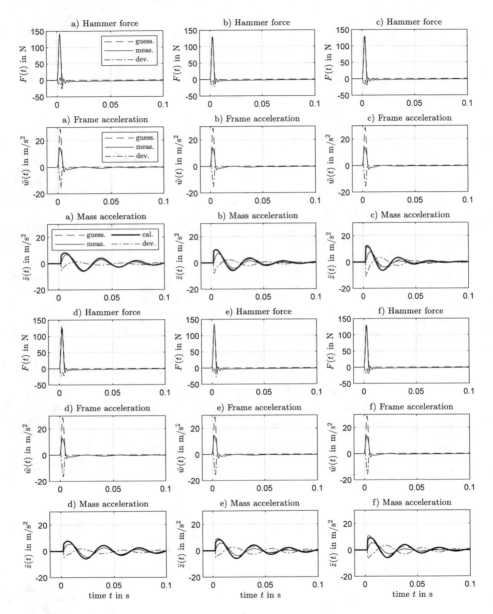

Fig. 5. Outcome, initially guessed from models, from measurement, from calibration, deviations between measurement and initially guessed/calibration of excitation force $F(t)$, velocities and accelerations $\dot{w}(t)$, $\dot{z}(t)$ and $\ddot{w}(t)$ and $\ddot{z}(t)$ of frame and mass in time domain for six damping cases a) to d), Table 2

calibration results of the stiffness. The authors observed that the higher passive damping force applied by the VCA leads to a shift in the system's eigenfrequency. With increasing gain in the active configuration, the stiffness calibration shows less deviation. After calibration, high deviation in both the calibrated damping coefficient and active gain compared to the initially guessed values remain.

6   Conclusion

The active approach for enhanced vibration isolation compared to passive isolation has been proven effective in experimental test and numerical simulation. Further, the investigation shows that with higher passive damping, the prediction of the dynamic outcome via the calibrated analytical model becomes less adequate. In case of data, resp. measurement uncertainty, the deviation of calibrated damping is up to 9% for highest applied damping, and up to 30% for stiffness. The stiffness prediction becomes more adequate in cases of active damping with only up to 8% at highest active damping. In case of model form uncertainty, the active damping cases lead to poor prediction quality of the calibrated passive damping coefficient and of the active gain, up to 50%. One reason is that the prediction by the LSM algorithm does not properly distinguish between the passive and the active model parameters when they are calibrated simultaneously. This problem occurs also when calibrating mass and stiffness properties would be calibrated simultaneously. In both problems, the functional relation in the model between passive and active damping parameters, and between inertia and stiffness, result from ambivalent parameter values within reasonable boundaries. Handling ambivalent calibration parameters is subject to further investigation.

Acknowledgment. This research is funded by the Deutsche Forschungsgemeinschaft (DFG, German Research Foundation) within the Sonderforschungsbereich (SFB, Collaborative Research Center) 805 "Control of Uncertainties in Load-Carrying Structures in Mechanical Engineering" – project number: 57157498.

References

1. Accel Technologies: Voice Coil Motor/VLR0113-0089-00A, data sheet, High-Tech Industrial Park, No. 1128 East Jiangxing Road, China (2020). www.accel-tec.com. Accessed 6 Nov 2020
2. Der Kiureghian, A., Ditlevsen, O.: Aleatory or epistemic? Does it matter? Struct. Saf. **31**(2), 105–112 (2009)
3. Eberhart, R.C., Kennedy, J.: A new optimizer using particle swarm theory. In: Proceedings of 6th International Symposium on Micro Machine and Human Science, pp. 39–43. IEEE, New Brunswick (1995)
4. Farajpour, I., Atamturktur, S.: Error and uncertainty analysis of inexact and imprecise computer models. J. Comput. Civ. Eng. **27**(4), 407–418 (2013). https://doi.org/10.1061/(ASCE)CP.1943-5487.0000233

5. Higdon, D., Gattiker, J., Williams, B., Rightley, M.: Computer model calibration using high-dimensional output. J. Am. Stat. Assoc. **103**(482), 570–583 (2008). https://doi.org/10.1198/016214507000000888
6. Irretier, H.: Basics in Vibration Technology (Grundlagen der Schwingungstechnik), Studium Technik (2000)
7. Kennedy, M.C., O'Hagan, A.: Bayesian calibration of computer models. J. Roy. Stat. Soc. Ser. B (Stat. Methodol.) **63**(3), 425–464 (2001). https://doi.org/10.1111/1467-9868.00294
8. Lenz, J., Platz, R.: Quantification and evaluation of parameter and model uncertainty for passive and active vibration isolation. In: IMAC-XXXVII A Conference and Exposition on Structural Dynamics, 28–31 January 2020, Orlando, pp. 135–147 (2020). Fl, USA (2019)
9. Li, S., Platz, R.: Observations by evaluating the uncertainty of stress distribution in truss structures based on probabilistic and possibilistic methods. ASME J. Verif. Valid. Uncert. Quantif. **2**(3), 031006–031006-9 (2017)
10. Markert, R.: Strukturdynamik (Structural Dynamics), Textbook. Shaker Verlag, Herzogenrath (2013)
11. Melzer, C.M., Platz, R., Melz, T.: Comparison of methodical approaches to describe and evaluate uncertainty in the load-bearing capacity of a truss structure. In: Fourth International Conference on Soft Computing Technology in Civil Engineering, Prague, Czech Republic, 1–4 September 2015
12. PCB Piezotronics, Inc.: ICP Impact Hammer Sensor 086C03, data sheet, 3425 Walden Avenue, Depew, NY, 14043-2495, USA (2010). www.pcb.com
13. PCB Piezotronics, Inc.: ICP Accelerometer 33B52, data sheet, 3425 Walden Avenue, Depew, NY, 14043-2495, USA (2016). www.pcb.com
14. PCB Piezotronics, Inc.: ICP Quarz Force Sensor 208C02, data sheet, 3425 Walden Avenue, Depew, NY, 14043-2495, USA (2019). www.pcb.com
15. Perfetto, S., Rohlfing, J., Infante, F., Mayer, D., Herold, S.: Test rig with active damping control for the simultaneous evaluation of vibration control and energy harvesting via piezoelectric transducers. J. Phys. Conf. Ser. **744** (2016)
16. Platz, R., Ondoua, S., Enss, G.C., Melz, T.: Approach to evaluate uncertainty in passive and active vibration reduction. In: IMAC–XXXII A Conference and Exposition on Structural Dynamics, 3–6 February, Orlando, FL, USA, pp. 345–352 (2014)
17. Platz, R., Enß, G.C.: Comparison of uncertainty in passive and active vibration isolation. In: IMAC–XXXIII A Conference and Exposition on Structural Dynamics, 2–5 February 2015, Orlando, FL, USA, pp. 15–25 (2015)
18. Platz, R., Melzer, C.: Uncertainty quantification for decision making in early design phase for passive and active vibration isolation. In: ISMA 2016 including USD 2016 International Conference on Uncertainty in Structural Dynamics, 19–21 September 2016, Leuven, Belgium, pp. 4501–4513 (2016)
19. Roy, C.J., Oberkampf, W.L.: A complete framework for verification, validation, and uncertainty quantification in scientific computing. Comput. Methods Appl. Mech. Eng. **200**(2528), 2131–2144 (2011)
20. Vandepitte, D., Moens, D.: Quantification of uncertain and variable model parameter in non-deterministic analysis. In: IUTAM Symposium on the Vibration Analysis of Structures With Uncertainties, St. Petersburg, Russia, 5–9 July 2009, pp. 15–24 (2009)
21. VDI 2062: Schwingungsisolierung – Vibration Insulation – Insulation elements, Part 2. Beuth Verlag Berlin, Verein Deutscher Ingenieure (2007)

22. VDI 2064: Aktive Schwingungsisolierung – Active vibration isolation. Beuth Verlag Berlin, Verein Deutscher Ingenieure (2010)
23. Vöth, S.: Dynamik schwingungsfähiger Systeme (Dynamics of vibrations systems). Vieweg & Sohn verlag, Wiesbaden (2006)
24. Zang, T.A., Hemsch, M.J., Hiburger, M.W., Kenny, S.P., Luckring, J.M., Maghami, P., Padula, S.L., Stroud, W.J.: Needs and opportunities for uncertainty—based multidisciplinary design methods for aerospace vehicles. The NASA STI Program 1, NASA Langley Research Center, Hampton, VA, Technical Report No. NASA/TM-2002-211462 (2002)

Reconstructing Stress Resultants in Wind Turbine Towers Based on Strain Measurements

Marko Kinne[1]([✉]), Ronald Schneider[1], and Sebastian Thöns[1,2]

[1] Bundesanstalt für Materialforschung und-prüfung (BAM), Berlin, Germany
marko.kinne@bam.de
[2] Lund University, Lund, Sweden

Abstract. Support structures of offshore wind turbines are subject to cyclic stresses generated by different time-variant random loadings such as wind, waves, and currents in combination with the excitation by the rotor. In the design phase, the cyclic demand on wind turbine support structure is calculated and forecasted with semi or fully probabilistic engineering models. In some cases, additional cyclic stresses may be induced by construction deviations, unbalanced rotor masses and structural dynamic phenomena such as, for example, the Sommerfeld effect. Both, the significant uncertainties in the design and a validation of absence of unforeseen adverse dynamic phenomena necessitate the employment of measurement systems on the support structures. The quality of the measurements of the cyclic demand on the support structures depends on (a) the precision of the measurement system consisting of sensors, amplifier and data normalization and (b) algorithms for analyzing and converting data to structural health information. This paper presents the probabilistic modelling and analysis of uncertainties in strain measurements performed for the purposes of reconstructing stress resultants in wind turbine towers. It is shown how the uncertainties in the strain measurements affect the uncertainty in the individual components of the reconstructed forces and moments. The analysis identifies the components of the vector of stress resultants that can be reconstructed with sufficient precision.

Keywords: Strain measurements · Bayesian updating of measurement uncertainties · Reconstruction of stress resultants

1 Introduction

The main components of a fixed offshore wind turbine are the rotor including the blades, the nacelle housing the generator, the tower, and the support structure [1]. The tower and the support structure are connected via a transition piece. In the design, the time-dependent behavior of a wind turbine system is determined using models describing the relevant physics and the turbine control system. Even though the models are highly sophisticated, they are subject to uncertainty. On the one hand, uncertainty is present in the model parameters. On the other hand, uncertainty is present in the models themselves (e.g. model uncertainty) as they do not necessarily include all influencing factors and/or processes. To reduce the uncertainty in the modelling and consequently in the estimates

© The Author(s) 2021
P. F. Pelz and P. Groche (Eds.): ICUME 2021, LNME, pp. 224–235, 2021.
https://doi.org/10.1007/978-3-030-77256-7_18

of the performance, operators can install monitoring systems to obtain data from the real turbine system. As an example, strain gauge rosettes can be applied at different elevations of the turbine tower to monitor its actual strain and stress state. If the rosettes are positioned appropriately around the circumference of a turbine tower, the actual time-dependent stress resultants (forces and moments) in this cross-section can be monitored. These load effects can, for example, be applied to updated predictions of the system's fatigue performance and improve decisions on inspection and maintenance actions as well as support decisions on a lifetime extension.

In this contribution, an approach to reconstruct stress resultants in a turbine tower cross-section from measured strains is presented. In this approach, a prior estimate of the uncertainty in strain measurements is determined based on a physical model of the measurement process [2–4]. Subsequently, the prior measurement uncertainty is updated with outcomes of actual strain measurements using Bayesian updating [2–4]. The updated measurement uncertainties together with material and modelling uncertainties are then considered in the reconstruction of the stress results. Their impact on the reconstructed stress resultants is investigated in a numerical example.

2 Modelling Uncertainty in Strain Measurements

2.1 Prior Probabilistic Model of the Measurement Uncertainty

Mechanical strains can be measured with strain gauges. Strain gauge configurations are commonly based on the Wheatstone bridge concept, through which small changes in electrical resistance can be measured. In a quarter-bridge configuration with one active strain gauge, an amplifier supplies voltage U_B [V] to the bridge circuit, amplifies the corresponding bridge output voltage U_A [mV] and determines the i th measured strain ε_M [μm/m] based on the following model [5]:

$$\varepsilon_M = \frac{4}{k}\frac{U_A}{U_B} \tag{1}$$

where k [-] is the batch-specific gauge factor provided by the manufacturer.

Strain measurements are subject to various uncertain influencing factors and hence the measured strain ε_M is not identical to the true mechanical strain ε [μm/m] – the measurand. In the following, the relation between ε and ε_M is modelled by the following process Eq. (2):

$$\varepsilon = B_p + \frac{f_{aa}}{c_k(\mathbf{X}_k, T)}\varepsilon_M + f_{az} + \varepsilon_T(\mathbf{X}_T, T) \tag{2}$$

wherein B_p [μm/m] describes the model uncertainty; f_{aa} [-] is the uncertain amplifying deviation factor; f_{az} [μm/m] is the uncertain amplifier zero deviation and T [°C] is temperature of the substrate, which is also measured. (Note that the uncertainty in the measured temperatures is neglected in the following. Also note that the values of f_{aa} and f_{az} depend on the measured value ε_M [2].) $c_k(\mathbf{X}_k, T)$ [-] is the correction coefficient of the gauge factor k, which is defined as [2]:

$$c_k(\mathbf{X}_k, T) = 1 + B_s + f_{s,v} + f_{s,q} + \alpha_k \cdot (T - 20\,°\text{C}) \tag{3}$$

where B_s [-] quantifies the model uncertainty associated with the gauge factor correction model; $f_{s,v}$ [-] is the gauge factor variation; $f_{s,q}$ [-] is the transverse strain correction factor and $\alpha_k \cdot (T - 20\,^\circ\text{C})$ [-] models the temperature variation of the gauge factor. The coefficient α_k [1/K] is an empirical quantity. The transverse strain correction factor $f_{s,q}$ is given by [5]:

$$f_{s,q} = \frac{q}{1 - q v_0} \left(\frac{\varepsilon_q}{\varepsilon_l} + v_0 \right) \tag{4}$$

wherein q [-] is the transverse sensitivity; v_0 [-] is Poisson's ratio of the material used in the experiments performed by the manufacture to determine the gauge factor and ε_q [μm/m] and ε_l [μm/m] are the actual strains perpendicular and parallel to the primary axis of the strain gauge. In Eq. (2), $\varepsilon_T(\mathbf{X}_T, T)$ [μm/m] is introduced to computationally compensate the temperature drift of the sensor. This quantity is referred to as the apparent strain and is defined as:

$$\varepsilon_T(\mathbf{X}_T, T) = \hat{\varepsilon}_T(T) + B_T \cdot (T - 20\,^\circ\text{C}) \tag{5}$$

where $\hat{\varepsilon}_T(T)$ [μm/m] is a batch-specific temperature-variation curve supplied by the manufacturer and $B_T \cdot (T - 20\,^\circ\text{C})$ [μm/m] is the model uncertainty of the temperature-variation curve.

The uncertain parameters of the process equation $\mathbf{X} = \left[B_p, f_{aa}, f_{az}, \mathbf{X}_k^{\mathrm{T}}, \mathbf{X}_T^T \right]^T$ with $\mathbf{X}_k = \left[B_s, f_{s,v}, \alpha_k \right]^T$ and $\mathbf{X}_T = [B_T]$ are modelled probabilistically [2], i.e. they are modelled as random variables. Their joint probability density function (PDF) is denoted by $p(\mathbf{x})$.

The process equation defined in Eq. (2) combined with the probabilistic model $p(\mathbf{x})$ of its parameters $\mathbf{X}$ constitutes the prior probabilistic model of the measurement uncertainty [2]. Based on this model, the prior distribution of the mechanical strain $p'(\varepsilon|\varepsilon_M, T)$ in function of the measurement ε_M and the temperature T is constructed using a Monte Carlo (MC) approach [2]. To this end, samples of the mechanical strain ε are generated by propagating the uncertainties in the parameters $\mathbf{X}$ through the process equation while the values of ε_M and T are kept fixed. Based on the samples of ε, a probabilistic model of ε is fitted (e.g. a normal distribution) [2]. $p'(\varepsilon|\varepsilon_M, T)$ quantifies the prior uncertainty in the measurand ε.

2.2 Bayesian Updating of the Measurement Uncertainty

The prior probabilistic model of the measurand $p'(\varepsilon|\varepsilon_M, T)$ determined based on the prior probabilistic process equation (Eq. (2)) is updated with actual strain measurements using a Bayesian approach. To this end, controlled experiments are performed. In each experiment, the true mechanical strain ε and the temperature T are kept fixed and a series of n strain measurements $\{\varepsilon_{M,i}\}_{i=1}^{n}$ is recorded with the measurement system modelled by the process equation. Based on the recorded strains, a likelihood function $L(\varepsilon|\varepsilon_M, T) \propto p(\varepsilon_M|\varepsilon, T)$ describing the measurement outcome is constructed.

It is assumed, that the relation between the true strain ε and the measured strain ε_M can be described by an additive measurement error e, i.e.

$$\varepsilon_M = \varepsilon + e \tag{6}$$

The probabilistic distribution of e at temperature T is denoted by $p(e|T)$. In the following, it is assumed that the measurement error e is normal distributed with zero mean and standard deviation σ_e. From Eq. (6), it follows that:

$$e = \varepsilon_M - \varepsilon \tag{7}$$

When the value of ε is known, the distribution of e corresponds to the distribution of ε_M shifted by ε. It follows, that for given ε the standard deviation of e is equal to the standard deviation of ε_M, which can be determined from the recorded strain measurements $\{\varepsilon_{M,i}\}_{i=1}^{n}$.

Once the parameters of the distribution $p(e|T)$ are known (i.e. the standard deviation σ_e), the likelihood function $L(\varepsilon|\varepsilon_M, T) \propto p(\varepsilon_M|\varepsilon, T)$ can be formulated as [6]:

$$L(\varepsilon|\varepsilon_M, T) = p(e = \varepsilon_M - \varepsilon|T) \tag{8}$$

Subsequently, the prior distribution $p'(\varepsilon|\varepsilon_M, T)$ of the measurand ε can be updated using Bayes' theorem:

$$p''(\varepsilon|\varepsilon_M, T) = \frac{L(\varepsilon|\varepsilon_M, T) \cdot p'(\varepsilon|\varepsilon_M, T)}{\int_{-\infty}^{\infty} L(\varepsilon|\varepsilon_M, T) \cdot p'(\varepsilon|\varepsilon_M, T)d\varepsilon} \tag{9}$$

where $p''(\varepsilon|\varepsilon_M, T)$ is the posterior (updated) distribution of the measurand ε, which quantifies the updated measurement uncertainty. $p''(\varepsilon|\varepsilon_M, T)$ depends on the measured strain ε_M and the temperature T and must be separately derived for each combination of the measured strain ε_M and temperature T [2].

3 Reconstructing Stress Resultants in Wind Turbine Towers

3.1 Inverse Mechanical Model (Relating a Strain State to Stress Resultants)

Figure 1(a) illustrates a cross-section of a tubular steel wind turbine tower. The blue coordinate system in Fig. 1(a) with origin at the center of the circular cross-section and coordinates (x, y, z) is the global reference coordinate system. Applying classical beam theory, the stress resultants in this cross-section consisting of axial force N_x, shear forces V_y and V_z, bending moments M_y and M_z, and torsional moment M_x can be reconstructed from strains measured by three strain rosettes $i = 1, 2, 3$ distributed evenly around the circumference of the section. The strain rosettes indicated as red dots in Fig. 1(a) are here assumed to be applied to the outer surface of the tower. Each rosette consists of three strain gauges a, b and c as illustrated in Fig. 1(b). At the position of each rosette $i = 1, 2, 3$, a local coordinate system with coordinates (ξ_i, η_i) is introduced. The ξ_i-axis points in tangential direction of the outer surface of the tube and the η_i-axis is aligned with the global x-axis. Figure 2 shows a top view of the cross-section including the positions of the three strain rosettes and the global and local coordinate systems.

It can be shown that the mechanical strains $\varepsilon_{a,i}$, $\varepsilon_{b,i}$ and $\varepsilon_{c,i}$ at position i can be transformed to the normal strains $\varepsilon_{\eta,i}$ and $\varepsilon_{\xi,i}$ and the shear strain $\psi_{\eta\xi,i}$ of an infinitesimal material element illustrated in Fig. 1(b) as follows:

$$\begin{bmatrix} \varepsilon_{\xi,i} \\ \varepsilon_{\eta,i} \\ \psi_{\xi\eta,i} \end{bmatrix} = \begin{bmatrix} \cos^2(\alpha) & \sin^2(\alpha) & 0.5 \cdot \sin(2\alpha) \\ \cos^2(\beta) & \sin^2(\beta) & 0.5 \cdot \sin(2\beta) \\ \cos^2(\gamma) & \sin^2(\gamma) & 0.5 \cdot \sin(2\gamma) \end{bmatrix}^{-1} \cdot \begin{bmatrix} \varepsilon_{a,i} \\ \varepsilon_{b,i} \\ \varepsilon_{c,i} \end{bmatrix} \tag{10}$$

(a)   (b)

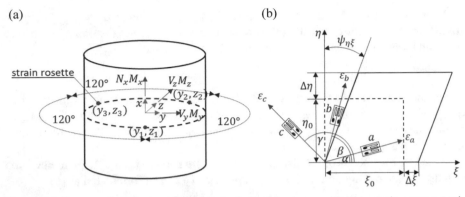

Fig. 1. (a) Cross-section along a tubular steel wind turbine tower with three strain rosettes positioned around the section's circumference and (b) illustration of the strain rosette on the surface of the tube consisting of three strain gauges a, b and c.

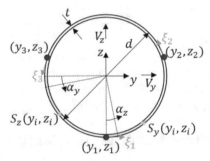

Fig. 2. Top view of the cross-section including the positions of the three strain rosettes and the global coordinate system with coordinates (x, y, z) and local coordinate systems with coordinates (ξ_i, η_i), $i = 1, 2, 3$. The local η_i-axis points in the same direction as the global x-axis.

The angles α, β and γ determine the orientation of the strain gauges (see Fig. 1(b)).

Assuming linear-elastic material behavior and a plane stress state, the normal stresses $\sigma_{\xi,i}$ and $\sigma_{\eta,i}$ and the shear stress τ_i at position i corresponding to $\varepsilon_{\xi,i}$, $\varepsilon_{\eta,i}$ and $\psi_{\xi\eta,i}$ can be determined as:

$$\begin{bmatrix} \sigma_{\xi,i} \\ \sigma_{\eta,i} \\ \tau_i \end{bmatrix} = \frac{E}{1 - v^2} \begin{bmatrix} 1 & v & 0 \\ v & 1 & 0 \\ 0 & 0 & 0.5 \cdot (1 - v) \end{bmatrix} \cdot \begin{bmatrix} \varepsilon_{\xi,i} \\ \varepsilon_{\eta,i} \\ 2\psi_{\xi\eta,i} \end{bmatrix} \tag{11}$$

where E and v are the Young's modulus and Poisson's ratio of steel.

Applying classical beam theory and noting that $\sigma_x(y_i, z_i) = \sigma_{\eta,i}$, the relation between the normal stress $\sigma_x(y_i, z_i)$ at position i with global coordinates (y_i, z_i) and the section force and moments N_x, M_y and M_z can be written as:

$$\sigma_x(y_i, z_i) = \sigma_{\eta,i} = \frac{N_x}{A} + \frac{M_y}{I_y} z_i - \frac{M_z}{I_z} y_i \tag{12}$$

where A is the cross-sectional area of the thin-walled tubular cross-section and I_y and I_z are the corresponding moment of inertia about y and z-axis. Since the normal stress $\sigma_x(y_i, z_i) = \sigma_{\eta,i}$ can be determined at each position $i = 1, 2, 3$ from the measured strains $\varepsilon_{a,i}$, $\varepsilon_{b,i}$ and $\varepsilon_{c,i}$, it is possible to formulate a system of linear equations with unknown N_x, M_y and M_z based on Eq. (12):

$$\begin{bmatrix} 1/A & z_1/I_Y & -y_1/I_z \\ 1/A & z_2/I_Y & -y_2/I_z \\ 1/A & z_3/I_Y & -y_3/I_z \end{bmatrix} \cdot \begin{bmatrix} N_x \\ M_y \\ M_z \end{bmatrix} = \begin{bmatrix} \sigma_{\eta,1} \\ \sigma_{\eta,2} \\ \sigma_{\eta,3} \end{bmatrix} = \begin{bmatrix} \sigma_x(y_1, z_1) \\ \sigma_x(y_2, z_2) \\ \sigma_x(y_3, z_3) \end{bmatrix} \tag{13}$$

This system of linear equations can be solved for N_x, M_y and M_z.

The shear stress $\tau(y_i, z_i)$ at position i can be expressed in terms of the shear forces V_y and V_z and the torsional moment M_x as:

$$\tau(y_i, z_i) = \frac{M_x}{2A_m t} - \frac{V_y S_z(y_i, z_i)}{I_z t} - \frac{V_z S_y(y_i, z_i)}{I_y t} \tag{14}$$

with

$$S_y(y_i, z_i) = -\int_0^{\alpha_z(y_i, z_i)} \left(\frac{d-t}{2}\right)^2 t\cos(\alpha_z)d\alpha_z$$

$$S_z(y_i, z_i) = -\int_0^{\alpha_y(y_i, z_i)} \left(\frac{d-t}{2}\right)^2 t\cos(\alpha_y)d\alpha_y \tag{15}$$

where $A_m = 0.25\pi(d-t)$, d is the outer diameter of the tubular cross-section, t is the wall thickness, and $S_y(y_i, z_i)$ and $S_z(y_i, z_i)$ are the first moments of area at position i in y and z-direction. The angles α_y and α_z in Eq. (15) are defined in Fig. 2. Note that the relation between $\tau(y_i, z_i)$ and M_x is based on Bredt's formula [7].

Based on Eq. (14), it is now possible to formulate a linear system of equation that relates the shear forces V_y and V_z and the torsional moment M_x to the shear stresses $\tau(y_i, z_i)$, $i = 1, 2, 3$ at the position of the strain rosettes:

$$\begin{bmatrix} 1/(2A_m t) & -S_z(y_1, z_1)/(I_z t) & -S_y(y_1, z_1)/(I_y t) \\ 1/(2A_m t) & -S_z(y_2, z_2)/(I_z t) & -S_y(y_2, z_2)/(I_y t) \\ 1/(2A_m t) & -S_z(y_3, z_3)/(I_z t) & -S_y(y_3, z_3)/(I_y t) \end{bmatrix} \cdot \begin{bmatrix} M_x \\ V_y \\ V_z \end{bmatrix} = \begin{bmatrix} \tau(y_1, z_1) \\ \tau(y_2, z_2) \\ \tau(y_3, z_3) \end{bmatrix} \tag{16}$$

This system of linear equations can be solved for V_y, V_z and M_x.

Equation (10), (11), (13) and (16) define a deterministic inverse model, which relates a strain state to stress resultants $\mathbf{R} = [N_x, V_y, V_z, M_x, M_y, M_z]^T$. In the following, this model is denoted by:

$$\mathbf{R} = f(\boldsymbol{\varepsilon}_1, \boldsymbol{\varepsilon}_2, \boldsymbol{\varepsilon}_3, \nu, E) \tag{17}$$

where $\mathbf{R} = [N_x, V_y, V_z, M_x, M_y, M_z]^T$ is the vector of stress resultants, $\boldsymbol{\varepsilon}_i = [\varepsilon_{a,i}, \varepsilon_{b,i}, \varepsilon_{c,i}]^T$ is the vector of mechanical strains determined from strains measured by strain gauges a, b and c at position $i = 1, 2, 3$.

3.2 Probabilistic Inverse Mechanical Model

The reconstruction of stress resultants from the strain measurements is subject to uncertainty. Uncertainty is present in the model itself (model uncertainty), the material properties and the measurements. To model the uncertainty in the (inverse) mechanical model itself, a random variable X_{str} with unit mean is introduced – in accordance with the probabilistic model code of the Joint Committee of Structural Safety (JCSS) [8] – which is multiplied to the model output, i.e.:

$$\mathbf{R} = X_{str} \cdot f(\varepsilon_1, \varepsilon_2, \varepsilon_3, v, E) \tag{18}$$

The uncertainty in the material properties is quantified by modelling the Young's modulus E and the Poisson's ratio v as random variables. This approach assumes that the spatial variability in the material properties is negligible. However, the model could be extended to account for the spatial variability by applying a random field approach.

The uncertainties in the mechanical strains ε_i, $i = 1, 2, 3$ corresponding to the strains measured by the strain gauges at the positions of the different strain rosettes are modelled as described in Sect. 2.

4 Numerical Example

In the following, the approach presented in Sect. 2 and 3 is illustrated in a numerical example. First, the prior distribution of the mechanical strain $p'(\varepsilon | \varepsilon_M, T)$ for a given measurement outcome $\varepsilon_M = 60$ μm/m and temperature $T = 20$ °C is constructed based on the process equation using MC simulation as described in Sect. 2.1 and the temperature-variation curve $\hat{\varepsilon}_s(T)$ given in Eq. (19). The simulation considers 10^5 samples of the parameters $\mathbf{X}$ of the process equation, which are defined in In this numerical example, the posterior mean and standard deviation are $\mu''_\varepsilon = 60.04$ μm/m and $\sigma''_\varepsilon = 1.15$ μm/m. The probability density function (PDF) of the posterior mechanical strain is shown together with the likelihood function and the prior distribution in Fig. 3.

Furth, a MC approach is applied to quantify the uncertainty in the reconstructed stress resultants. To this end, a constant strain state at the positions of the three strain rosettes is determined based on damage equivalent shear forces and bending moments, which were determined for the cross-section at the interface between the tower and the transition piece of an offshore wind turbine [12]. The computed strains are applied as the measurement outcomes. Strictly, an analysis of the posterior measurement uncertainty for each measured strain ε_M and temperature T as described in Sect. 2 has to be performed. In this numerical example, however, we adopt a simplified approach, in which we assume that (a) the temperature is T = 20 °C, (b) the posterior standard deviation σ''_ε determined above can be applied to quantify the uncertainty in the strain measurement regardless of the measured strain ε_M and (c) individual measurements are statistically independent. Based on these assumptions, we generate 10^5 independent and identical distributed (i.i.d.) samples of the mechanical strain $\varepsilon_{j,i}$, j = a, b, c and i = 1, 2, 3 at the position of the strain rosettes from a normal distribution with mean equal to $\varepsilon_{M,j,i}$, j = a, b, c and i = 1, 2, 3 and standard deviation equal to σ''_ε. In this way, we implicitly model the error in the strain measurements to be additive and normal distributed with zero mean

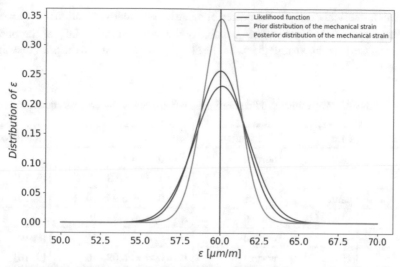

Fig. 3. Prior distribution $p'(\varepsilon|\varepsilon_M, T)$ of ε, likelihood function $L(\varepsilon|\varepsilon_M, T)$ and posterior distribution $p''(\varepsilon|\varepsilon_M, T)$ of ε for $\varepsilon_M = 60\ \mu$m/m and $T = 20\ °$C.

and standard deviation σ''_ε. In addition, 10^5 samples of the Young's modulus E, the Poisson's ratio v and the model uncertainty X_{str} are generated. The probabilistic model of the parameters of the inverse mechanical model and the value of the angles α, β and γ defining the orientation of the strain gauges in each strain rosette (see Fig. 1(b)) are given in Table 2.

It is assumed that ε is normal distributed. The resulting prior mean and standard deviation of ε are $\mu'_\varepsilon = 60.1\ \mu$m/m and $\sigma'_\varepsilon = 1.73\ \mu$m/m.

$$\hat{\varepsilon}_T(T) = -31.8 + 2.77T - 6.55 \cdot 10^{-2}T^2 + 3.28 \cdot 10^{-4}T^3 - 3.26 \cdot 10^{-7}T^4 \quad (19)$$

Second, a hypothetical laboratory experiment is performed as the basis for updating the prior measurement uncertainty. In the experiment, the mechanical strain and the temperature have fixed values $\varepsilon = 60\ \mu$m/m and $T = 20\ °$C. The mean and standard deviation of the measured strains are assumed to be $\mu_{\varepsilon_M} = 60\ \mu$m/m and $\sigma_{\varepsilon_M} = 1.55$ μm/m. As described in Sect. 2.2, the standard deviation σ_e of the measurement error e in Eq. (6) is equal to σ_{ε_M} if ε has a fixed value. Based on this, the likelihood function $L(\varepsilon|\varepsilon_M, T)$ is constructed as described in Sect. 2.2.

Third, the prior distribution $p'(\varepsilon|\varepsilon_M, T)$ of the mechanical strain ε is updated with $\varepsilon_M = 60\ \mu$m/m and $T = 20\ °$C to the posterior distribution $p''(\varepsilon|\varepsilon_M, T)$ according to Eq. (9). Given that $p'(\varepsilon|\varepsilon_M, T)$ and $L(\varepsilon|\varepsilon_M, T)$ have the functional form of a normal distribution, the posterior distribution $p''(\varepsilon|\varepsilon_M, T)$ also has this functional form. It can be shown that the posterior mean and variance μ''_ε and σ''_ε are given by [11]:

$$\mu''_\varepsilon = \frac{\mu'_\varepsilon \sigma_e^2 + \varepsilon_M \sigma'^2_\varepsilon}{\sigma_e^2 + \sigma'^2_\varepsilon} \qquad\qquad \sigma''_\varepsilon = \sqrt{\frac{\sigma_e^2 \sigma'^2_\varepsilon}{\sigma_e^2 + \sigma'^2_\varepsilon}} \quad (20)$$

In this numerical example, the posterior mean and standard deviation are $\mu_\varepsilon'' = 60.04$ μm/m and $\sigma_\varepsilon'' = 1.15$ μm/m. The probability density function (PDF) of the posterior mechanical strain is shown together with the likelihood function and the prior distribution in Table 1.

Table 1. Probabilistic and deterministic parameters of the process equation

Parameter	Unit	Distribution	Values[a]	References
B_p	μm/m	normal	$\mu = 0, \sigma = 1$	[2]
f_{aa}	–	uniform	$\mu = 1, \sigma = 1.73 \cdot 10^{-4}$	[9, 10]
f_{az}	μm/m	uniform	$\mu = 0, \sigma = 6.93 \cdot 10^{-1}$	[9, 10]
B_s	–	normal	$\mu = 0, \sigma = 4.38 \cdot 10^{-4}$	[2]
$f_{s,v}$	–	normal	$\mu = 0, \sigma = 7 \cdot 10^{-3}$	[9, 10]
α_k	1/°C	normal	$\mu = 0, \sigma = 6.99 \cdot 10^{-3}$	[9, 10]
B_T	μm/(m K)	normal	$\mu = 0, \sigma = 1.19$	[9, 10]
q	–	deterministic	$5 \cdot 10^{-4}$	[5]
v_0	–	deterministic	0.285	[5]
$\varepsilon_q/\varepsilon_l$	–	deterministic	0.3	

[a] μ and σ are the mean and standard deviation of the corresponding parameter

Furth, a MC approach is applied to quantify the uncertainty in the reconstructed stress resultants. To this end, a constant strain state at the positions of the three strain rosettes is determined based on damage equivalent shear forces and bending moments, which were determined for the cross-section at the interface between the tower and the transition piece of an offshore wind turbine [12]. The computed strains are applied as the measurement outcomes. Strictly, an analysis of the posterior measurement uncertainty for each measured strain ε_M and temperature T as described in Sect. 2 has to be performed. In this numerical example, however, we adopt a simplified approach, in which we assume that (a) the temperature is $T = 20\,°C$, (b) the posterior standard deviation σ_ε'' determined above can be applied to quantify the uncertainty in the strain measurement regardless of the measured strain ε_M and (c) individual measurements are statistically independent. Based on these assumptions, we generate 10^5 independent and identical distributed (i.i.d.) samples of the mechanical strain $\varepsilon_{j,i}, j = a, b, c$ and $i = 1, 2, 3$ at the position of the strain rosettes from a normal distribution with mean equal to $\varepsilon_{M,j,i}$, $j = a, b, c$ and $i = 1, 2, 3$ and standard deviation equal to σ_ε''. In this way, we implicitly model the error in the strain measurements to be additive and normal distributed with zero mean and standard deviation σ_ε''. In addition, 10^5 samples of the Young's modulus E, the Poisson's ratio v and the model uncertainty X_{str} are generated. The probabilistic model of the parameters of the inverse mechanical model and the value of the angles α, β and γ defining the orientation of the strain gauges in each strain rosette (see Fig. 1(b)) are given in Table 2.

Table 2. Probabilistic model of the parameters of the inverse mechanical model and values of the angles α, β and γ defining the orientation of the strain gauges in each strain rosette.

Parameter	Unit	Distribution	Values	References
E	N/mm^2	lognormal	$\mu = 2.1 \cdot 10^5, \sigma = 0.03 \cdot \mu$	[8]
ν	–	lognormal	$\mu = 0.3, \sigma = 0.03 \cdot \mu$	[8]
X_{str}	–	lognormal	$\mu = 1, \sigma = 0.03 \cdot \mu$	[13]
d	mm	deterministic	7000	
t	mm	deterministic	70	
α	°	deterministic	45	
β	°	deterministic	90	
γ	°	deterministic	135	

The histograms of the reconstructed bending moments M_y and M_z and shear forces V_y and V_z are shown in Fig. 4 (Note that the values of the reconstructed section forces and moments are here normalized by their mean value). In addition, Table 3 summarizes the coefficient of variations of the reconstructed section forces and moments, which are estimated based on their samples. The uncertainties in the bending moments are significantly lower than the uncertainties in the shear forces.

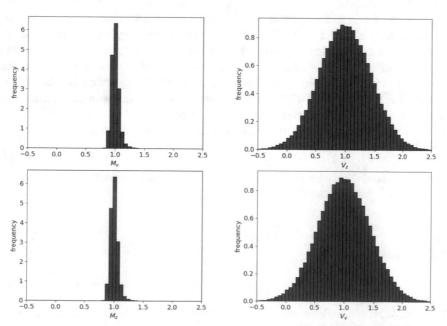

Fig. 4. Histograms of the reconstructed bending moments M_y and M_z and shear forces V_y and V_z. The values of the reconstructed stress resultants are normalized by their mean values.

Table 3. Coefficient of variation of the reconstructed stress resultants

Shear forces	Coefficient of variation	Bending moments	Coefficient of variation
V_y	0.37	M_y	0.07
V_z	0.37	M_z	0.07

5 Summary and Concluding Remarks

In the first part of this contribution, a probabilistic model of strain measurements is formulated. The model is based on a process equation of the measurement process. It is applied to quantify a prior estimate of the measurement uncertainty. Bayesian methods are then applied to update the prior measurement uncertainty with outcomes of actual strain measurements. This approach takes all available information on the measurement process into account. The prior measurement uncertainty is based on the probabilistic formulation of the physical measurement process including model uncertainties associated with the applied physical model. The likelihood function is constructed based on a model of the observation-based measurement uncertainty and used to update the prior measurement uncertainty. By updating the prior measurement uncertainty, an improved estimate of the uncertainty in the strain measurement is obtained.

In the second part of the paper, an inverse mechanical model for reconstructing stress resultants in a cross-section of a wind turbine tower is presented. Finally, the methods for modelling strain measurements and the inverse mechanical model are applied in a numerical study considering a strain state derived by applying bending moments and shear forces to a turbine tower cross-section. In addition to the measurement uncertainties, the reconstruction is affected by the uncertainties in the material properties and the uncertainty in the mechanical model itself. The numerical study shows that the bending moments can be determined with the highest precision while the uncertainty in the shear forces is higher. The precision of the reconstruction depends mostly on the accuracy of the reconstructed stresses, which itself depend on the measurement uncertainty, its dependencies and the size of the measured strains.

Acknowledgment. This work was supported by the German Ministry for Economic Affairs and Energy (BMWi) through grant 03SX449Z.

References

1. Böttcher, J.: Handbuch Offshore-Windenergie: rechtliche, technische und wirtschaftliche Aspekte: Walter de Gruyter (2013)
2. Thöns, S.: Monitoring based condition assessment of offshore wind turbine support structures. IBK Bericht, vol. 345 (2012)
3. Thöns, S., Faber, M.H., Rücker, W.: On the utilization of monitoring data in an ultimate limit state reliability analysis, pp. 1762–1769

4. Thöns, S., Faber, M.H., Rücker, W.: Life cycle cost optimized monitoring systems for offshore wind turbine structures. IRIS Industrial Safety and Life Cycle Engineering: Technologies/Standards/Applications, pp. 75–90 (2013)
5. Keil, S.: Beanspruchungsermittlung mit Dehnungsmessstreifen, Cuneus (1995)
6. Simoen, E., Papadimitriou, C., Lombaert, G.: On prediction error correlation in Bayesian model updating. J. Sound Vib. **332**(18), 4136–4152 (2013)
7. Schnell, W., Gross, D., Hauger, W.: Technische Mechanik: Band 2: Elastostatik. Springer-Verlag, Heidelberg (2013)
8. JCSS Joint Committee on Structural Safety, "Probabilistic Model Code" 2001-2015
9. TML, TML WFLA-6–17 strain gauge test data (2008)
10. Verein Deutscher Ingenieure/Verband der Elektrotechnik Elektronik Informationstechnik, "VDI/VDE 2635 Part 1", Experimental structure analysis- Metallic bonded resistance strain gages - Characteristics and testing conditions, Verein Deutscher Ingenieure (2007)
11. Gelman, A., Carlin, J.B., Stern, H.S., Dunson, D.B., Vehtari, A., Rubin, D.B.: Bayesian Data Analysis. CRC Press, Cambridge (2013)
12. Wölfel Engineering GmbH + Co. KG, Fatigue Parameters of Arkona Becken OWF wind turbines for the MISO project (2020)
13. Tarp-Johansen, N., Madsen, P.H., Frandsen, S.T.: Calibration of partial safety factors for extreme loads on wind turbines

Mastering Uncertain Operating Conditions in the Development of Complex Machine Elements by Validation Under Dynamic Superimposed Operating Conditions

Thiemo Germann$^{(\boxtimes)}$ (iD), Daniel M. Martin (iD), Christian Kubik (iD), and Peter Groche (iD)

Institut für Produktionstechnik und Umformmaschine, Technische Universität Darmstadt, Otto-Berndt-Str. 2, 64287 Darmstadt, Germany
`{germann,daniel.martin,kubik,groche}@ptu.tu-darmstadt.de`

Abstract. Machine elements produced in large quantities undergo several development cycles and can be adapted from generation to generation. Thus, experiences from real operation can be taken into account in further development. This is not possible for innovative investment goods such as special purpose machines, as these are usually individual items. Therefore, functionality and quality of newly developed components must be assured by previous investigations.

Conventional methods are inadequate at this point, as they cannot represent the actual, complex operating conditions in the later application. A reliable statement about the behavior of the system through a comprehensive validation in laboratory tests under standardized conditions is not achievable in this way due to a multitude of diversified load cases.

In previous work, a method was developed to allow testing of machine elements in the laboratory under detuned operating conditions. For this purpose, disturbance variables are applied to the system using paraffin wax phase change actuators in order to simulate real operation states and to analyze the behavior of the machine element under these conditions. The investigated disturbance variables are fluctuations and asymmetries of the operating load through superimposed temperature gradients. Complex interactions between the machine element and the adjacent components or the overall system can thus be taken into account.

The functionality of the methodology has been developed and briefly demonstrated so far. This paper presents the next level within the development process of the methodology. The necessary components are explained in detail and an AI black box evaluation tool is discussed. This work is based on a test bench that applies dynamically changing states of detuning under superimposed disturbances. Additionally, energy efficiency and performance of the test setup is advanced. As presented, the method opens up the possibility of validating new machine elements in the laboratory under realistic conditions.

Keywords: Product development · Uncertainty · AI · Machine elements

© The Author(s) 2021
P. F. Pelz and P. Groche (Eds.): ICUME 2021, LNME, pp. 236–251, 2021.
https://doi.org/10.1007/978-3-030-77256-7_19

1 Introduction

Investment goods are by definition productive assets within a company, used for the production of goods or services within the production process. Typically, they have a much longer period of usage than the goods they produce [1]. A characteristic example for an innovative investment good is the servo mechanical press with the integration of a fully electrically driven shaft. It changes the usability significantly due to the increased operating speed [2]. Due to their significant impact on functionality and quality, investment goods are key-elements of production lines. A subsequent impact on technical, ecological and economic key figures is notable.

Innovative investment goods are unique items. Therefore, the practical knowledge of the product usage is confined. Additionally, investment goods and their accompanying processes are highly complex. This complexity along with the small quantities causes very high manufacturing costs resulting in a significant development effort. Since the manufacturer covers the total development costs, the case of failure often involves an existential risk. For this reason, and due to the high level of complexity, there is a need for functional reliability over the entire service life and within differing use cases.

As there are numerous stakeholders, the above-mentioned points are further complicated. These stakeholders (e.g. manufacturer, buyer, operator) must be involved in the development process of investment goods [3]. Different objectives, interests and varying social, economic conditions lead to a complex development environment.

However, due to its complexity the typical development of innovative investment goods is separated into the development and selection of numerous individual machine elements (ME). Those ME are often characterized by their uniqueness and development for this single purpose. Due to the influence of the specific installation and the actual operating conditions, the performance of the ME or modules is, on the one hand, highly dependent on the other parts of the investment good. On the other hand, the forces and energies acting on the structure of an investment good depend on the behavior of the embedded ME and modules. These interactions result in a high degree of complexity.

For a proper validation, experimental tests under environmental conditions and disturbances, as in later operation conditions, are necessary. Today, costly and numerous emulations of operating solutions are performed for machine tools or its subsystems. Still, the experience of different operating conditions and their overlay are only available for the next generation of the ME (compare Fig. 1). The idle development process directly combines the laboratory results with the usage experience to create an entirely satisfactory product (green process, Fig. 1).

However, due to their novelty and the aforementioned challenges, the development of innovative investment goods is a long and iterative process with high uncertainty. Initial designs are usually based on models and inputs derived from individual subsystem data. Additionally, inputs from testing under idealized conditions not closely reflecting the real application are used. Thus, unpredictable time-consuming and costly modifications are necessary when hidden interdependencies between the innovative ME and interacting components of the investment good impede the required performance.

Still, there are some methods to tackle the challenges of a complex design process. Typical tools are e.g. the Design Structure Matrix (DSM), the Quantification of Margins and Uncertainty (QMU) or the SCRUM method.

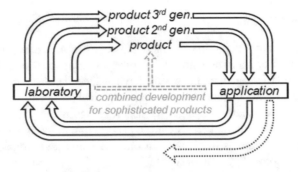

Fig. 1. Trade-off within the development process

QMU uses virtual models to quantitatively describe system behavior. It uses nominal values (M) and uncertainties (U) of components (compare Fig. 2). In general, QMU is a promising approach but it is not able to deal with the described challenges in the development of investment goods in an acceptable time frame [4].

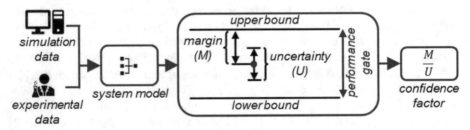

Fig. 2. Typical procedure of a QMU based development process, according to [4]

The DSM is a model-based method to uncover and handle complex interdependencies. It allows structuring the development process for qualitatively captured dependencies [5]. This is available for some standardized ME such as screws [6]. However, for innovative ME in the early development of investment goods this requirement is typically not fulfilled. A quantitative description of their behavior has to consider geometrical influences, material properties and the actual operating conditions. Due to deviations between ideal and real conditions, this methodology is close to inapplicable.

A modern approach to keeping up with uncertainty and unknown interdependencies are agile and adaptive methods such as the SCRUM methodology. It is a software development approach to incorporate user experience into the development process. In this process, intermediate results are used for evaluation, stakeholder's feedback and definition of the next steps to eliminate uncertainty within the validation process [7]. Methodological challenges continue to hinder the use of SCRUM for innovative investment goods. As the method is used to change the requirements during the design process, conflicts with some of the stakeholders are foreseeable. Subdividing a physical object is challenging. In addition, materiality makes it difficult to create and evaluate a prototype after each sprint, and to make changes to a designed product when new requirements arise. Obviously, it is not possible to offer customers a complex investment good, as for example the 3D Servo Press [2], with unfinished machine tools to tryout.

Summarizing, existing tools and methodologies for reliable development processes are not fully applicable, due to the complexity, novelty and characteristics of investment goods. Therefore, an approach for the validation and verification process is needed that handles the high level of unknown interdependencies and uncertainty. Thus, it is possible to ensure a sound base for the functional reliability of the investment goods and for the limitation of the development risks of the involved stakeholders.

The research described in this contribution therefore consistently continues the methodology developed in previous work [8]. The functionality of the methodology has so far been demonstrated at defined, static operating points. This contribution presents a test bench, which applies dynamically changing states of detuning under superimposed disturbances. Additionally, the energy efficiency and performance of the test setup is further developed. The successful application is presented and an evaluation method consisting of a black box AI model is discussed.

2   Approach

2.1   Model Representation – Bringing the Application to the Laboratory

Investment goods consist of a multitude of ME. Typically, several of these ME are individually designed for a unique investment good. Validation and verification within the design process is done separately. A laboratory environment with defined and standardized conditions is common. Especially for innovative ME, the effect of altered changing conditions on their behavior is unknown, since changing interface conditions or stochastic variations within the embedding structure are highly complex to simulate. In assembled investment goods, these changing interface conditions of individual MEs can reach a critical level that jeopardizes the investment good's overall functionality. Predicting these complex interface conditions is central to ensuring the products functionality. Therefore, superimposing thermal and mechanical loads have to be impinged in the laboratory. Additionally, it is of essential meaning to consider the complete field of possibly occurring operating points caused by the machine interfaces and the clamping conditions. Therefore, the test bench must be equipped with actuators that offer the option of simulating gradual changes in installation conditions and replicating dimensional changes due to tolerances or thermal loads. Furthermore, these actuators need to be compact enough to avoid larger design changes for their integration. Therefore, the paraffin wax phase change actuator (PCA) is the means of choice.

2.2   Paraffin Wax Phase Change Actuators for Detuning

Paraffin wax is a product of the petroleum industry, consisting of linear hydrocarbon strings of 20 to 40 carbon molecules. It is a phase change material with a significant increase of volume when changing from solid to liquid phase. Due to the small compression modulus it presents a suitable material for PCA. The activation of the paraffin wax is done by applying heat [9].

The used paraffin wax, SIGMA-ALDRICH paraffin wax mp. 58 – 62 °C, has a free volume extension of approx. 15% by heating from room temperature (25 °C) to the target temperature $\theta_t = 75$ °C (compare Fig. 3), a compression module of 1035 MPa [10] and a thermal conductivity of 0.371 W/mK.

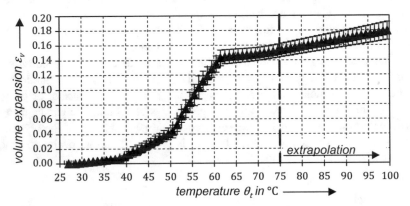

Fig. 3. Dilatation ϵ_V (change in volume relative to the initial volume) of SIGMA-ALDRICH paraffin wax mp. 58–62 °C, according to [10]

The necessary characteristics of the actuators are the provision of sufficient actuating forces under thermal loads as well as a compact and robust housing. Neither rapid responses nor big displacements are needed. Therefore, a deep drawn sheet metal housing (dual phase steel 1.0936) consisting of two cups is used. The inner cup is filled with paraffin wax and a metallurgical bond via laser welding joins the actuator. A cutting seal by a brass sheet and the inner cups chamfer seals the PCA (compare Fig. 4) [11].

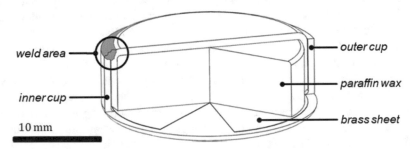

Fig. 4. Design of the PCA in a cross-sectional view, according to [11]

A radially applied thermal current activates the PCA. Due to the low thermal conductivity of wax, a delayed reaction is typical. This leads to a continuous increase of the axial compression force until the whole filled paraffin wax has reached the target temperature θ_t, which also determines the achievable maximum force. Forces of up to 60 kN have been proven at a working temperature of $\theta_t = 80$°C [4]. The front faces transmit the force and are also responsible for the displacement. A displacement increases the inner volume and is thus accompanied by a loss of force. A linear force-displacement-characteristic therefore follows. To avoid irreversible housing deformations, the axial displacement is limited to $w \leq 0, 1$ mm. Additionally, a slight hysteresis behavior is visible (compare Fig. 5). The described material behavior shows a very good repetition accuracy.

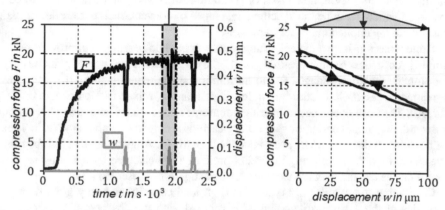

Fig. 5. Force characterization (left) and force-displacement-characteristic (right) of a typical PCA for a target temperature of $\theta_t = 63°C$; the displacement w is the controlled increase of the PCAs axial height due to the paraffin wax volume expansion in the characterization setup.

The use case within this work does not need a high dynamic; a slow but continuously increasing level of actuating force is rather needed to evaluate the ME behavior through the parameter field of clamping conditions. Nevertheless, higher actuator dynamics are achievable by thermal structures within the actuators housing, which shorten the thermal paths within the PCA or by additives within the paraffin wax.

2.3 Evaluation of Detuned Behavior

The paraffin wax PCAs are positioned between the ME and embedding laboratory environment. By activating the PCAs, a detuning as of a variation within the installation conditions can be simulated. Adjusting the PCA activation level allows for a wide variety of disturbances to emulate even more situations. Thus, complex disturbances in the component tests can be simulated by displacement fields that are close to the application. These small disturbances can have a significant impact on the functionality of key components as joints or bearings and subsequently on the over-all system behavior.

The resulting findings can be classified into two key areas. The first area focuses on the ME. Sensitivity with respect to specific disturbances can be evaluated and thus critical operation conditions are identified. This knowledge enables improvements at an early stage of the development process when modifications are still associated with comparatively low effort and expense. The second area affects the embedding structure. Necessary adjustments to the installation conditions can be deduced and even measures to improve the basic structure can be derived.

2.4 Backlash Free High-Load Bearing

The high machine flexibility of the 3D Servo Press is accompanied by challenging requirements for the MEs, especially for the bearings. The higher loads experienced in comparison with conventional forming presses can no longer be carried by pure roller bearings. For this reason, new types of combined roller and plain bearings have been developed which combine the specific advantages of conventional, pure roller or plain bearings for the application in servo presses.

Roller bearings have a lower starting torque than plain bearings, but provide poorer damping properties. At high operating loads, the economic advantages of plain bearings predominate the disadvantages due to installation space requirements. However, these are subject to backlash. In addition, in low speed ranges, mixed friction occurs in plain bearings. The developed special bearing counteracts these disadvantages. This is done by extending a plain bearing by two roller bearings arranged on either side of it. These can be either cylindrical roller bearings or ball bearings. Angular contact ball bearings have the advantage of allowing the radial preload to be changed by applying an axial force. For this reason, they are used for the investigations presented here.

Combining the two bearing types, the high load capacity and very good damping properties of the plain bearings can be combined with the absence of backlash and the good starting properties of the roller bearings [12]. In the range of low speeds and loads, the roller bearings carry most of the load and keep the bearing shaft centered in the bearing shell, thus avoiding mixed friction [13]. An increase in the operating load causes a displacement of the bearing shaft in the plain bearing shell and thus a deflection of the roller bearings. This in turn increases the load ratio of the plain bearing. An increase in speed also causes a higher plain bearing force, which leads to a reduction in the eccentricity of the shaft and relieves the roller bearings. Thus, high loads can be transmitted while avoiding backlash during start-up. Combined roller and plain bearings therefore offer great potential for meeting the challenging requirements of bearing arrangements for servo presses [14].

Both measurements and simulations confirm the load- and speed-dependent functional transition between the two bearing types [13, 15]. With increasing load, the shaft displacement initially increases more strongly until the plain bearing takes over the load-bearing components of the roller bearing and the increase in displacement decreases with increasing press force [12].

2.5 Test Bench

In preliminary work, a test bench was developed enabling the condition monitoring of a sensory equipped bearing during operation. This test bench enables continuous testing and evaluation with regard to defined technological and economic criteria as well as comparison with conventional bearings.

The developed sensor-equipped bearing and the test bench have been presented in [15]. The operating force is provided by a hydraulic actuator and absorbed by two support bearings which hold the shaft (compare Fig. 6 (1)) in the axis of the electric motor that generates the rotational movement. The test bearing (2, 3) is mounted in a housing (11), which transmits the force to the bearing. To detect the shaft displacement path in the plain bearing, the test bearing is equipped with two eddy current sensors (10) placed 90° apart from each other, which measure the position of the shaft relative to the housing. Three temperature sensors measure the temperature on the outside of the roller bearing outer rings (8) as well as the outside temperature of the plain bearing shell (9). In order not to influence the lubricant film in the plain bearing, the lubricant temperature is not measured directly in the plain bearing, but only in the oil tank. In addition, the operating force is recorded by means of a piezo force transducer. The oil supply pressure is measured via a manometer attached to the inlet.

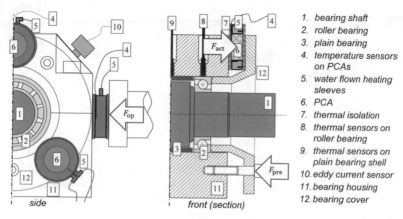

1. bearing shaft
2. roller bearing
3. plain bearing
4. temperature sensors on PCAs
5. water flown heating sleeves
6. PCA
7. thermal isolation
8. thermal sensors on roller bearing
9. thermal sensors on plain bearing shell
10. eddy current sensor
11. bearing housing
12. bearing cover

Fig. 6. Test bench of the combined roller and plain bearing

The distribution of the force flow F_{pb}/F_{rb} is determined by the stiffness of the roller bearings and the radial displacement $\vec{e}$ of the shaft assuming a constant stiffness of the rolling elements. Knowing the individual bearing forces, the operating behavior and the service life of the bearings can be estimated [14]. The compliance of the bearing is an important parameter to characterize the mechanical behavior of the bearing and, especially with regard to the use in a press, it is important for a model-based control of the ram movement. The total compliance δ of the combined bearing is determined by the displacement of the shaft in the direction of the operating force, and the operating force with e_x/F_{op}. In order to validate the bearing behavior at varying operating conditions, full-rotation and pivoting tests can be carried out [16]. Additionally, the damping properties of the bearing combination have been evaluated using punch tests [8]. To investigate the influence of detuned conditions, a test setup has been generated using PCAs to apply disturbance variables [8]. This procedure is adopted in the present work.

For ecological reasons, the actuator temperature control is based on water-flow actuator sleeves. The water is heated with a water boiler and pumped by an electrically driven pump in the test stand. Each actuator is coated by its own sleeves and is thermally isolated to the outer clamping to avoid energy losses as well as undesired heating of the bearing. There are four groups of actuators with eight actuators in total. Three actuators are positioned around each bearing cover and two separated PCAs are located on the operating force side of the bearings housing. Each group of actuators can be controlled separately and is supplied with hot or cold water.

2.6 Evaluation via AI

In order to quantify the influence of different input variables on the behavior of an engineering system, Machine Learning (ML) models can be used for predicting characteristic quantities. In general, the procedure of building a ML model to evaluate process states consists of several steps, which base on the heuristic process model Knowledge Discovery in Databases (KDD) that considers the acquisition, preprocessing and transformation of given data sets as well as the training and validation of the ML model [17]. One of the most efficient supervised ML model approaches is the multiple regression [18]. Multiple regression models allow to quantify the influence of several independent variables on one output parameter. Here, the output is represented by a weighted sum of the input parameters, allowing the influence of the input parameters to be quantified. In the simplest case, a multiple regression is linear and describes the correlation between input and output as follows:

$$y = \beta_0 + \sum_{i=1}^{p} \beta_i x_i + \epsilon .$$

The predicted output y is a weighted sum of its p input parameters x_i (called features). β_i represent the learned feature weights or coefficients. The first weight in the sum β_0 is called the intercept and is not multiplied with a feature. ϵ is an error term according to a Gaussian distribution. To find the optimal weight for each feature, a common approach is the least squares method to find the best fit for a data set

$$\hat{\beta} = \arg \min_{\beta_0, \, \dots, \, \beta_p} \sum_{i=1}^{n} \left(y^{(i)} - \left(\beta_0 + \sum_{j=1}^{p} \beta_j x_j^{(i)} \right) \right)^2 .$$

In order to finally apply the multiple regression approach to predict the compliance of the bearing, in this work the procedure for implementing the model as shown in Fig. 7 is used.

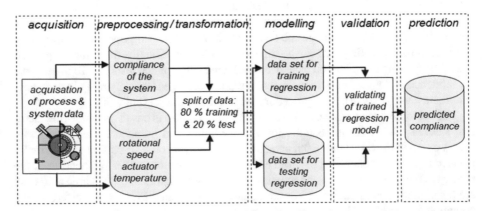

Fig. 7. Procedure for predicting the bearings compliance using a multiple regression model

From the bearing system, process data are acquired using four actuator temperatures, the rotational speed and system data using of the compliance of the system. Based on the time series features are derived and split into a training and a test data set. To quantify the performance of the model, the root mean square error (RMSE) value between the actual and predicted compliance of the bearing is determined.

3 Application

In previous work [8], an approach for the validation of novel ME in investment goods was presented, which allows the emulation of realistic operating conditions in laboratory tests. Therefore, potential uncertainty influences were identified that can arise at interfaces between the ME and surrounding system, whereupon these were specifically manipulated in experiments. Paraffin wax PCAs were used, compare Sect. 2.3. The approaches are demonstrated uses the example of a novel combined roller and plain bearing used in the drive train of a newly developed flexible press (c.f. [2]).

Thereupon, on the one hand, the test setup is optimized within the scope of the present work. The PCAs are activated by heating sleeves through which a working medium (here water) flows, which that can be heated or cooled via a peripheral circuit. This modified control of the actuators allows both heating and active cooling. This allows different load situations to be generated one after the other in a targeted manner. Figure 8 shows the system for controlling the temperature of the actuators in a flow diagram. On the other hand, the design of an experimental design is presented, which simultaneously examines a maximum number of disturbance variables with a relatively small number of test series. An evaluation via AI is also presented, to evaluate individual disturbance variable influences from these predefined experimental results.

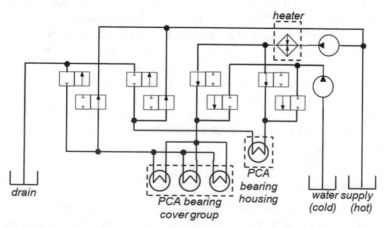

Fig. 8. PCA control: Schematic representation of the heating (shown)/cooling system; for simplicity only one bearing cover PCAs and one PCA on the operating force side are illustrated.

In [8], for comparability the same initial conditions are generated by a warm-up run before each test, until approximately stationary temperatures are reached at the bearing points. Since heat transfer is not completely prevented by the used thermal isolation between PCA and bearing, the actual conditions may differ for each test run. Additionally, surrounding work floor conditions influence the experimental conditions significantly. In addition, the time required to carry out tests increases to an unmanageable extent when a higher number of possible disturbance and manipulated variables have to be investigated. Therefore, this procedure is now abandoned and instead the different initial conditions are taken into account by a continuously recording of all temperatures not only during the tests, but also in the intermediate periods.

The question is addressed whether the findings on the properties of the bearing can be reconstructed when operating conditions are approached from different initial conditions. The experimental procedure is described in the following section.

3.1 Test Procedure

In order to demonstrate the feasibility of the approach, the load distribution on the different bearing types and the total compliance of the combined roller and plain bearing are selected as output variables for the following considerations. Initially, only tests in full-rotation mode will be carried out. During a series of tests, several speed levels are approached. All data is measured continuously both during a test series and in between in order to resolve time-dependent effects. In order to reduce the test effort, test series are carried out without a defined time interval and without a previously performed warm-up phase. In order to identify and isolate the influence of individual input variables, an AI-based evaluation method is then applied.

Control variables for the system under investigation include speed, operating force, the preload of the rolling bearings and the oil feed pressure. In the pivoting operation mode, the pivoting angle and frequency determine the operating point. Possible interference effects result from the bearing temperatures, a reduction in the preload force (shown in the test by expansion of the PCAs attached to the bearing cover) or an asymmetrical application of force (possible in presses due to tilting of the ram or as a result of transverse forces). The latter is achieved in the test by applying different forces to the actuators mounted on the operating force side. From the possible manipulated variables, only a few selected ones are considered in the investigations presented here. The operating mode investigated is full-rotation mode. The total operating force F_{op} is 10 kN, the oil pressure at the infeed is 10 bar. The shaft speed is varied in 5 steps between 10 and 400 rpm. The roller bearings are preloaded by the bearing covers in a defined manner in the initial state (cold actuators).

3.2 Exemplary Results

The compliance of the tested bearing is considered as a decisive property for the dimensioning of the surrounding system. Compliance values are not only important for the characterization of the mechanical behavior of the drivetrain of presses but they are also necessary for a model based control of the ram movement. Figure 9 shows the compliance of the combined bearing determined during a representative series of measurements under different disturbance influences for different speeds. Tests under symmetrical operating load are shown in black. Tests in which the PCAs attached to the force side apply different forces due to different temperatures, resulting in an asymmetrical load, are shown in blue dashed lines. Experiments with a one-sided loss of roller bearing preload due to increased PCA temperatures at one of the bearing covers (see Fig. 6) are marked with triangles. The results from tests with a decreased preload on both sides are plotted with circles. Normal preload conditions are labeled with diamonds.

In general, the compliance decreases with increasing speed due to the higher effectiveness of the plain bearing. The influence of the detuning decreases with increasing speeds as the values of different measurements converge at high speeds. The influence of the roller bearing preload is higher at low speeds due to the lower plain bearing load. Under symmetrical load, a one- or two-sided loss of preload leads to an increase in compliance. In the case of asymmetrical operating load, this is not observed. In the present case, the compliance values are even slightly lower, which can be explained by a tilting of the shaft and the associated solid body contact in the plain bearing shell.

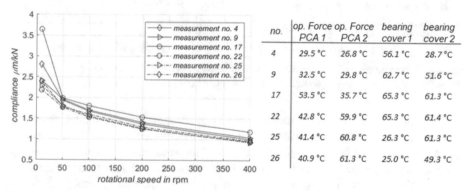

no.	op. Force PCA 1	op. Force PCA 2	bearing cover 1	bearing cover 2
4	29.5 °C	26.8 °C	56.1 °C	28.7 °C
9	32.5 °C	29.8 °C	62.7 °C	51.6 °C
17	53.5 °C	35.7 °C	65.3 °C	61.3 °C
22	42.8 °C	59.9 °C	65.3 °C	61.4 °C
25	41.4 °C	60.8 °C	26.3 °C	61.3 °C
26	40.9 °C	61.3 °C	25.0 °C	49.3 °C

Fig. 9. Compliance of the combined bearing for different operating conditions of a representative test series and corresponding averaged PCA temperatures

These observations are consistent with the results from [8], although the tests conducted there were carried out individually with prior warm-up and thus under stationary, reproducible conditions, whereas the investigations presented here involved dynamically variable operating conditions with uneven starting conditions. This reduces the effort required to validate the bearing behavior under realistic operating conditions. It is thus possible to investigate a much larger number of possible uncertainty influences. Since a large number of measurements are conducted in a short time and a correspondingly large amount of data is generated, the AI-based approach presented below is used to evaluate the complex relationships.

In order to quantify the influence of the expansion of the wax actuators as a result of their temperature on the performance of the investigated bearing, a regression model for predicting the compliance of the bearing is implemented according to Fig. 7. In the first step, the acquisition of temperature signals in each actuator as well as the rotational speed are conducted for different operating points. The temperatures of the four actuator groups are used as input parameters, with the three actuators on each of the two bearing covers being combined being averaged. These four temperatures are also averaged over a defined period of time prior to the actual acquisition of the data sets. Since the PCAs achieve an approximately stationary force after 20 min at constant temperature, the averaging interval is set to 20 min. Taking into account five rotational speeds and 27 test series, there are 135 operating points with different speeds and actuator temperatures. The acquired time series are averaged for each rotational speed stage, resulting in a total of 175 predictor variables and 35 labeled responds given by the compliance of the system.

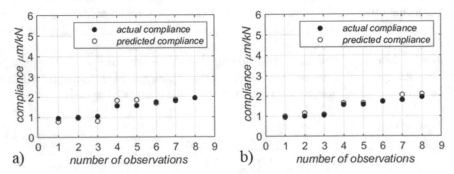

Fig. 10. Representation of the measured and predicted compliances of the test data using linear regression model (left) and a quadratic regression model (right)

To predict this compliance, both a linear and a quadratic regression model are derived considering the rotational speeds and the actuator temperatures. Finally, the transformed data set is randomly split into a training and a test data set to validate the model. Thereby, 25 operating points are used for training and ten operating points for testing the model. Figure 10(b) shows the results for the randomly selected operating points predicted by the regression model. While the linear regression model shows a deviation quantified by the RMSE of $1.7236 \cdot 10^{-9}$ m/N, the quadratic regression model performs slightly better with a value of $1.2284 \cdot 10^{-9}$ m/N. This is confirmed by the results shown in Fig. 10(a) for compliance of the bearing for different operating points. From the experimental results, an approximately quadratic response for the compliance of the bearing system can be seen, which accordingly justifies the quadratic performance of the regression model. The results show that both regression models are able to predict the compliance of the system as a function of actuator temperatures and rotational speeds. The performance of the models quantified by the RMSE, the maximum error as well as the coefficient of determination (R^2) is shown in Table 1.

Table 1. Statistical evaluation of the prediction results

	Linear model	Quadratic model
Root Mean Square Error (RMSE)	$1.7236 \cdot 10^{-9}$ m/N	$1.2284 \cdot 10^{-9}$ m/N
absolute maximal error	$2.7581 \cdot 10^{-9}$ m/N	$2.4865 \cdot 10^{-9}$ m/N
coefficient of determination (R^2)	78.57%	89.11%

4   Conclusion

Within this work, a consequent advancement of the methodology presented in [8] is promoted. So far, only explicit targeted states could be approached and an initial preheating to steady state conditions has been necessary. A cool down completed each run to secure reproducibility. Since only a few influencing variables and their combination can be investigated in manageable investigation periods, this procedure is not suitable for an efficient validation process. Considering the large number of influencing variables (e.g. speed, oil pressure etc.) and disturbances (e.g. bearing temperatures, asymmetrical load, etc.) a complete test cycle is not feasible. Additionally, no time-dependent disturbance overlays and dynamically changing conditions can be investigated.

Thus, it is a considerably more efficient test procedure to run through randomly selected operating points in sequence. This approach automatically incorporates dynamic processes and their resulting influence on system behavior into the test data. The test bench has been modified to implement this more efficient test method. In addition to an improved data recording, the concept for applying disturbances was revised. With the aid of water-flown sleeves, the paraffin wax PCA can be selectively heated or cooled. Four independent actuator groups (three PCAs on each bearing cover and two separate PCAs on the bearing housing) have been implemented on the test rig.

On the one hand, the dynamic application of disturbances as well as the investigation of different superimposed states allow a considerable reduction of the necessary test cycles. On the other hand, there is also a complex superposition of results, which makes the evaluation of the isolated effects of applied disturbances considerably more difficult. For the determination of results and the identification of necessary improvement measures, a methodology was discussed, which trains an algorithm by means of an AI black-box model. In this way, influences of individual values could be determined from the superimposed test results. This has been demonstrated by the compliance of the combined bearing.

The goal of this work was to show the applicability of the presented approach. The results show that the method works for the application example shown. The next step is to optimize the presented test setup by the implementation of additional measuring systems to record the PCA forces and displacements in order to improve the understanding of the observed relationships. The accuracy of the evaluation algorithm can be increased by more training data. Subsequently, the procedure has to be transferred to more complex investment goods and their components.

Acknowledgement. The results of this paper are outcomes of the Collaborative Research Centre SFB 805 (projects B2 and T6). The German Research Foundation (DFG) funds both projects. The authors want to thank for funding the projects.

References

1. Nieschlag, R., Dichtl, E., Hörschgen, H.: Marketing: Ein entscheidungstheoretischer Ansatz. Duncker und Humblot, Berlin (1972)
2. Groche, P., Scheitza, M., Kraft, M., et al.: Increased total flexibility by 3D servo presses. CIRP Ann. **59**, 267–270 (2010). https://doi.org/10.1016/j.cirp.2010.03.013
3. Thommen, J.-P., Achleitner, A.-K., Gilbert, D., et al.: Allgemeine Betriebswirtschaftslehre. Springer Gabler, Wiesbaden (2017)
4. Zhou, W., Qiu, N., Zhang, N., et al.: Research on steady-state characteristics of centrifugal pump rotor system with weak nonlinear stiffness. TFAMENA **42**, 87–102 (2018). https://doi.org/10.21278/TOF.42306
5. Koga, T., Aoyama, K.: Design process guide method for minimizing loops and conflicts. JAMDSM **3**, 191–202 (2009). https://doi.org/10.1299/jamdsm.3.191
6. Pai, N., Hess, D.: Experimental study of loosening of threaded fasteners due to dynamic shear loads. J. Sound Vib. **253**, 585–602 (2002). https://doi.org/10.1006/jsvi.2001.4006
7. Klein, T., Reinhart, G.: Towards agile engineering of mechatronic systems in machinery and plant construction. Procedia CIRP **52**, 68–73 (2016). https://doi.org/10.1016/j.procir.2016.07.077
8. Groche, P., Sinz, J., Germann, T.: Efficient validation of novel machine elements for capital goods. CIRP Ann. **69**, 125–128 (2020). https://doi.org/10.1016/j.cirp.2020.03.004
9. Ogden, S., Klintberg, L., Thornell, G., et al.: Review on miniaturized paraffin phase change actuators, valves, and pumps. Microfluid. Nanofluid. **17**, 53–71 (2014). https://doi.org/10.1007/s10404-013-1289-3
10. Mann, A.: Gestaltungsrichtlinien für geschlossene Paraffinaktoren zur gezielten Beeinflussung von Systemen hoher Steifigkeit, Berichte aus Produktion und Umformtechnik, vol. 120. Shaker, Düren (2019)
11. Mann, A., Germann, T., Ruiter, M., et al.: The challenge of up scaling paraffin wax actuators. Mater. Des. **190**, (2020). https://doi.org/10.1016/j.matdes.2020.108580
12. Groche, P., Sinz, J.: Innovation durch Kombination: Wälz-Gleit-Lagerung für Servopressen, MM Maschinenmarkt 2016 (2016)
13. Groche, P., Sinz, J., Felber, P.: Kombinierte Wälz-Gleitlager - Anforderungsgerechter Funktionsübergang, Konstruktionspraxis Spezial (2018)
14. Sinz, J., Groche, P.: Effekte der Größenskalierung auf die Funktionsfähigkeit kombinierter Wälz-Gleitlager. In: 13. VDI-Fachtagung: Gleit- und Wälzlagerungen 2019: Gestaltung - Berechnung - Einsatz, vol. 2348, Schweinfurt, Deutschland (2019)
15. Sinz, J., Niessen, B., Groche, P.: Combined roller and plain bearings for forming machines: design methodology and validation. In: Schmitt R., Schuh G. (eds), Advances in Production Research: Proceedings of the 8th Congress of the German Academic Association for Production Technology (WGP), pp 126–135. Springer Nature (2018)
16. Sinz, J., Knoll, M., Groche, P.: Operational effects on the stiffness of combined roller and plain bearings. Proc. Manuf. **41**, 650–657 (2019). https://doi.org/10.1016/j.promfg.2019.09.054
17. Fayyad, U., Piatetsky-Shapiro, G., Smyth, P.: From data mining to knowledge discovery in databases. AI Mag. **17**, 37–54 (1996). https://doi.org/10.1609/aimag.v17i3.1230
18. Hastie, T., Tibshirani, R., Friedman, J.: The Elements of Statistical Learning: Data Mining, Inference, and Prediction, Second Springer Series in Statistics. Springer-Verlag, New York (2009)

On Uncertainty, Decision Values and Innovation

Sebastian Thöns[1,2](✉), Arifian Agusta Irman[3], and Maria Pina Limongelli[4]

[1] Faculty of Engineering LTH, Lund University, Lund, Sweden
sebastian.thoens@kstr.lth.se
[2] German Federal Institute for Materials Research and Testing, Berlin, Germany
[3] Rambøll, Copenhagen, Denmark
[4] Politechnico di Milano, Milan, Italy

Abstract. This paper contains a description, an alignment and a joint approach for technology readiness development with a three phases support of decision value analyses. The three phases are separated into the decision value forecasting, decision value analysis and the technology value quantification supporting the technological concept formulation and experimental testing, the prototype development and the technology qualification and operation. Decision value forecasting allows technology development guidance by technology performance requirements and the value creation even before the technology development is started. This approach is exemplified with load, damage and resistance information based integrity management of a structure and the ranking of the different strategies. The results can be used to guide a technology screening for matching with performance characteristics in terms of precision, cost and employability. Moreover, the first estimate of value creation of the technology for stakeholders, business models and market evaluation is provided.

Keywords: Innovation · Technology readiness · Decision analysis

1 Introduction

Decision theory has been introduced from economic sciences to built environment engineering by Benjamin and Cornell (1970) based on the works of Raiffa and Schlaifer (1961). In recent years, many studies have been published on topic of value quantification of structural health information (SHI) for built environment systems (e.g., Pozzi and Der Kiureghian (2011) and Thöns (2018)) also in conjunction with the COST Action TU1402 (www.cost-tu1402.eu and https://en.wikipedia.org/wiki/Value_of_structural_health_information).

The SHI value quantification in the frame of the COST Action TU1402 has resulted in the scientific evidence of a high SHI value for built environment systems and its boundaries, an enhanced accessibility of the value of information analyses and guidelines for scientific utilisation, engineering and infrastructure owner usage. A scientific potential of guiding the technology development with a SHI value quantification has been identified.

© The Author(s) 2021
P. F. Pelz and P. Groche (Eds.): ICUME 2021, LNME, pp. 252–263, 2021.
https://doi.org/10.1007/978-3-030-77256-7_20

Technology readiness has been introduced by the USA National Aeronautics and Space Administration NASA (Sadin, Povinelli et al. (1989)) and has since penetrated technological management in various organisations such as e.g. military organisations and the European Space Agency ESA. Technology readiness levels have been defined for the European research and innovation program Horizon 2020 since 2009 (see e.g., Héder (2017)).

The technology development is subdivided in a stepwise technology readiness process starting with the observation, concept formulation and experimental testing and (Technology Readiness Levels - TRLs 1 to 4) followed by technology demonstration and prototype development (TRLs 5 to 7) and the technology qualification and operation (TRL 8 and 9), see e.g. Héder (2017) and Table 1.

Table 1. European Technology Readiness Levels according to Héder (2017)

TRL 1	Basic principles observed
TRL 2	Technology concept formulated
TRL 3	Experimental proof of concept
TRL 4	Technology validated in lab
TRL 5	Technology validated in relevant environment
TRL 6	Technology demonstrated in relevant environment
TRL 7	System prototype demonstration in operational environment
TRL 8	System complete and qualified
TRL 9	Actual system proven in operational environment

A technology development has the potential for innovation and innovation scaling when the technology performs and creates value for market stakeholders as a premise for the development of a new market according to the Disruptive Innovation Theory (e.g. Bower and Christensen (1995) and Christensen (1997)). However, the TRL development accounts solely for technological steps and not for technology value quantification.

This paper focuses on the utilisation of decision theory - originating from the field of economic management science – along the technology development for information acquirement system innovation guidance. The paper starts out with summarising the decision analytical formulation for built environment systems and how a value quantification is performed (Sect. 2). An approach to align the technology development with different types of decision analyses is developed in Sect. 3. The first step of this approach namely decision value forecasting is described and exemplified in Sect. 4. The paper closes with a summary and conclusions highlighting the potential for innovation guidance and pointing to further research.

2   Decision Analytical Formulation

Decision analyses encompass the modelling of built environment systems, information about the system performance and actions to modify the system states or system performance.

System models are used to assess and predict the behaviour of real-world systems subjected to exposures and disturbances, which influence the component and the system states. Information is based on observations of the physical world from which data can be extracted, indicators for evaluating the system states and performance can be derived and information to adapt and update the system states and system performance can be obtained. Information in turn facilitates closer to reality predictions of the system performance and implies that the adaptation with information is solely on the side of the models and will not affect the real-word system performance.

An action - as a physical system change - influences the physical world system performance and the modelled system performance. In this sense, actions can be used for enforcing a coherence of the modelled system performance and the physical world system performance. For planning of actions, an enhanced system knowledge by adaption with information may be used.

The information i_i supported decision analytical formulation for built environment systems consists of models for the information and integrity management and for the system performance composed of system states X_l and associated utilities $u(\ldots)$. It is distinguished between probabilistic models for information outcomes $Z_{i,j}$ and system states and decision variables relating to information and actions a_k (Fig. 1).

Following Benjamin and Cornell (1970), the decision analytical objective functions are formulated for a prior analysis without additional information, a posterior analysis with known additional information and a pre-posterior analysis with predicted information. In a prior and posterior decision analysis, the actions constitute the decision variables maximizing the prior or posterior expected utilities (U_{prior} and U_{Post}, respectively) based on the expected utility theorem, see e.g. Von Neumann and Morgenstern (1947). A pre-posterior decision analysis facilitates to optimise the expected value of the utility also by the choice of the information acquirement strategy (with the index i).

The expected value of the utility does not constitute per se a value. Only in relation to a threshold, a value can be quantified as described in Raiffa and Schlaifer (1961) by interrelating decision theory to economical concepts. Following the original formulation, the value of information has been quantified as the difference of the expected utilities stemming from a decision analysis without and with information constituting the base the and the enhancement scenario. This original formulation is extended here with varying the base and enhancement scenarios for more comprehensiveness (Thöns and Kapoor (2019)).

The value of a predicted action can be quantified by subtracting the expected system performance utility without or with an implemented action, U_{SP} and $U_{SP}(a)$, respectively, from an expected system performance utility with a predicted action U_{Prior}, see Eq. (1). When the optimal action (sets) a_1 and a_2 are not identical, then the action (set) value in relation to another action (set) can be quantified as the difference between two prior decision analyses, $U_{Prior}(a_{k_1})$ and $U_{Prior}(a_{k_2})$, see Eq. (2). The value contains the action costs and consequences.

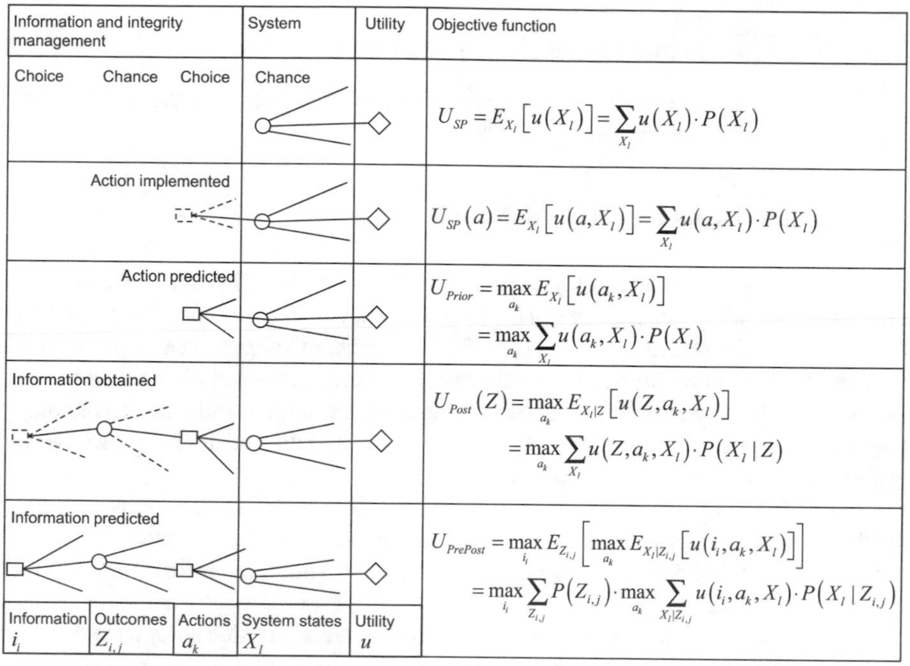

Information and integrity management			System	Utility	Objective function
Choice	Chance	Choice	Chance		$U_{SP} = E_{X_I}\left[u(X_I)\right] = \sum_{X_I} u(X_I) \cdot P(X_I)$
Action implemented					$U_{SP}(a) = E_{X_I}\left[u(a, X_I)\right] = \sum_{X_I} u(a, X_I) \cdot P(X_I)$
Action predicted					$U_{Prior} = \max_{a_k} E_{X_I}\left[u(a_k, X_I)\right]$ $= \max_{a_k} \sum_{X_I} u(a_k, X_I) \cdot P(X_I)$
Information obtained					$U_{Post}(Z) = \max_{a_k} E_{X_I|Z}\left[u(Z, a_k, X_I)\right]$ $= \max_{a_k} \sum_{X_I} u(Z, a_k, X_I) \cdot P(X_I \mid Z)$
Information predicted					$U_{PrePost} = \max_{i_i} E_{Z_{i,j}}\left[\max_{a_k} E_{X_I|Z_{i,j}}\left[u(i, a_k, X_I)\right]\right]$ $= \max_{i_i} \sum_{Z_{i,j}} P(Z_{i,j}) \cdot \max_{a_k} \sum_{X_I|Z_{i,j}} u(i, a_k, X_I) \cdot P(X_I \mid Z_{i,j})$
Information i_i	Outcomes $Z_{i,j}$	Actions a_k	System states X_I	Utility u	

Fig. 1. Decision analytical formulation for quantification of the expected utilities by a system performance analysis (U_{SP}), a prior decision analysis (U_{Prior}), a posterior decision analysis (U_{Post}) and a pre-posterior decision analysis ($U_{PrePost}$).

$$V_{SP}^{Prior}(a_k) = U_{Prior} - U_{SP} \quad \text{and} \quad V_{SP}^{Prior}(a_k, a) = U_{Prior} - U_{SP}(a) \tag{1}$$

$$V_{Prior,a_{k_2}}^{Prior,a_{k_1}}(a_{k_1}, a_{k_2}) = U_{Prior}(a_{k_1}) - U_{Prior}(a_{k_2}) \tag{2}$$

The predicted value of an information can be quantified in analogy to the action value as the expected utility difference with an enhancement scenario containing predicted, i.e. pre-posterior, information and a base scenario excluding this predicted information. The base scenario can be of the types of a prior, a posterior and a pre-posterior decision analysis. For the latter, the information acquirement strategy sets i_1 and i_2 are exclusive. The value contains the information costs:

$$V_{Prior}^{PrePost}(i_i) = U_{PrePost} - U_{Prior} \quad \text{and} \quad V_{Post}^{PrePost}(i_i, Z) = U_{PrePost} - U_{Post}(Z) \tag{3}$$

$$V_{PrePost,i_2}^{PrePost,i_1}(i_1, i_2) = U_{PrePost}(i_1) - U_{PrePost}(i_2) \tag{4}$$

The value of information and actions can be quantified by using the base scenarios of the action value quantification and the enhancements scenarios of the information value quantification. Both, the information costs and the action costs and consequences are included in the value quantification:

$$V_{SP}^{PrePost}(i_i, a_k) = U_{PrePost} - U_{SP} \quad \text{and} \quad V_{SP}^{PrePost}(i_i, a) = U_{PrePost} - U_{SP}(a) \tag{5}$$

$$V^{\text{PrePost,1}}_{\text{PrePost,2}}\left(i_1, i_2, a_{k_1}, a_{k_2}\right) = U_{\text{PrePost}}\left(i_2, a_{k_2}\right) - U_{\text{Prior}}\left(i_1, a_{k_1}\right) \qquad (6)$$

The decision value can be divided with the expected utility of the base scenario resulting in a normalised decision value $\bar{V}$.

3 Decision Value and Technology Readiness

Three types of decision value analyses are distinguished namely (1) value forecasting, (2) value analysis and (3) technology value quantification, which are temporally aligned with the technology development phases (Fig. 2). The decision value forecasting analysis is performed solely with a probabilistic built environment system performance analysis and the consideration of a base scenario constituting the conventional and known technology. The information and action modelling exploits characteristics of the built environment system performance model.

The decision value analysis is performed with a probabilistic and experimentally verified technology performance for (a) the quantification of the current decision value and (b) potentials for decision value optimisation and (c) boundaries for optimality such as e.g. decision rules.

The technology value quantification constitutes a decision value analysis in the operational environment and with consideration of technology production boundaries. For the stages of a decision value analysis and the technology value quantification there are many studies such as e.g. Thöns (2018), Long, Döhler et al. (2020) and Thöns and Stewart (2020).

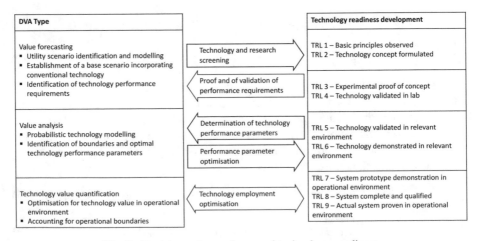

Fig. 2. Decision value analyses and technology readiness.

In relation to the TRLs 1 and 2, the decision value forecasting phase is temporally located before the technology development starts. With a base scenario and the utility scenario modelling as well as the identification of technology performance requirements, a basic principle and conceptual technology screening can be performed. With experimental proof and validation, the technology performance parameters (at TRLs 3 and 4) can be validated against the performance requirements for the desired decision value.

Further experimental testing in a relevant environment (TRLs 5 and 6) will improve the performance parameter optimisation and development of probabilistic technology performance models to be integrated in a decision value analysis. The decision value analysis will reveal conditions and potentials for utility gains, which can be used as an input for the further technology development. A technology value quantification is performed to fully represent the technology performance in an operational environment. The technology value quantification can be performed with different base scenarios constituting different conventional technology approaches.

4 Decision Value Forecasting

The described decision value forecasting approach is introduced. For this purpose, a decision scenario is formulated consisting of a built environment system performance model for which information acquirement technology is to be developed facilitating an efficient information and integrity management.

The information value forecasting is developed with a built environment system performance model distinguishing the complementary intact ($X_1 = S$) and failure state ($X_2 = F$) with the limit state function (7), g_F. The resistance R, the damage D (with its capacity transformation function t_D) and the loading S are based on models, for which the precision is known with the respective model uncertainties M. The limit state equation is representative for a non-redundant built environment system subjected to a dominating failure mode under deterioration (JCSS (2001–2015)).

$$X_1 = S : \quad g_F = M_R \cdot R(1 - t_D \cdot M_D \cdot D) - M_S \cdot S > 0, \quad X_2 = F : \quad g_F \leq 0 \quad (7)$$

The consequences of the intact and failure state are modelled to calculate the expected value of the utility with a system performance (SP) analysis.

Information in its fundamental meaning is about the knowledge of system states (see Raiffa and Schlaifer (1961)). Progressing this fundamental concept, the limit state model as introduced with Eq. (7) and the fact that information can be forecasted with the help of realisations of the model uncertainties is applied (see e.g. Thöns (2018)). By introducing model uncertainty realisation thresholds η, indication events can be discretised (see Eq. (8) with one threshold and two complementary indication events for use in conjunction with Eq. (7)). The thresholds can be defined and/or optimised for system performance in relation to availability, failure and/or utility probabilities (see Agusta (2020)).

$$Z_{i,1} : \ P(Z_{i,1}) = \int_{-\infty}^{\eta} f_M(m)dm; \quad Z_{i,2} : \ P(Z_{i,2}) = \int_{\eta}^{\infty} f_M(m)dm \quad (8)$$

The pre-posterior and the posterior probabilities of the system states are calculated with the truncated and normalised or just truncated model uncertainty distributions $f_{M\,|^{t_M}_{-\infty}}$ and $f_{M\,|^\infty_{t_M}}$, respectively. Additionally, the random variable M_U can be multiplied to the respective model uncertainty to account for a limited precision of the information.

Actions can be introduced as engineering actions and/or utility actions modifying the system state probabilities and/or the system state utilities, respectively.

4.1 Exemplary Study of Decision Value Forecasting

The decision value forecasting approach takes basis in the system state Eq. (7), which is representative for a built environment system subjected to a dominating failure mode under deterioration (JCSS (2001–2015), Fig. 3). The resistance is without damage and in analogy to a structural design process calibrated to a target failure probability of $P_T = 10^{-5}$. Such target represents a the reliability of typical engineering structure subjected to moderate consequences of failure and normal costs of safety measures (see e.g. ISO 2394 (2015)). The model uncertainties are adjusted in conjunction with Part 3.09 of the Probabilistic Model Code of the Joint Committee on Structural Safety (JCSS (2001–2015)) with higher model uncertainties for the loading and the damages. Failure consequence, i.e. a negative utility $u(F)$, is normalised. A utility $u(S|m)$ for a possible service life extension is assigned in case the structural reliability is high due to compliance with the threshold η_1 and its calibration to a target reliability.

Variable	Distribution	Expected Value	St. dev.
S	Weibull	3.5	0.10
R	Lognormal	$P(F\,|\,D=0.0)=P_T$	0.10
D	Normal	2.0	0.10
t_D	Det.	0.1	-
M_R	Lognormal	1.0	0.05
M_S	Lognormal	1.0	0.10
M_D	Lognormal	1.0	0.20
P_T	Det.	10^{-5}	-
$u(S\,|\,m_S \leq \eta_1)$ or $u(S\,|\,m_D \leq \eta_1)$ or $u(S\,|\,m_R \geq \eta_1)$	Det.	0.001	-
$u(F)$	Det.	-1.0	-

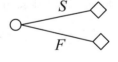

Fig. 3. System performance model (Eq. (7)) and part of decision tree (see Fig. 1)

Information is modelled by exploiting the characteristics of model uncertainties namely that the system state behaviour of a constructed built environment system represents a realisation of the model uncertainty (Thöns (2018), Agusta and Thöns (2018)).

The model uncertainty thresholds are introduced for a higher or equal (target) failure probability than required, η_1, and for the optimal action being repair, η_2. In this way the thresholds are optimised to comply with the decision rules $r(\ldots)$ of repairing (denoted with a_1) only for a $Z_{i,3}$ indication informing low structural reliability (Eq. (9)). The indication $Z_{i,1}$ informs about a high reliability, the indication $Z_{i,2}$ of a reliability as expected:

$$r\big[Z_{i,1}, Z_{i,2}, Z_{i,3}\big] = [a_0, a_0, a_1] \tag{9}$$

Equation (10) shows the probabilities of indications calculation for the discretisation of the load model uncertainties. Note that the discretisation for the resistance model uncertainty requires a different order of the integration boundaries in conjunction with the threshold determination rules in Figs. 3 and 4.

The information may be subjected to a finite precision expressed with a generic, Normal distributed information uncertainty M_U with a coefficient of variation of 5% (Fig. 4). The information has a cost of $c(i_i) = 0.0015$ adjusted to similar consequence and cost ratios in e.g. Thöns (2018).

$$P\big(Z_{1,1}\big) = \int_{-\infty}^{\eta_1} f_{M_S}(m_S)dm_S; \quad P\big(Z_{1,2}\big) = \int_{\eta_1}^{\eta_2} f_{M_S}(m_S)dm_S; \quad P\big(Z_{1,3}\big) = \int_{\eta_2}^{\infty} f_{M_S}(m_S)dm_S \tag{10}$$

The repair action will lead to damage of zero, $D(a_1) = 0.0$, and costs of $c(a_1) = 0.01$ see e.g. Thöns (2018).

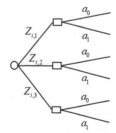

Variable	Distribution	Expected Value	St. dev.
$D(a_1)$	Det.	0.0	-
$c(a_1)$	Det.	0.01	-
M_U	Normal	1.0	0.05
$c(i_i)$	Det.	0.0015	-
η_1	Det.	$P(F\mid m=\eta_1)=P_T$	-
η_2	Det.	$a_{opt}=a_1$	-

Fig. 4. Information and integrity management model and part of decision tree (see Fig. 1)

The information and integrity management model will be used to predict and to pre-posteriorly and posteriorly update the probabilities of failure and survival. For example, the posteriorly updated probability of failure with a $Z_{1,1}$ indication subjected to the information uncertainty M_U is calculated with the threshold-truncated and normalised distribution $M_S\big|_{-\infty}^{\eta_1}$:

$$F|Z_{1,1}: \quad g_{F|Z_{1,1}} = M_R \cdot R(1 - t_D \cdot M_D \cdot D) - M_U \cdot M_S\big|_{-\infty}^{\eta_1} \cdot S \le 0 \tag{11}$$

With the decision value analysis, the indication dependent posterior values and the probabilities of the information and integrity management strategies are forecasted. For information about the load, both the $Z_{1,1}$ and $Z_{1,2}$ indication, denoting a behaviour better or as expected, respectively, lead to a positive posterior decision value (Fig. 5). The $Z_{1,3}$ indication requires a repair (see decision rules in Eq. (9)) and leads to a negative value. The indication $Z_{1,2}$ has a significantly higher probability than the other indications. The influence of the information precision is not very pronounced as only the $Z_{1,1}$ and $Z_{1,2}$ probabilities and the $Z_{1,2}$ posterior value are slightly influenced.

For resistance information, the most probable $Z_{2,2}$ indication lead to low or even negative posterior relative decision value with consideration of the information precision. The low probability indication $Z_{2,1}$ leads to high decision value. The influence of the information uncertainty is rather pronounced affecting the probabilities of the indications $Z_{2,2}$ and $Z_{2,3}$.

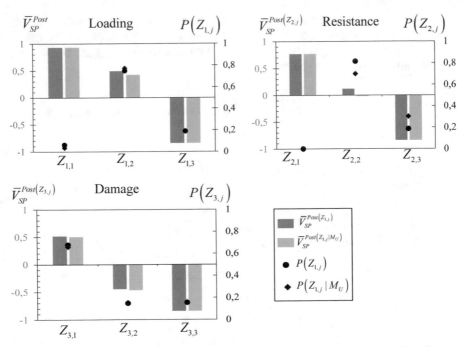

Fig. 5. Posterior decision value of information and integrity management and perfect and imperfect indication probabilities for loading, resistance and damage information

For damage information, the indication $Z_{3,1}$ has the highest probability followed by approximately equal probabilities of the $Z_{3,2}$ and $Z_{3,3}$ indications. The information uncertainty has minor influence both on the indication probabilities and the values.

The pre-posterior, i.e. the predicted, value of information and integrity management has been calculated by summing the product of the indication probabilities and the posterior values (Table 2). Positive values are calculated for load and damage information with and without consideration of the information precision. The information precision significantly influences the value of the integrity management with load information.

The damage information value is less influenced, which is attributed to the damage-resistance transfer function. The influence of the information precision is lower for the integrity management with damage information. Resistance information do not lead to a positive value.

The decision value between the strategies $\bar{V}_{\mathrm{PrePost,1}}^{\mathrm{PrePost,2}}$ is quantified with the load information strategy (Table 2) as a basis. Damage information and with consideration of its uncertainties leads to the highest value $\bar{V}_{\mathrm{PrePost,1}}^{\mathrm{PrePost,2}}$ explicitly quantifying the second best alternative in the ranking of the information and integrity management strategies.

Table 2. Thresholds and pre-posterior value of information

Strategy	Description	Thresholds		$\bar{V}_{\mathrm{SP}}^{\mathrm{PrePost}}$	$\bar{V}_{\mathrm{PrePost,1}}^{\mathrm{PrePost,2}}$
		η_1	η_2		
Perfect Information (PI)					
1	Load	0.84	1.11	0.32	–
2	Resistance	1.29	0.95	-0.05	-0.37
3	Damage	1.09	1.20	0.16	-0.16
Imperfect Information (II)					
1	Load	0.83	1.09	0.18	–
2	Resistance	1.31	0.97	-0.29	-0.47
3	Damage	1.08	1.19	0.13	-0.05

5   Summary and Conclusions

This paper contains a description, an alignment and a joint approach for technology readiness development with a three phases support by decision value analyses. The decision value analyse are divided into the decision value forecasting, value decision value analysis and the technology value quantification phases. The technology readiness development levels (TRLs) are seperated into technological concept formulation and experimental testing (TRLs 1 to 4), the prototype development (TRLs 5 to 7) and the technology qualification and operation (TRLs 8 and 9).

The decision value forecasting approach relies solely on a built environment system performance model. Decision value forecasting facilitates support for the first phases of technology readiness development, i.e. for technological concept formulation and experimental testing. With an exemplary decision value forecasting analysis, it was found that for a built environment system, load or damage information acquirement systems with a high precision should be developed. Resistance information acquirement system should not be developed as the forecasted decision value is negative. In the context of business model and technology markets, a forecast of the achievable technology value for the structural information and integrity management has been provided in the order of 13% to 32%.

The alignment of technology development, innovation and decision value analyses requires more systematic research and applications to substantiate and detail their interrelations and to demonstrate the targeted support of innovation decisions with case studies.

The approach has been written in the context of technology development, which may, however, be composed of a technological and algorithmic readiness level development (see e.g. Limongelli, Orcesi et al. (2018)).

References

Agusta, A.: Structural Integrity and Risk Management based on Value of Information and Action Analysis. Ph.D. thesis (2020)

Agusta, A., Thöns, S.: Structural monitoring and inspection modeling for structural system updating. In: Sixth International Symposium on Life-Cycle Civil Engineering (IALCCE), Ghent, Belgium, 28–31 October 2018 (2018)

Benjamin, J.R., Cornell, C.A.: Probability, Statistics and Decision for Civil Engineers. McGraw-Hill, New York (1970). ISBN: 070045496

Bower, J.L., Christensen, C.M.: Disruptive technologies: catching the wave. Harv. Bus. Rev. **73**(1), 43–53 (1995)

Christensen, C.M.: The Innovator's Dilemma: When New Technologies Cause Great Firms to Fail, University of Illinois at Urbana-Champaign's Academy for Entrepreneurial Leadership (1997)

Héder, M.: From NASA to EU: the evolution of the TRL scale in public sector innovation. Innov. J. **22**(2), 1–23 (2017)

ISO 2394: General Principles on Reliability for Structures ISO 2394 (2015)

JCSS: Probabilistic Model Code, JCSS Joint Committee on Structural Safety. ISBN: 978-3-909386-79-6 (2001–2015)

Limongelli, M.P., Orcesi, A., Vidovic, A.: The indicator readiness level for the classification of research performance indicators for road bridges. In: The Sixth International Symposium on Life-Cycle Civil Engineering, IALCCE 2018 GHENT, Ghent University, Belgium (2018)

Long, L., Döhler, M., Thöns, S.: Determination of structural and damage detection system influencing parameters on the value of information. Struct. Heal. Monitor. (2020). https://doi.org/10.1177/1475921719900918

Pozzi, M., Der Kiureghian, A.: Assessing the value of information for long-term structural health monitoring. In: Health Monitoring of Structural and Biological Systems 2011, San Diego, California, United States, 7–10 March 2011 (2011)

Raiffa, H., Schlaifer, R.: Applied Statistical Decision Theory. New York, Wiley (2000). ISBN: 047138349X (1961)

Sadin, S.R., Povinelli, F.P., Rosen, R.: The NASA technology push towards future space mission systems. Acta Astronautica **20**(C), 73–77 (1989). https://doi.org/10.1016/0094-5765(89)90054-4

Thöns, S.: On the value of monitoring information for the structural integrity and risk management. Comput.-Aid. Civil Infrastruct. Eng. **33**(1), 79–94 (2018). https://doi.org/10.1111/mice.12332

Thöns, S., Kapoor, M.: Value of information and value of decisions. In: 13th International Conference on Applications of Statistics and Probability in Civil Engineering (ICASP), Seoul, Korea, 26–30 May 2019

Thöns, S., Stewart, M.G.: On the cost-efficiency, significance and effectiveness of terrorism risk reduction strategies for buildings. Struct. Saf. **85**, (2020). https://doi.org/10.1016/j.strusafe.2020.101957

Neumann, V., Morgenstern, O.: Theory of Games and Economical Behavior. Princeton University Press, Princeton (1947)

Assessment of Model Uncertainty in the Prediction of the Vibroacoustic Behavior of a Rectangular Plate by Means of Bayesian Inference

Nikolai Kleinfeller[1(✉)], Christopher M. Gehb[1], Maximilian Schaeffner[1], Christian Adams[1], and Tobias Melz[1,2]

[1] Mechanical Engineering Department, Research Group System Reliability, Adaptive Structures and Machine Acoustics SAM, Technical University of Darmstadt, Otto-Berndt-Straße 2, 64287 Darmstadt, Germany
nikolai.kleinfeller@sam.tu-darmstadt.de
[2] Fraunhofer Institute for Structural Durability and System Reliability LBF, Bartningstr. 47, 64289 Darmstadt, Germany

Abstract. Designing the vibroacoustic properties of thin-walled structures is of particularly high practical relevance in the design of vehicle structures. The vibroacoustic properties of thin-walled structures, e.g., vehicle bodies, are usually designed using finite element models. Additional development effort, e.g., experimental tests, arises if the quality of the model predictions are limited due to inherent model uncertainty. Model uncertainty of finite element models usually occurs in the modeling process due to simplifications of the geometry or boundary conditions. The latter highly affect the vibroacoustic properties of a thin-walled structure. The stiffness of the boundary condition is often assumed to be infinite or zero in the finite element model, which can lead to a discrepancy between the measured and the calculated vibroacoustic behavior. This paper compares two different boundary condition assumptions for the finite element (FE) model of a simply supported rectangular plate in their capability to predict the vibroacoustic behavior. The two different boundary conditions are of increasing complexity in assuming the stiffness. In a first step, a probabilistic model parameter calibration via Bayesian inference for the boundary conditions related parameters for the two FE models is performed. For this purpose, a test stand for simply supported rectangular plates is set up and the experimental data is obtained by measuring the vibrations of the test specimen by means of scanning laser Doppler vibrometry. In a second step, the model uncertainty of the two finite element models is identified. For this purpose, the prediction error of the vibroacoustic behavior is calculated. The prediction error describes the discrepancy between the experimental and the numerical data. Based on the distribution of the prediction error, which is determined from the results of the probabilistic model calibration, the model uncertainty is assessed and the model, which most adequately predicts the vibroacoustic behavior, is identified.

Keywords: Bayesian inference · Model uncertainty · Simply supported rectangular plates

P. F. Pelz and P. Groche (Eds.): ICUME 2021, LNME, pp. 264–277, 2021.
https://doi.org/10.1007/978-3-030-77256-7_21

1 Introduction

The design of the vibroacoustic behavior of technical structures is of particularly high practical relevance especially in the design of vehicle structures [1]. In an early design phase, finite element (FE) models of the vehicle structures are usually built up to predict the vibroacoustic behavior, which is described in general by the natural frequencies and the associated mode shapes of the structure [1]. If the FE model predictions are limited due to inherent uncertainty, there may be additional development effort in a later design phase because the model predictions do not correspond to the real behavior of the developed structure [1]. Parameter uncertainty exists if the distributions of the model parameters are unknown [2]. Further, model uncertainty occurs if the model has been simplified, e.g., neglecting complexity of the technical structure by assuming simplified geometries or boundary conditions [2]. Boundary conditions highly affect the vibroacoustic behavior of a thin-walled structure, e.g., a rectangular plate [3]. In an experimental model, ideal boundary conditions cannot be realized. There are different ways to model the boundary conditions of the rectangular plate in the modeling process. Consequently, competing models including model uncertainty are present and the model that predicts the reality most adequately needs to be identified.

In this context, the first step is to calibrate model parameters of the competing models under consideration so that they can evolve their full prediction potential. Kennedy et al. [4] define the calibration procedure as the matching of the model predictions to observed data by identifying the unknown distributions of the model parameters. Thus, model parameter calibration adjusts the model parameters to physical observations of experimentally observed data. A deterministic model parameter calibration solves an optimization problem, e.g., [5], to achieve a best fit for the unknown model parameters based on a defined calibration criterion. Consequently, there is no information about the model parameter uncertainty after the calibration procedure. One possible way to take model parameter uncertainty into account during the calibration procedure is the application of a probabilistic model parameter calibration by means of Bayesian inference [6]. Then, the unknown model parameters are defined as random variables and their distributions are determined using the information included in the observed data. In this context, Goller et al. [6] conclude that a model never describes reality exactly. Therefore, there are no true values of the model parameters and there is always a discrepancy between the model predictions and the observed data, which is called the prediction error. Because of this fact, the model uncertainty can be assessed by evaluating the remaining prediction error after the calibration procedure [6].

This paper aims to assess two different FE models in their capability to predict the vibroacoustic behavior, i.e. the first six natural angular frequencies and mode shapes, of a simply supported rectangular plate, which is examined as an experimental model in a test stand. The two FE models differ in their modeling of the simply supported boundary conditions. A probabilistic model parameter calibration by means of Bayesian inference is performed so that the model parameter uncertainty is reduced and the FE models reach their maximum potential of predicting the vibroacoustic behavior observed in the experimental model. The model uncertainty of the two FE models is assessed based on the prediction error, which is a measure for the remaining model uncertainty after the calibration procedure.

This paper is organized as follows. In Sect. 2, the test stand, the test specimen and the corresponding FE models are presented. In Sect. 3, a probabilistic model parameter calibration by means of Bayesian inference is introduced. Section 4 shows the results for the probabilistic model calibration for the two FE models and the model uncertainty is assessed by the prediction error. The conclusions are given in Sect. 5.

2   Experimental and Finite Element Models

The data of the experimental model are obtained on a test stand of the research group SAM called SAMple test stand (System reliability, Adaptive structures, and Machine acoustics test stand for Primary Laboratory Experiments) [7]. The test stand, which is located inside a semi-anechoic room, is shown in Fig. 1(a). Here, a surrounding truss structure carries a scanning laser Doppler vibrometer (SLDV). The test specimen is an aluminium rectangular plate and is screwed to the top of an acoustic box. A detailed illustration of the test specimen with plate length a and plate width b is shown in Fig. 1(b). The corresponding design parameters are listed in Table 1. The simply supported boundary conditions are realized in the experimental model according to an approach of Robin et al. [8]. Therefore, the edges of the rectangular plate are bonded to thin blades with a certain blade thickness t_{bl}. This blades are slotted and clamped between two brackets. Thus, a defined blade length l_{bl} is adjusted, as illustrated in Fig. 1(c).

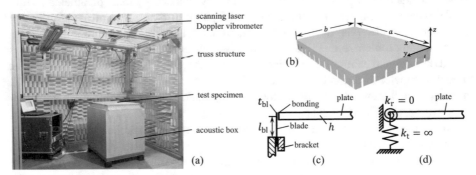

Fig. 1. (a) SAMple test stand in the semi-anechoic room [7]; (b) illustration of the simply supported rectangular plate and the corresponding design parameters; (c) mechanism of the construction to achieve simply supported boundary conditions in the experimental model; (d) model M_1 with ideal simply supported boundary condition;

The rectangular plate is excited by an automatic impact hammer with an integrated sensor for force measurement. The vibration velocities are acquired in the time domain by an SLDV measurement at defined measurement points on the surface of the test specimen. The transfer functions are determined and an experimental modal analysis is performed using a single degree-of-freedom (SDOF) analysis.

In this paper, the first six ($N_m = 6$) natural frequencies of the rectangular plate up to 300 Hz are considered which are significantly involved in the low frequency excitation of the fluid inside of the acoustic box shown in Fig. 1(a). The vibrational behavior is

Table 1: Nominal design parameters of the simply supported rectangular plate

Parameter	Nomenclature	Value	Unit
Plate length	a	870.0	mm
Plate width	b	620.0	mm
Plate thickness	h	5.0	mm
Blade thickness	t_{bl}	0.5	mm
Blade length	l_{bl}	25	mm

measured at $N_0 = 402$ points on the surface of the plate to ensure sufficient spatial resolution. Thus, the natural angular frequencies $\omega_r^{(e)}$ and the complex mode shapes vectors $\underline{\psi}_r^{(e)}$ with $\underline{\psi}_r^{(e)} \in \mathbb{R}^{N_0}$, $r = 1, \ldots N_m$ of the rectangular plate are obtained. Here, the superscript (e) denotes the observed data of the experimental model. In order to take into account variation due to assembly in the experimental data the test specimen is reassembled and measured three times. This involves unfasten the bolts between the blades and the brackets and lifting out the rectangular plate. Then, the rectangular plate reassembled. Due to high assembly effort, a limited set ($N_s = 3$) of experimental data

$$\mathbf{D} = \left\{ \omega_{1n}^{(e)} \ldots \omega_{N_m n}^{(e)}, \underline{\psi}_{1n}^{(e)} \ldots \underline{\psi}_{N_m n}^{(e)} \right\}_{n=1}^{N_s}, n = 1, \ldots, N_s \text{ is available for this paper.}$$

In order to predict the vibroacoustic behavior of the experimental model, two competing FE models M_1 and M_2 are set up in the software ANSYS (release 19.2). For both models M_1 and M_2, the geometry of the rectanglar plate is discretized using a structuted mesh with 8-node shell elements (SHELL281) and an element size of approximately $4 \cdot 10^3$ m. This leads to a total number of 5353 FE nodes and to a sufficient spatial resolution for vibroacoustic analysis. The models M_1 and M_2 differ in the modeling of the simply supported boundary conditions. For model M_1, the boundary conditions of the simply supported rectangular plate are assumed by an ideal simply supporting modeling approach [3]. The nodes at the edges of the plate are directly connected to the ground. Consequently, the vertical stiffness k_t is infinite and the rotational stiffness k_r is set to zero as shown in Fig. 1(d). For model M_2, the boundary conditions of the simply supported rectangular plate are modelled by linear spring elements (COMBIN14), whose vertical stiffness k_t and rotational stiffness k_r can arbitrarily be defined. It is expected that the measured vibroacoustic behavior can be better predicted using model M_2 because it includes a more detailed modeling approach of the boundary conditions and provides two more parameter to be calibrated. Nevertheless, model M_1 is easier to implement and is used more often. An assessment of the model uncertainty should show which model is best suited to predict the bavior of the experimental model. The natural angular frequencies $\omega_r^{(n)}$ and the mode shapes $\underline{\psi}_r^{(n)}$ of both FE models M_1 and M_2 are obtained by a numerical modal analysis using a BLOCK-LANCZOS algorithm. The superscript (n) denotes the data of the numerical FE models. It is generally of interest to calibrate the unknown model parameters, which cannot be measured directly. The plate length a, plate width b and plate thickness h of the plate can be measured and, therefore assumed to be well-known. Thus, for the model M_1, the parameters of model M_2, the parameters Young's modulus E, mass density ρ and two additional unknown model parameters the vertical stiffness k_t and the rotational stiffness k_r are considered in the calibration procedure. Table 2 summarizes the unknown model parameters for model M_1 and model M_2.

Table 2. Unknown model parameters with lower and upper bounds

Parameter	Model M_1	Model M_2
E	$U\left(\left[6 \cdot 10^9, 8 \cdot 10^9\right] \mathrm{Nm}^{-2}\right)$	$U\left(\left[6 \cdot 10^9, 9 \cdot 10^9\right] \mathrm{Nm}^{-2}\right)$
ρ	$U\left(\left[2.4 \cdot 10^3, 2.9 \cdot 10^3\right] \mathrm{kgm}^{-3}\right)$	$U\left(\left[2.4 \cdot 10^3, 2.9 \cdot 10^3\right] \mathrm{kgm}^{-3}\right)$
k_t	---	$U\left(\left[10^4, 10^7\right] \mathrm{Nm}^{-1}\right)$
k_r	---	$U\left(\left[0, 10^2\right] \mathrm{Nm}\right)$

It is further assumed that no prior knowledge about the distribution of the unknown model parameters is available. Consequently, the unknown model parameters are assumed to be uniformly distributed (U) between the lower and upper bounds, which are given in Table 2. The respective lower and upper bounds of the model parameters are best guesses so that the calibration is not restricted. A uniform distribution is a common choice if no information about the model parameters is available [9].

The introduced experimental model and the FE models are embedded within a probabilistic model parameter calibration by means of Bayesian inference, which is described in the following section. Within this framework, the unknown model parameters to be calibrated are defined as random variables and are combined in the vector of unknown parameters.

$$\boldsymbol{\theta} = \begin{cases} (E, \rho) \text{ for } M_1, \\ (E, \rho, k_t, k_r) \text{ for } M_2. \end{cases} \tag{1}$$

3   Probabilistic Parameter Calibration by Means of Bayesian Inference

The aim of a probabilistic parameter calibration is the identification of the unknown distributions of the model parameters. The Bayesian Theorem [9] describes the posterior probability density as

$$p(\boldsymbol{\theta}|\mathbf{D}, M_i) = c^{-1} p(\mathbf{D}|\boldsymbol{\theta}, M_i) p(\boldsymbol{\theta}|M_i). \tag{2}$$

The prior probability density $p(\boldsymbol{\theta}|M_i)$ quantifies the prior probability of a specific parameter set $\boldsymbol{\theta}$ of Eq. (2) for a model M_i. The likelihood $p(\mathbf{D}|\boldsymbol{\theta}, M_i)$ quantifies the probability that a specific model evaluation of the model M_i with the parameter set $\boldsymbol{\theta}$ corresponds to the experimental model data $\mathbf{D}$. The parameter c, the total probability or evidence, is typically not computable with reasonable effort and is only normalizing the result anyway [10]. To avoid this effort, methods for sampling Eq. (2) can be chosen. In this paper, a Transitional Markov Chain Monte Carlo (TMCMC) algorithm by Ching et al. [11] is used to solve Eq. (2). In the context of TMCMC, Eq. (2) is not solved directly, but it is assumed that the posterior probability

$$p_j \propto p(\mathbf{D}|\boldsymbol{\theta}, M_i)^{q_j} p(\boldsymbol{\theta}|M_i) \tag{3}$$

of a calculation step $j = 0, \ldots, m$ is proportional to the product of the prior probability $p(\boldsymbol{\theta}|M_i)$ and the likelihood $p(\mathbf{D}|\boldsymbol{\theta}, M)^{q_j}$. Thus, the problem is solved using $m > 1$ calculation steps. The power q_j is defined as

$$q_j \in [0, 1], q_0 = 0 < q_1 < \ldots < q_m = 1 \tag{4}$$

and is used to control the transition of the prior distribution to the posterior distribution. Thus, the samples are gradually moved from the prior ($j = 0$, $q_0 = 0$) to the posterior distribution ($j = m$, $q_m = 1$). During each step, the parameter space is resampled with a certain number of N_s points. The TMCMC terminates with the condition $q_m = 1$. The TMCMC overcomes the weaknesses of methods, which are working with Markov Chain Monte Carlo (MCMC), e. g., the reduced statistical efficiency, the presence of the burn-in period, and the inefficiency of the calibration procedure for a high number of model parameters [12]. More detailed information can be found in [11, 12].

The likelihood $p(\mathbf{D}|\boldsymbol{\theta}, M_i)$ includes the discrepancy between the FE model outputs and the experimental data, which is termed the prediction error [6, 13]. A formulation for the likelihood $p(\mathbf{D}|\boldsymbol{\theta}, M_i)$ based on modal properties is proposed by Vanik et al. [14] and is also adopted by Goller et al. [6]. It is assumed based on the principle of maximum entropy that the prediction error follows a normal distribution [15]. Additionally to modal properties the likelihood used for this contribution also contains the mass m of the rectangular plate as a model output. The likelihood of the data set $\mathbf{D}$ for a model parameter set $\boldsymbol{\theta}$ and a model M_i

$$p(\mathbf{D}|\boldsymbol{\theta}, M_i) = c_1 \exp\left(-\frac{1}{2}\sum_{n=1}^{N_s}\sum_{r=1}^{N_m}\left(\varepsilon_r^{-2} e_{\omega_{rn}^2}^2 + \delta_r^{-2} e_{\psi,rn}^2\right) + \sigma^{-2} e_{\text{mass}}^2\right) \tag{5}$$

is then composed using the prediction error of the natural angular frequencies

$$e_{\omega_{rn}^2} = \omega_{rn}^{2(e)} - \omega_r^{2(n)}, \tag{6}$$

the prediction error of the mode shapes

$$e_{\psi,rn} = 1 - \frac{\left|\underline{\psi}_{rn}^{(e)T} \underline{\psi}_r^{(n)}\right|^2}{\underline{\psi}_r^{(n)T} \underline{\psi}_r^{(n)}} \tag{7}$$

and the prediction error of the mass.

$$e_{\text{mass}} = \frac{m^{(e)}}{m^{(n)}} - 1 \tag{8}$$

The factor c_1 is described in detail in [14] and just normalizes the likelihood. The prediction error variances of the natural angular frequencies ε_r^2 and of the mode shapes δ_r^2 are estimated from the three observed data sets. Consequently, individual values of ε_r^2 and δ_r^2 are used for the first six mode shapes. It is assumed that the prediction error variance of the mass has a constant value of $\sigma^2 = 10^{-2}$. The prediction error of the mode shapes $e_{\psi,m}$ in Eq. (7) corresponds to the modal assurance criterion (MAC). In vibroacoustics, the agreement of the numerically and experimentally obtained mode shapes is usually assessed by the MAC [16]. The MAC describes the linear correlation between the vector of the mode shapes of the experimental model $\underline{\psi}_r^{(e)}$ and those of the numerical model $\underline{\psi}_r^{(n)}$. In Eq. (7), the vector of the mode shapes for the experimental $\underline{\psi}_r^{(e)}$ and for the numerical model $\underline{\psi}_r^{(n)}$ must have equal length and take into account equal spatial coordinates. To match the number of points of the experimental model, only the closest node of the FE mesh in relation to a measurement point is considered. Thus, the vector of the mode shapes of the FE model is reduced from 5353 to 402 points. The results of the probabilistic model parameter calibration are shown in the following section.

4   Calibration Results and Assessment of the Model Uncertainty

A probabilistic model parameter calibration by means of TMCMC is performed for both models M_1 and M_2, whereby a number of $N = 1000$ is used to cover the parameter space. The computations are performed on a standard desktop PC with a single model evaluation taking approximately 20 s. In this paper, only the results for the natural angular frequencies are shown and discussed in detail since the results of the mode shapes lead to similar conclusions. Nevertheless, the natural angular frequencies as wells as the mode shapes are used to compute the likelihood following Eq. (5).

Figure 2 depicts the posterior distributions of the calibrated model parameters Young's modulus E (a) and mass density ρ (b) for model M_1 as histogramms.

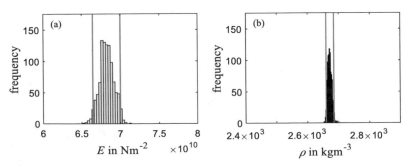

Fig. 2. Posterior distributions with the 95% interpercentile intervals (solid lines) of the calibrated model parameters Young's modulus E (a) and mass density ρ (b) for model M_1. The limits of the abscissae correspond to the lower and upper bounds of the prior distributions of the unknown model parameters.

As expected, the distributions of the model parameters E and ρ are narrowed down by the information contained in the observed data set **D**. The lower and upper bounds of the prior distributions are represented by the limits of the abscissa in Fig. 2. The relations between the limits of the abscissae and the 95% interpercentile intervalls of the posterior distributions show the reduction of model parameter uncertainty due to the probablistic model calibration. The model parameter ranges are reduced by approximetly 82% for the Young's modulus E and 95% for the mass density ρ.

The effect of the probabilistic model parameter calibration procedure on the accuracy of the model prediction of the model M_1 is shown in Fig. 3. Here, the histograms of the prior and posterior distributions of the first six ($N_m = 6$) natural angular frequencies $\omega_r^{(n)}$, the 95% interpercentile intervals of the posterior distribution as well as the corresponding observed data $\mathbf{D} = \left\{ \omega_1^{(e)} \ldots \omega_{N_m n}^{(e)} \right\}_{n=1}^{N}$, $N = 3$, are plotted. As expected, the distributions of the posterior model predictions using calibrated model parameters are narrowed down in comparison to the distribution of the prior model predictions using non-calibrated model parameters. For the second to the sixth natural angular frequency (b)–(f), the posterior model predictions are closely distributed around the observed data values.

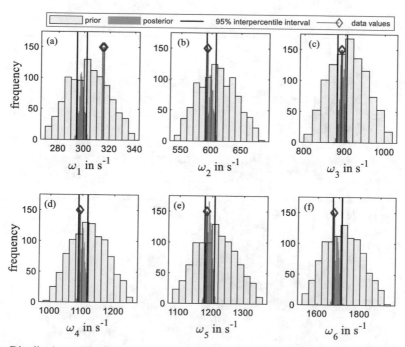

Fig. 3. Distributions of the first six natural angular frequencies of the simply supported rectangular plate predicted with the model M_1 using non-calibrated (prior) and calibrated (posterior) model parameters. The 95% interpercentile intervals of the posterior distributions are plotted as solid lines. The corresponding values of the observed data are plotted as stems.

In the case of the first natural angular frequency ω_1 (a), a deviation between the observed data values and the posterior model predictions still remains, which can not be further reduced by the calibration procedure. This leads to the fact, that the observed data values are not part of the posterior distribution of the model predition of the first natural angular frequency ω_1.

Figure 4 depicts the posterior distributions of the calibrated model parameters Young's modulus E (a), mass density ρ (b), vertical stiffness k_t (c), and rotational stiffness k_r (d) for model M_2 as histogramms. The distributions of the model parameters Young's modulus E, mass density ρ and rotational stiffness k_r are also narrowed down by the information contained in the observed data sets **D**. The vertical stiffness k_t is still distributed over the entire range even after the calibration. The data **D** do not contain the necessary information to extend the knowledge of the vertical stiffness k_t due to the calibration procedure. The lower and upper bounds of the prior distributions are represented by the limits of the abscissae in Fig. 4. The relations between the limits of the abscissae and the 95% interpercentile intervals of the posterior distributions visualize the reduction of model parameter uncertainty for the model M_2 due to the probablistic model calibration. The model parameter ranges are reduced by approximetly 71% for the Young's modulus E, 94% for the mass density ρ and 84% for the rotational stiffness k_r.

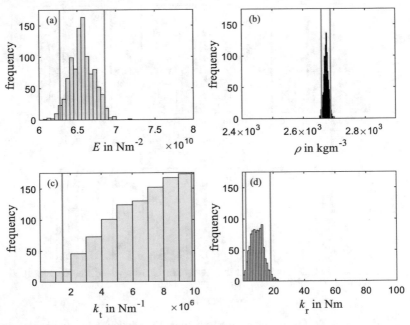

Fig. 4. Posterior distributions with the 95% interpercentile intervals (solid lines) of the calibrated model parameters Young's modulus E (a), mass density ρ (b), vertical stiffness k_t (c) and rotational stiffness k_r (d) for model M_2. The limits of the abscissae correspond to the lower and upper bounds of the prior distributions of the unknown model parameters.

For the vertical stiffness k_t, a reduction of the model parameter ranges cannot be achieved by the probabilistic model parameter calibration procedure.

The effect of the probabilistic model parameter calibration procedure on the accuracy of the model prediction of the model M_2 is shown by Fig. 5. Here, the histograms of the prior and posterior distributions of the first six ($N_m = 6$) natural angular frequencies $\omega_r^{(n)}$, the 95% interpercentile intervals of the posterior distribution as well as the corresponding observed data $\mathbf{D} = \left\{ \omega_1^{(e)} \dots \omega_{N_m n}^{(e)} \right\}_{n=1}^{N}$, $N = 3$, are plotted. The distributions of the posterior model predictions using the calibrated model parameters are narrowed down in comparison to the distribution of the prior model predictions using non-calibrated model parameters. For the first to the sixth natural angular frequency see Fig. 5(a)–(f), the posterior model predictions are closely distributed around the observed data values. Consequently, the observed data values are always part of the posterior distribution of the model preditions of the first six natural angular frequencies $\omega_r^{(n)}$. It can be concluded that the preditction quality of the models M_1 and M_2 has improved due to the probabilistic model parameter calibration procedure.

Table 3 summarizes the results of the calibrated model parameters for the models M_1 and M_2 by means of the 95% interpercentiles and the mean value of the distributions.

Table 3. Posterior uncertainty of the calibrated model parameters for the model M_1 and M_2

Parameter	Model M_1			Model M_2		
	95% interpercentile		mean	95% interpercentile		mean
	min	max		min	max	
E in Nm^{-2}	$6.64 \cdot 10^{10}$	$7.00 \cdot 10^{10}$	$6.82 \cdot 10^{10}$	$6.27 \cdot 10^{10}$	$6.85 \cdot 10^{10}$	$6.56 \cdot 10^{10}$
ρ in kgm^{-3}	$2.66 \cdot 10^3$	$2.68 \cdot 10^3$	$2.67 \cdot 10^3$	$2.66 \cdot 10^3$	$2.69 \cdot 10^3$	$2.67 \cdot 10^3$
k_t in Nm^{-1}	---	---	---	$1.47 \cdot 10^6$	$9.84 \cdot 10^6$	$6.58 \cdot 10^6$
k_r in Nm	---	---	---	1.94	18	9.3

Finally, the model uncertainty of the models M_1 and M_2 is assessed based on the prediction error of the first six natural angular frequencies of the rectangluar plate. Goller et al. [6] conclude in their contribution that the distribution of the prediction error after the calibration procedure is a measure for the remaining model uncertainty, which can not be further reduced by adjusting the model parameter.

In Fig. 6 the distributions of the prediction error of the natural angular frequencies $e_{\omega_r^2}$ averaged over all three observed data values are plotted for the first six natural angular frequencies ($N_m = 6$) and for the models M_1 and M_2, respectively. The prediction error $e_{\omega_r^2}$ is calculated using Eq. (6) and the model predictions are based on calibrated model parameters. If the model predictions of the natural angular frequencies $\omega_r^{(n)}$ match the observed data values on average, the prediction error $e_{\omega_r^2}$ is distributed around zero and consequently, the averaged prediction error $\bar{e}_{\omega_r^2} = 0$. Table 4 shows the averaged prediction errors $\bar{e}_{\omega_r^2}$ of the first six natural angular frequencies for model M_1 and model M_2. The averaged prediction errors $\bar{e}_{\omega_r^2}$ are plotted as stems in Fig. 6.

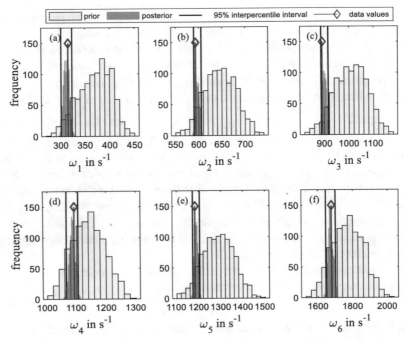

Fig. 5. Distributions of the first six natural angular frequencies of the simply supported rectangular plate predicted with the model M_2 using non-calibrated (prior) and calibrated (posterior) model parameters. The 95% interpercentile intervals of the posterior distributions are plotted as solid lines. The corresponding values of the observed data are plotted as stems.

Table 4. Averaged prediction error $\bar{e}_{\omega_r^2}$ in s^{-2} of the first six natural angular frequencies $\omega_r^{(n)}$ of the simply supported rectangular plate for the models M_1 and M_2

	$\bar{e}_{\omega_1^2}$	$\bar{e}_{\omega_2^2}$	$\bar{e}_{\omega_3^2}$	$\bar{e}_{\omega_4^2}$	$\bar{e}_{\omega_5^2}$	$\bar{e}_{\omega_6^2}$
M_1	$-1.017\cdot10^4$	$0.828\cdot10^4$	$0.307\cdot10^4$	$2.401\cdot10^4$	$2.355\cdot10^4$	$5.952\cdot10^4$
M_2	$-0.205\cdot10^4$	$0.554\cdot10^4$	$1.835\cdot10^4$	$-1.356\cdot10^4$	$1.211\cdot10^4$	$-1.427\cdot10^4$

It can be concluded that except for the third natural angular frequency $\omega_3^{(n)}$, the model M_2 has a smaller averaged prediction errors $\bar{e}_{\omega_r^2}$ than the model M_1. Consequently, the model M_2 leads to a better prediction of the vibroacoustic behavior of the simply supported rectangluar plate for the first six natural angular frequencies. A possible option to close the remaining shift of the prediction error is the definition and calibration of a discrepancy function according to Kennedy and O'Hagan [4]. This can be done by a Gaussian process as shown by Feldmann et al. [17].

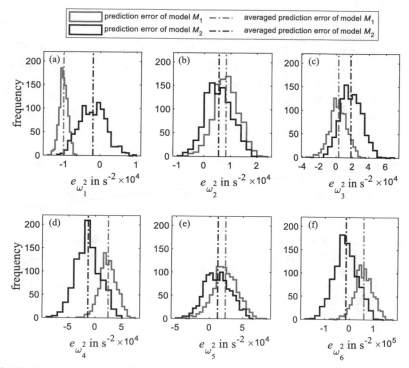

Fig. 6. Distributions of the prediction error $e_{\omega_r^2}$ after the calibration procedure for the first six natural angular frequencies of the simply supported rectangular plate using model M_1 and model M_2 respectively. The averaged prediction errors $\bar{e}_{\omega_r^2}$ are plotted as stems.

5   Conclusions

This paper presents the assessment of the model uncertainty in the prediction of the vibroacoustic behavior of a simply supported rectangular plate. For this purpose, a probabilistic model calibration via Bayesian inference is performed and the model uncertainty is assessed by the remaining prediction error after the calibration procedure. The test specimen is a simply supported rectangular plate which is investigated at a test stand. The rectangular plate is excited by an automatic impact hammer and the vibration velocities are obtained by a SLDV at specific measurement points on the surface of the plate. The vibroacoustic behavior is described by the natural angular frequencies and the corresponding mode shapes of the simply supported rectangular plate. Consequently, the vibroacoustic behavior of the experimental model is obtained by an experimental modal analysis using the measured transfer functions. Two competing FE models for the prediction of the vibroacoustic behavior of the simply supported rectangular plate are set up, which differ in the modeling of the simply supported boundary conditions. Both models are embedded in a probabilistic model calibration procedure via Bayesian inference.

The aim of the calibration is the reduction of the model parameter uncertainty by the identification of the distribution of the unknown model parameters. The measurement data used for the calibration are the results of the experimental modal analysis. The likelihood for the Bayesian inference consists of three parts, which consider the natural angular frequencies, the corresponding mode shapes as well as the mass of the simply supported rectangular plate. For both FE models, a significant reduction of the lower and upper bounds of the unknown model parameters is achieved by means of probabilistic model calibration. It is concluded that the model parameter uncertainty is reduced for both models. For assessment of the model uncertainty involved in the FE models, the remaining prediction error, which is a measure for the model uncertainty, is analyzed after the calibration. It is shown that a more detailed modeling of the boundary conditions leads to better calibration results. For future work, the experimental data base has to be extended for the calibration procedure in order to better take into account the scattering of the vibroacoustic behavior occurring in reality due to the assembling process and to validate the prediction of the vibroacoustic behavior against independent measurement data.

Acknowledgements. The authors like to thank the Deutsche Forschungsgemeinschaft (DFG, German Research Foundation) for funding this project within the Sonderforschungsbereich (SFB, Collaborative Research Center) 805 "Control of Uncertainties in Load-Carrying Structures in Mechanical Engineering" – project number: 57157498. The authors also acknowledge the financial support for the 3D scanning laser Doppler vibrometer from the DFG under the grant number INST 163/520–1.

References

1. Luegmair, M., Schmid, J.D.: Challenges in vibroacoustic vehicle body simulation including uncertainties. SAE Technical Paper 2020-01-1571 (2020)
2. Mallapur, S., Platz, R.: Uncertainty quantification in the mathematical modelling of a suspension strut using Bayesian inference. Mech. Syst. Signal Proc. **118**, 158–170 (2019)
3. Bapat, A.V., Suryanarayan, N.: A theoretical basis for the experimental realization of boundary conditions in the vibration analysis of plates. J. Sound Vibr. **163**(3), 463–478 (1993)
4. Kennedy, M.C., O'Hagan, A.: Bayesian calibration of computer models. J. Roy. Stat. Soc. Ser. B (Stat. Methodol.) **63**(3), 425–464 (2001)
5. Wang, X., Lin, S.I., Wang, S.: Dynamic friction parameter identification method with LuGre model for direct-drive rotary torque motor. Math. Probl. Eng. **2016**, 1–8 (2016)
6. Goller, B., Schuëller, B.I.: Investigation of model uncertainties in Bayesian structural model updating. J. Sound Vibr. **330**, 6122–6136 (2011)
7. Adams, C., Bös, J., Slomski, E.M., Melz, T.: An experimental investigation of vibrating plates in similitude and the possibility to replicate the responses using sensitivity-based scaling laws. In: 6th Conference on Noise and Vibration emerging methods, Ibiza, pp. 175574-1–175574-12 (2018)
8. Robin, O., Chazot, A.D., Boulandet, R., Michau, M., Berry, A., Atalla, N.: A plane and thin panel with representative simply supported boundary conditions for laboratory vibroacoustic tests. Acta Acustica United with Acustica **102**(1), 403–425 (2016)
9. Smith, R.C.: Uncertainty Quantification: Theory, implementation and application Computational Science & Engineering, 12th edn. Society for Industrial and Applied Mathematics, Philadelphia (2014)
10. Green, P.L., Worden, K.: Modelling friction in a nonlinear dynamic system via Bayesian inference. In: Allemang, R., et al. (eds.) Special Topics in Structural Dynamics, Vol. 6, pp. 543–553. Springer, New York (2013)

11. Ching, J., Chen, Y.-C.: Transitional Markov Chain Monte Carlo method for Bayesian model updating, model class selection, and model averaging. J. Eng. Mech. **133**(7), 816–832 (2007)
12. Betz, W., Papaioannou, I., Straub, D.: Transitional Markov Chain Monte Carlo: observations and improvements. J. Eng. Mech. **142**(5), 04016016 (2016)
13. Prajapat, K., Ray-Chaudhuri, S.: Prediction error variances in Bayesian model updating employing data sensitivity. J. Eng. Mech. **142**(12), 04016096 (2016)
14. Vanik, M., Beck, J., Au, S.-K.: Bayesian probabilistic approach to structural health monitoring. J. Eng. Mech. **126**, 738–745 (2000)
15. Jaynes, E.T.: Probability Theory: The logic of science. Cambridge University Press, Cambridge (2003)
16. Allemang, R.J.: The modal assurance criterion – twenty years of use and abuse. Sound Vibr. **1**(August), 14–21 (2003)
17. Feldmann, R., Platz, R.: Assessing model form uncertainty for a suspension strut using Gaussian Processes. In: 3rd ECCOMAS Thematic Conference on Uncertainty Quantification in Computational Sciences and Engineering, Crete, pp. 569–582 (2019)

Optimization Under Uncertainty

Detection of Model Uncertainty in the Dynamic Linear-Elastic Model of Vibrations in a Truss

Alexander Matei[✉] and Stefan Ulbrich

Department of Mathematics, Research Group Optimization,
Technical University of Darmstadt, Dolivostraße 15, 64293 Darmstadt, Germany
matei@mathematik.tu-darmstadt.de

Abstract. Dynamic processes have always been of profound interest for scientists and engineers alike. Often, the mathematical models used to describe and predict time-variant phenomena are uncertain in the sense that governing relations between model parameters, state variables and the time domain are incomplete. In this paper we adopt a recently proposed algorithm for the detection of model uncertainty and apply it to dynamic models. This algorithm combines parameter estimation, optimum experimental design and classical hypothesis testing within a probabilistic frequentist framework. The best setup of an experiment is defined by optimal sensor positions and optimal input configurations which both are the solution of a PDE-constrained optimization problem. The data collected by this optimized experiment then leads to variance-minimal parameter estimates. We develop efficient adjoint-based methods to solve this optimization problem with SQP-type solvers. The crucial test which a model has to pass is conducted over the claimed true values of the model parameters which are estimated from pairwise distinct data sets. For this hypothesis test, we divide the data into k equally-sized parts and follow a k-fold cross-validation procedure. We demonstrate the usefulness of our approach in simulated experiments with a vibrating linear-elastic truss.

Keywords: Model uncertainty · Optimum experimental design · Sensor placement · Optimal input configuration · k-fold cross-validation

1 Introduction

In science and technology, dynamic processes are often described by time-variant mathematical models. However, the accurate prediction of the motion and behavior of technical systems is still challenging. Due to our incomplete knowledge of the internal relations between model parameters, state variables and the time domain, the user frequently encounters model uncertainty [20]. In [12] we developed an algorithm to identify this model uncertainty using parameter estimation, optimal experimental design and classical hypothesis testing. It is the aim of this paper to extend this approach to dynamic models. In the following, we adopt the

© The Author(s) 2021
P. F. Pelz and P. Groche (Eds.): ICUME 2021, LNME, pp. 281–295, 2021.
https://doi.org/10.1007/978-3-030-77256-7_22

framework described in [12] and extend it to meet mathematical models which are comprised of time-variant partial differential equations (PDE).

There is abundant literature on the assessment of descriptive and predictive qualities of dynamic models. Most common are techniques like residual analysis [27,29] and interval simulation [25], maximum likelihood methods [29] and Bayesian model updating [31]. Our approach comes from a frequentist perspective and offers an alternative: we minimize the extent of data uncertainty by optimizing the experimental design and employ a k-fold cross-validation to test the model's fitness and consistency. The validation is hereby performed via a classical hypothesis test in the parameter space.

A subproblem that needs to be solved in our approach to detect model uncertainty is the PDE-constrained optimal experimental design (OED) problem where the PDE is time-dependent. We specifically focus on experiments where sensors need to be positioned and inputs must be chosen in order to achieve a maximum information gain for the estimated values of the model parameters. Optimal sensor placement has been addressed within the PDE-context in [1,2,23] and optimal input configuration has been extensively analyzed for both linear and nonlinear ordinary differential equations in various engineering applications [6,17,21,22,28]. However, in these cases the problem dimension is small compared to a (discretized) time-variant PDE and thus, gradient-based optimization with a sensitivity approach, as suggested by [4] and [19], works fine. In our case, this approach is no longer computationally tractable. Our framework meets high-dimensionality by employing efficient adjoint techniques in a sequential quadratic programming (SQP) solver scheme.

This paper is organized as follows. In Sect. 2 and Sect. 3, we introduce our model equations and briefly present the concepts of parameter estimation and OED followed by efficient solution techniques for the OED problem. Then, in Sect. 4 we show how our algorithm to detect model uncertainty is adapted to the dynamic setting. Section 5 contains numerical results for the OED problem applied to vibrations of a truss and the application of our algorithm to detect model uncertainty. We end the paper with concluding remarks.

2   Model Equations of Transient Linear Elasticity and Their Discretization

Let $G \subset \mathbb{R}^2$ be a bounded Lipschitz domain with sufficiently smooth boundary $\partial G = \Gamma_D \cup \Gamma_F \cup \Gamma_N$ where $\Gamma_D, \Gamma_F, \Gamma_N$ are pairwise disjoint and non-empty. Furthermore, let $(0, T)$, with $T > 0$, be an open and bounded time interval. We consider the parameter-dependent equations of motion for the linear-elastic body G of mass density $\varrho > 0$ and weak damping constant $a > 0$, see [15, Sec. 7.2]:

$$\varrho\,\partial_{tt}^2 y + a\varrho\,\partial_t y - \text{div } \sigma(y,\partial_t y,p) = 0, \quad \text{in } (0,T) \times G,$$
$$y = 0, \quad \text{on } (0,T) \times \Gamma_{\text{D}},$$
$$\sigma(y,\partial_t y,p) \cdot n = 0, \quad \text{on } (0,T) \times \Gamma_{\text{F}},$$
$$\sigma(y,\partial_t y,p) \cdot n = u, \quad \text{on } (0,T) \times \Gamma_{\text{N}}, \qquad (1)$$
$$y(0,\cdot) = 0, \quad \text{in } G,$$
$$\partial_t y(0,\cdot) = 0, \quad \text{in } G.$$

We include Rayleigh damping in our modeling by the generalized law of Hooke:

$$\sigma(y,\partial_t y,p) = \mathcal{C}(p)\big(\varepsilon(y) + b\varepsilon(\partial_t y)\big),$$

where $b > 0$ is the strong damping constant, $\varepsilon(y) = \frac{1}{2}\big(\nabla y^\top + \nabla y\big)$ is the linearized strain and $\mathcal{C} : \varepsilon \mapsto p_1 \cdot \text{trace}\,(\varepsilon)\,I + 2p_2 \cdot \varepsilon$ is the fourth order elasticity tensor, see also [7]. The parameters in this PDE are the well known Lamé constants $p = (\lambda_{\text{L}}, \mu_{\text{L}}) \in \mathbb{R}_+^2$. It is evident from (1) that the displacement $y : (0,T) \times G \mapsto \mathbb{R}^2$ is caused by the traction $u : (0,T) \times \Gamma_{\text{N}} \mapsto \mathbb{R}^2$ alone.

After adopting the weak formulation of (1) according to [15, Sec. 7.2] and [9] we perform a finite-dimensional approximation of this weak formulation known as the Galerkin ansatz. We employ standard quadratic finite elements for the elastic body G for the space discretization. Then the finite element approximation leads to the (high-dimensional) second-order ordinary differential equation

$$M\partial_{tt}^2 y(t) + C(p)\partial_t y(t) + A(p)y(t) - Nu(t) = 0, \qquad (2)$$

with the stiffness matrix $A(p)$, the mass matrix M and the boundary mass matrix N. For the Rayleigh damping term, we introduce the damping matrix

$$C(p) := aM + bA(p), \qquad (3)$$

where $a, b > 0$ are the damping constants as before.

We want to use a numerical time-update scheme with a predefined step size Δt to solve (2). Therefore, we rewrite (2) in the form

$$Ma_n + C(p)v_n + A(p)d_n - Nu_n = 0,$$

with the acceleration vector $a_n = \partial_{tt}^2 y(t_n)$, the velocities $v_n = \partial_t y(t_n)$ and the displacements $d_n = y(t_n)$ for time steps t_n, where $n = 1,\dots,n_{\text{t}}$, respectively. The implicit Newmark method is suitable to solve this equation. It can be implemented in the following way. First, we choose constants $\beta_{\text{N}} = \frac{1}{4}$, $\gamma_{\text{N}} = \frac{1}{2}$ for stability reasons, see [15,30], and define with them other constants $\alpha_1,\dots,\alpha_6$ as depicted in [30, Sec. 6.1.2]. Then the iteration scheme reads as follows:

$$a_{n+1} = \alpha_1(d_{n+1} - d_n) - \alpha_2 v_n - \alpha_3 a_n,$$
$$v_{n+1} = \alpha_4(d_{n+1} - d_n) + \alpha_5 v_n + \alpha_6 a_n,$$
$$[\alpha_1 M + \alpha_4 C(p) + A(p)]\,d_{n+1} = Nu_{n+1} + M(\alpha_1 d_n + \alpha_2 v_n + \alpha_3 a_n) \qquad (4)$$
$$+ C(p)(\alpha_4 d_n - \alpha_5 v_n - \alpha_6 a_n)\,.$$

This scheme can be written in matrix form:

$$L(p)y - Fu = 0,$$

where $y = (y_1, \ldots, y_{n_t})^\top$ are the states, $u = (u_1, \ldots, u_{n_t})^\top$ are the boundary forces at all time points, $y_n = (a_n, v_n, d_n)^\top$ and $u_n = (u_{n,x}, u_{n,y})^\top$. The matrices L and F have the block form

$$L(p) = \begin{bmatrix} Q(p) & & & \\ P(p), X(p) & & & \\ & \ddots & \ddots & \\ & & P(p), X(p) \end{bmatrix}, \qquad F = \begin{bmatrix} E_0 & & & \\ & E_1 & & \\ & & \ddots & \\ & & & E_1 \end{bmatrix},$$

where

$$Q(p) = \begin{bmatrix} M, C(p), A(p) \\ I \\ I \end{bmatrix}, \quad X(p) = \begin{bmatrix} I, 0, -\alpha_1 I \\ 0, I, -\alpha_4 I \\ 0, 0, D(p) \end{bmatrix}, \quad E_0 = \begin{bmatrix} N, \\ 0, \\ 0 \end{bmatrix}, \quad E_1 = \begin{bmatrix} 0, \\ 0, \\ N \end{bmatrix},$$

with $D(p) := \alpha_1 M + \alpha_4 C(p) + A(p)$ and

$$P(p) = \begin{bmatrix} \alpha_3 I, & \alpha_2 I, & \alpha_1 I \\ -\alpha_6 I, & -\alpha_5 I, & \alpha_4 I \\ -\alpha_3 M + \alpha_6 C(p), & -\alpha_2 M + \alpha_5 C(p), & -\alpha_1 M - \alpha_4 C(p) \end{bmatrix}.$$

For the following optimization problems let $Y = Z = \mathbb{R}^{n_y}$ be the state space and let

$$U_{\text{ad}} := \left\{ u \in H^1(0, T; \Gamma_N) : [u(t)](x) = c(t) \text{ and } u_{\min} \le u \le u_{\max} \right\},$$

be the space of admissible inputs with $n_y = n_d n_t$ being the product of the space dimension after discretization n_d and the number of time steps n_t. Furthermore, let $e : Y \times \mathbb{R}_+^2 \times U_{\text{ad}} \to Z$ be an operator defining the *state equation* as

$$e(y, p, u) := L(p)y - Fu = 0 \tag{5}$$

and denote its unique solution by $y(p, u)$. We assume the operator $\partial_y e(y, p, u)$ to be continuously invertible such that we can use the Implicit Function Theorem to define a mapping $p \mapsto y(p, u)$. Its derivatives $s_i := \partial_{p_i} y(p, u)$ for $i = 1, 2$ are computed by solving

$$\partial_y e(y(p, u), p, u)s_i + \partial_{p_i} e(y(p, u), p, u) = 0,$$

which in our setting is equivalent to

$$L(p)s_i + \partial_{p_i} L(p)y(p, u) = 0, \qquad i = 1, 2. \tag{6}$$

Thus, the sensitivity variable $s := [s_1, s_2] \in Y \times Y$ depends on the solution of the state equation and on the parameters, i.e., $s_i = s_i(y(p, u), p_i)$. Equations (6) are solved by rewriting them using the iteration scheme (4).

For the input space U_{ad} we employ a time discretization with linear finite elements and denote by M_T the mass matrix and by A_T the stiffness matrix in the time domain.

3   Lamé-Parameter Estimation and the Optimal Experimental Design Problem

Given a set of experimental data, we are concerned with an accurate estimation of the Lamé-parameters $p = (\lambda, \mu) \in \mathbb{R}_+^2$ which are part of the model equations. The measurements are taken at selected points on the discretized free boundary part Γ_F of the elastic body G with specified sensor types. We denote by n_s the number of available sensors. In order to compare the output of the model equations, i.e., the state, with experimental data we introduce a nonlinear *observation operator* $h \colon Y \to \mathbb{R}^{n_s n_t}$ that maps components of the state to quantities that are actually measured during the experiment at all n_t time steps.

Within the framework of optimal experimental design, we introduce binary weights $\omega \in \{0, 1\}^{n_s}$ for all sensor locations and types. These weights operate as a selection tool, i.e., $\omega_k = 1$ if, and only if, sensor k is used at its specified location. Since the position of these sensors and their usage throughout the experiments stay the same, the values of ω are copied n_t times and summarized in the diagonal matrix $\Omega \in \mathbb{R}^{n_z \times n_z}$, where $n_z = n_s n_t$. In addition, each sensor has a fixed operating precision, i.e., standard deviation, which we associate by the variable $\sigma_{pr} \in \mathbb{R}^{n_s}$. We again summarize n_t copies of σ_{pr} in a diagonal matrix $\Sigma \in \mathbb{R}^{n_z \times n_z}$.

The data $z \in \mathbb{R}^{n_z}$ is used to estimate the parameters $p \in \mathbb{R}_+^2$ by solving a least-squares problem:

$$\min_{p \in \mathbb{R}_+^2} \frac{1}{2} r(z, y(p, u))^\top \Omega \Sigma^{-2} r(z, y(p, u)), \tag{7}$$

where $r(z, y(p, u)) := h(y(p, u)) - z$ are the residuals and $y(p, u)$ is the unique solution of (5) for given p and u. Since the measurements are random variables $z = z^* + \varepsilon$ with unknown true values z^* and noise ε, so are the parameters. We model the noise to be Gaussian, i.e., $\varepsilon \in \mathcal{N}(0, \Omega^{-1} \Sigma^2)$. In a first order approximation, like in a Gauss-Newton solver scheme, the parameters are also Gaussian with unknown mean p^* and covariance matrix C, see [8, 19]. Then the confidence region of the parameters with a fixed confidence level $1 - \alpha$, where $\alpha \in (0, 1)$, is given by

$$K(p^*, C, \alpha) := \left\{ p \in \mathbb{R}_+^2 : (p - p^*)^\top C^{-1} (p - p^*) \le \chi_2^2(1 - \alpha) \right\}.$$

We assume that the solution $\bar{p}$ of (7) for given z and ω, emerging from the Gauss-Newton algorithm, is sufficiently close to p^*. Thus, we make the assumption that for a given data set, $\bar{p}$ is a fairly good approximation of p^*. Then the covariance matrix C can be approximated by employing the Gauss-Newton scheme as well and it has the following form [8]:

$$C_{GN} = \left[s(y(p, u), p)^\top \partial_y h(y(p, u))^\top \Omega \Sigma^{-2} \partial_y h(y(p, u)) s(y(p, u), p) \right]^{-1}.$$

We aim at minimizing the confidence region where the estimated parameters $\bar{p}$ lie:

$$K(\bar{p}, C_{GN}, \alpha) = \left\{ p \in \mathbb{R}_+^2 : (p - \bar{p})^\top C_{GN}^{-1} (p - \bar{p}) \le \chi_2^2(1 - \alpha) \right\} \to \min.$$

The reduction of the size of the confidence ellipsoid K is equivalent to reducing the "size" of the covariance matrix. This is realized by choosing best sensor locations, determined by the weights ω, and by finding optimal inputs u. In practice, there are various design criteria Ψ that measure the "size" of a matrix C, see [11]. In this paper we decide to use the E-criterion which is related to the maximal expansion of K:

$$\Psi(C) = \Psi_{\mathrm{E}}(C) = \lambda_{\max}(C) \sim \mathrm{diameter}(K)^2.$$

We add a cost term $P_\varepsilon(\omega)$ to penalize the number of used sensors and a regularizer $R(u) := u^\top (M_\mathrm{T} + A_\mathrm{T})u$ to the objective function. Moreover, we relax the binary restriction on ω to employ gradient-based solution techniques for the following optimal experimental design problem.

Definition 1. *Let $p \in \mathbb{R}_+^2$ be an estimate of p^* and let $\kappa, \beta > 0$ be fixed. Furthermore, choose $\varepsilon \in (0, 1]$. Then we call $(\bar{\omega}, \bar{u})$ an optimal design of an experiment with the linear-elastic body G if it is the solution of*

$$\min_{\omega, u, y, s} \ \Psi(C_{\mathrm{GN}}(\omega, y, s)) + \kappa \cdot P_\varepsilon(\omega) + \beta \cdot R(u), \tag{8}$$

where (u, y, s) are subject to the equality constraints

$$\begin{aligned} L(p)y - Fu &= 0, \\ L(p)s_i + \partial_{p_i} L(p)y &= 0, \end{aligned} \tag{9}$$

for $i = 1, 2$ and (ω, u) satisfy the inequality constraints

$$\omega \in [0, 1]^{n_\mathrm{s}}, \quad u \in U_{\mathrm{ad}}. \tag{10}$$

The penalty term $P_\varepsilon(\omega)$ is a smooth approximation of the l_0-"norm". It ensures sparse solutions in ω for suitable choices of κ but does not lead to $\{0, 1\}$-valued weights yet. To achieve the latter, we adopt a continuation strategy as described in [1, 2].

Note, that the penalty parameter κ must not be chosen too large since the matrix C_{GN} becomes singular if too many weights ω are switched to zero. We refer to [19] for more details on lower bounds for the sum of the weight variables.

In practice, problem (8)–(10) is solved using its reduced formulation, i.e., by eliminating the equality constraints (9) and inserting $y(p, u)$ and $s(y(p, u), p)$ into the objective function.

3.1 Derivative and Adjoint Computation

Let $J(\omega, u, y, s_1, s_2)$ be the objective function in (8). We show how the derivative of the reduced objective function $\hat{J}(\omega, u)$, where the solutions $y(p, u)$ and $s(y(p, u), p)$ of (9) have been inserted into J, with respect to the inputs u is efficiently computed. To do so, we follow a standard Lagrangian view of the optimization problem (8)–(10). For simplicity, we ignore the inequality constraints (10) and still denote by $\partial_y \Psi$ the derivative of Ψ with respect to y even though we used the Clarke directional derivatives in the case of $\Psi = \Psi_{\mathrm{E}}$, cf. [13]. Let $\mu, \lambda_1, \lambda_2 \in Y^*$ be Lagrange multipliers and let the Lagrangian be defined as

$$\mathcal{L}(\omega, u, y, s_1, s_2, \mu, \lambda_1, \lambda_2) := J(\omega, u, y, s_1, s_2) + \langle \mu, L(p)y - Fu \rangle_{Y^*, Y}$$

$$+ \sum_{i=1}^{2} \langle \lambda_i, L(p)s_i + \partial_{p_i} L(p)y \rangle_{Y^*, Y}.$$

The adjoint equations follow from $\partial_y \mathcal{L} = \partial_{s_i} \mathcal{L} = 0$ for $i = 1, 2$:

$$L(p)^\top \mu + [\partial_{p_1} L(p)]^\top \lambda_1 + [\partial_{p_2} L(p)]^\top \lambda_2 + \partial_y \Psi = 0,$$
$$L(p)^\top \lambda_1 + \partial_{s_1} \Psi = 0, \qquad (11)$$
$$L(p)^\top \lambda_2 + \partial_{s_2} \Psi = 0.$$

The fact that the matrix $L(p)$ is transposed on the left hand side of (11) leads to an iteration scheme backwards in time. We demonstrate this for the second and third adjoint equations in order to obtain λ_i whereby adopting ideas from [18, Sec. 5.4]. Let $\lambda = \lambda_i$ and $r := \partial_{s_i} \Psi$ for $i \in \{1, 2\}$. Note that $r = (r_1, \ldots, r_{n_t})^\top$ and $r_n = (r_n^d, 0, 0)$ since the velocities and accelerations do not enter Ψ. In the terminal point t_{n_t} we have to solve

$$X(p)^\top \lambda_{n_t} = r_{n_t},$$

or equivalently $\lambda_{n_t}^a = 0$, $\lambda_{n_t}^v = 0$ and

$$-\alpha_1 \lambda_{n_t}^a - \alpha_4 \lambda_{n_t}^v + D(p)\lambda_{n_t}^d = r_{n_t}^d.$$

For other time points t_n, $n \neq 1$ the current iterate is obtained from the one which is a step forward in time:

$$[X(p)^\top, P(p)^\top] \begin{pmatrix} \lambda_n \\ \lambda_{n+1} \end{pmatrix} = r_n,$$

or equivalently

$$\lambda_n^a = \alpha_3 \lambda_{n+1}^a + \alpha_6 \lambda_{n+1}^v + [\alpha_3 M - \alpha_6 C(p)] \lambda_{n+1}^d,$$
$$\lambda_n^v = \alpha_2 \lambda_{n+1}^a + \alpha_5 \lambda_{n+1}^v + [\alpha_2 M - \alpha_5 C(p)] \lambda_{n+1}^d,$$
$$-\alpha_1 \lambda_n^a - \alpha_4 \lambda_n^v + D(p)\lambda_n^d = r_n^d - \alpha_1 \lambda_{n+1}^a - \alpha_4 \lambda_{n+1}^v + [\alpha_1 M + \alpha_4 C(p)] \lambda_{n+1}^d.$$

In order to obtain the adjoint variable at the initial time point t_1 we solve

$$[Q(p)^\top, P(p)^\top] \begin{pmatrix} \lambda_1 \\ \lambda_2 \end{pmatrix} = r_1,$$

or equivalently

$$M\lambda_1^a = \alpha_3 \lambda_2^a + \alpha_6 \lambda_2^v + [\alpha_3 M - \alpha_6 C(p)] \lambda_2^d,$$
$$C(p)\lambda_1^a + \lambda_1^v = \alpha_2 \lambda_2^a + \alpha_5 \lambda_2^v + [\alpha_2 M - \alpha_5 C(p)] \lambda_2^d,$$
$$A(p)\lambda_1^a + \lambda_1^d = r_1^d - \alpha_1 \lambda_2^a - \alpha_4 \lambda_2^v + [\alpha_1 M + \alpha_4 C(p)] \lambda_2^d.$$

The matrix vector product $q := [\partial_{p_i} L(p)]^\top \lambda_i$ is computed likewise using the iteration scheme.

Finally, the full derivative of the reduced objective function $\hat{J}(\omega, u)$ with respect to the inputs u is given by

$$\frac{\mathrm{d}\hat{J}}{\mathrm{d}u} = -F^\top \mu + 2\beta(M_\mathrm{T} + A_\mathrm{T})u,$$

where $\mu \in Y^*$ is the adjoint variable obtained from (11).

3.2   Computational Remarks

In order to solve (8)–(10) we employ an SQP algorithm with BFGS updates [10] for the Hessian H_k of the Lagrangian. We modify the update formula in the following way:

$$H_0 = \begin{bmatrix} I \\ & 2\beta(M_\mathrm{T} + A_\mathrm{T}) \end{bmatrix},$$

$$H_{k+1} = H_k - \frac{H_k d^k (H_k d^k)^\top}{(d^k)^\top H_k d^k} + \frac{r^k (r^k)^\top}{(r^k)^\top d^k},$$

where d^k is the current step, y^k is the difference between gradients of the Lagrangian at the new and old iterate and

$$r^k := \begin{cases} y^k & \text{if } (y^k)^\top d^k \geq 0.2(d^k)^\top H_k d^k, \\ \theta y^k + (1 - \theta) H_k d^k & \text{otherwise,} \end{cases}$$

with $\theta = \frac{0.8(d^k)^\top H_k d^k}{(d^k)^\top H_k d^k - (y^k)^\top d^k}$. After every tenth iteration we reset the Hessian to H_0 to avoid matrix filling and to ensure a gradient descent with respect to ω from time to time.

4   Detection of Uncertainty in Dynamic Models

We adopt the algorithm presented in [12] and describe the main differences when applied to a time-variant model $\mathcal{M}$ of a dynamic process. In general, we presuppose that a valid model should reproduce *all* measurements obtained with *all* admissible inputs at *all* sensor locations with the *same* set of parameters. Our approach is summarized in Algorithm 1.

First, initial (or artificial) data z_{ini} is needed for an appropriate guess p_{ini} of the parameter values. Having fixed these parameters, one can solve the OED problem (8)–(10) to obtain best sensor positions $\bar{\omega}$ and optimal input configurations $\bar{u}$, see lines 02 and 03.

Algorithm 1. (Detection of Model Uncertainty, adapted from [12])

Input: Model $\mathcal{M}$, test level TOL (e.g. 5%), number of test scenarios n_{tests}.
Output: Does $\mathcal{M}$ need to be rejected? YES (1) or NO (0).

01: Initialize $i := 1$.
02: Generate initial data z_{ini} in all feasible sensor locations.
03: Calculate p_{ini} from (7) using z_{ini}. Fix $p = p_{\text{ini}}$ and solve (8)–(10). Obtain $(\bar{\omega}, \bar{u})$.
04: Acquire data z in optimal sensor locations $\bar{\omega}$ for inputs close to the optimum $\bar{u}$.
05: Check whether measurement errors are Gaussian. Otherwise go to line 04 or exit.
06: Divide z into one calibration set z_{cal} and one validation set z_{val}.
07: Calculate $(p_{\text{cal}}, C_{\text{cal}})$ using z_{cal}. Likewise, obtain p_{val} using z_{val}.
08: Determine $\alpha_{\min} \in (0, 1)$, such that p_{val} lies on boundary of $K(p_{\text{cal}}, C_{\text{cal}}, \alpha_{\min})$.
09: **if** $\alpha_{\min} \geq$ TOL **then**
10:   **if** $i < n_{\text{tests}}$ **then**
11:     $i := i + 1$. Go to line 06.
12:   **else**
13:     **return** 0.
14:   **end if**
15: **else if** $\alpha_{\min} <$ TOL **then**
16:   **return** 1.
17: **end if**

The acquisition of experimental data z in line 04 is done at the optimal sensor locations and for inputs close to the optimum. Since we assume that the true values of the model parameters remain the *same* for all inputs $u \in U_{\text{ad}}$, we can ensure that our data are truly divers by performing measurements for different input values within a small neighborhood of the optimum $\bar{u}$. Evidently, the size of the confidence ellipsoid K stays small because of continuity of the objective function (8) with respect to the inputs.

Recall, that for time-variant systems each measurement at a given time depends on the past. Since the order of the data is important, the splitting of z into one calibration and one validation set must not happen over the time axis. Since our methodology is different from forecasting [5] we do not allow such splittings over the time domain.

We perform the division regarding the *different inputs* in a k-fold cross-validation manner [16]. Divide the data into k groups where each group is distinguished by the input for which it was collected in the whole time domain. We then use $k - 1$ groups for calibration and the remaining group for validation. When repeating this procedure we run through all k possible combinations.

For the validation itself, we perform a classical hypothesis test from line 08 onward as documented in [12]. The threshold TOL is identical to the error of the first kind. It is common to set a 5% limit to this error. The $\alpha_{\min}$ which is computed in line 08 is the p-value of the statistical test. This is the smallest test level for which the null hypothesis can only just be rejected.

There is no need to account for the problem of multiple testing here, since we are using a k-fold cross-validation manner to divide the data which ensures pairwise disjoint validation sets.

5   Numerical Results for Simulated Vibrations of a Truss

We employ a 2D-truss consisting of nine beams and six connectors with about 5 000 spatial degrees of freedom in order to exemplify the application of Algorithm 1. The Dirichlet boundary Γ_D is located at the two outer top connectors and the Neumann boundary Γ_N on the bottom left connector, see Fig. 1. We use pairs of strain gauges as sensors that can measure either the axial deflection or the displacement caused by bending of the beams, see [14] and [26]. The strain gauges are located on the upper and lower boundaries of the beams, indicated as black bullets and connecting lines in the figure, which are part of the free boundary Γ_F of the body G. Each strain gauge measures the relative displacement of two adjacent nodes: $\varepsilon_u = y_{N1} - y_{N2}$ and $\varepsilon_\ell = y_{N3} - y_{N4}$, see Fig. 1a. For simplicity, we compute the square of the axial deflection $h_a(y)$ and the square of the displacement caused by bending $h_b(y)$:

$$
h_a(y) = \frac{1}{4} \left\| \varepsilon_u + \varepsilon_\ell \right\|^2, \qquad\qquad h_b(y) = \frac{1}{4} \left\| \varepsilon_u - \varepsilon_\ell \right\|^2.
$$

Thus, the overall observation operator h consists of h_a and h_b *at all time points* and we create for each such sensor five weight variables. These additional weights shall give the experimenter information about which pairs of strain gauges are more important than others. The discretization of the truss allows for 117 sensors in total. Hence, we have $n_s = 117 \times 2 \times 5 = 1\,170$ weight variables.

Throughout our numerical simulations we use pure stiffness damping, i.e., $a = 0$ in (3). This is promising to provide better resemblance with actual experimental data, see [3] and [24]. The accuracy of our sensors is fixed to $\sigma_{\mathrm{pr},k} = 10\,\mu\mathrm{m}$ for $k = 1, \ldots, n_s$.

We simulated vibrations of the truss for $n_t = 600$ time steps with a step size of $\Delta t = 5$ ms. Thus, three seconds were simulated in total and the solution of the PDE (1) involves about 3 000 000 degrees of freedom. Initially, there were all 117 pairs of strain gauges used measuring both the axial deflection and the displacement caused by bending with maximum weight, respectively. We also use a constant maximally feasible force as a starting point for the inputs u. The excitation forces u act solely on the Neumann boundary Γ_N.

Since we were not able to conduct real experiments, all the data was simulated, i.e., generated on the computer with random numbers. Thus, line 05 in Algorithm 1 became obsolete. We assume the beams of the *real truss* $\mathcal{R}$ to have an equal cross-sectional area in the displacement-free state except for two beams having a 5% and a 7% smaller diameter, respectively. For the detection of model uncertainty it is not important to know which beams differ from the standard diameter. However, *our model* $\mathcal{M}$ operates on the assumption that all

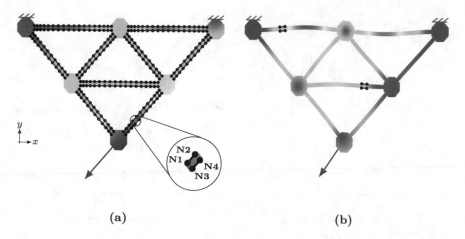

(a)   **(b)**

Fig. 1. (a) snapshot of the truss with all possible locations for the strain gauges marked as bullets with connecting lines and the excitation force displayed by a red arrow, (b) snapshot after problem (8)–(10) has been solved with displayed optimal positions for strain gauges and optimal excitation

beams have the same cross-sectional area. This directly impacts the mathematical terms in the mass, damping and stiffness matrices, see (2), since a model with different cross-sectional beam areas would induce other finite element terms. It is our aim in this section to show that Algorithm 1 successfully detects model uncertainty in $\mathcal{M}$ when compared to $\mathcal{R}$.

Since we only simulate experiments, we skipped line 02 in Algorithm 1 and adopted textbook values for p_{ini}, namely, the well-known Lamé-constants for steel $\lambda_{\mathrm{L}} = 121\,154$ N/mm^2 and $\mu_{\mathrm{L}} = 80\,769$ N/mm^2. These are the values which we use to generate all measurements from the real truss $\mathcal{R}$. Problem (8)–(10) is solved after about 80 iterations with an overall computation time of about 8 h on an AMD EPYC 48 × 2.8 GHz machine. The design criterion decreased by ≈99% which means that the maximal expansion of the confidence ellipsoid decreased by ≈98% compared to the initial design, see Fig. 2. The final design employs only two pairs of strain gauges that measure the axial deflection, the upper with weight two the lower with weight five, cf. Fig. 1b.

Let $\bar{u}$ be the optimal input force obtained from solving (8)–(10). For the application of the hypothesis test in Algorithm 1, consider the following perturbed inputs:

$$u_1(t) = \bar{u}(t) + \delta_1, \qquad\qquad u_2(t) = \bar{u}(t) + 4\sin(t/(2\pi)),$$
$$u_3(t) = \bar{u}(t) + 4\cos(t/(2\pi)), \qquad u_4(t) = \bar{u}(t) + \delta_2(t),$$
$$u_5(t) = \bar{u}(t) + \delta_3(t), \qquad\qquad u_6(t) = \bar{u}(t) \cdot (1 + 0.06\sin(t/(2\pi))),$$
$$u_7(t) = \bar{u}(t) \cdot (1 + 0.06\cos(t/(2\pi))), \quad u_8(t) = \bar{u}(t) + \delta_4,$$

where $\delta_1, \delta_4 \sim \mathcal{N}(0, 4 \cdot I)$ and $\delta_2(t), \delta_3(t) \sim \mathcal{N}(0, 4t/n_t \cdot I)$ for all $t \in \{t_1, \ldots, t_{n_t}\}$ with equal time step size Δt as introduced before. With these inputs we generate

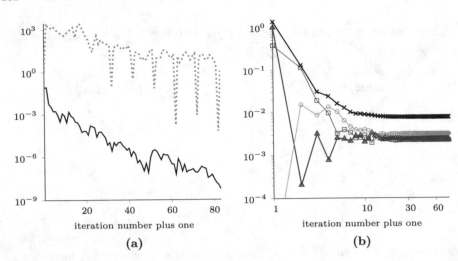

Fig. 2. Optimization results for problem (8)–(10): (a) first order optimality (——) and norm of the step (......), (b) objective function (—×—), design criterion (—▲—), penalty (—■—) and regularization value (—⊝—)

eight different data sets and perform an 8-fold cross-validation. We use seven sets for calibration and one set for validation in line 06 of Algorithm 1. Thus, we conducted eight different hypothesis tests, four of which are shown in Table 1. It is clearly seen, that the model $\mathcal{M}$ does not pass any test when a threshold of 5% is applied to α_{min}. According to our assumption that a valid model should reproduce *all* measurements conducted with *all* admissible inputs with the *same* set of parameters, this is a significant indication of model uncertainty.

Table 1. Excerpt of the results for the hypothesis tests from Algorithm 1

#	p_{cal}	p_{val}	$\|p_{cal} - p_{val}\|$	$\lambda_{min}(C)^{-1}$	α_{min} in %
1	$4.588 \cdot 10^{-5}$	$1.206 \cdot 10^5$	$1.253 \cdot 10^5$	1.073	$\ll 0.001$
	$1.130 \cdot 10^5$	$7.909 \cdot 10^4$			
2	$4.562 \cdot 10^{-5}$	$1.206 \cdot 10^5$	$1.253 \cdot 10^5$	1.073	$\ll 0.001$
	$1.130 \cdot 10^5$	$7.909 \cdot 10^4$			
3	$4.340 \cdot 10^{-5}$	$1.206 \cdot 10^5$	$1.253 \cdot 10^5$	1.073	$\ll 0.001$
	$1.130 \cdot 10^5$	$7.909 \cdot 10^4$			
4	$6.167 \cdot 10^{-5}$	$1.206 \cdot 10^5$	$1.253 \cdot 10^5$	1.073	$\ll 0.001$
	$1.130 \cdot 10^5$	$7.909 \cdot 10^4$			

6   Conclusion

In this paper we showed that our algorithm to detect model uncertainty, which was first presented in [12], is applicable to dynamic models. We efficiently solved the OED problem with time-dependent PDE-constraints using modified BFGS-updates and adjoint methods within an SQP solver scheme. Thus, in finding optimal sensor positions and optimal inputs we were able to significantly reduce the size of the confidence region of the estimated model parameters. By an 8-fold cross-validation using hypothesis tests in the parameter space, we demonstrated on simulations of vibrations in a truss that our algorithm is able to detect inaccuracies of the linear-elastic model which is deficient in the geometrical description of the truss. It is the object of further investigation to show that our algorithm detects other forms or kinds of model uncertainty as well.

Acknowledgements. This research was funded by the German Research Foundation (DFG) – project number 57157498 – CRC 805 within the subproject A3. We would like to thank the DFG for funding.

The authors would also like to thank Philip Kolvenbach for providing efficient finite element code.

References

1. Alexanderian, A., Petra, N., Stadler, G., Ghattas, O.: A-optimal design of experiments for infinite-dimensional Bayesian linear inverse problems with regularized ℓ_0-sparsification. SIAM J. Sci. Comput. **36**(5), A2122–A2148 (2014)
2. Alexanderian, A., Petra, N., Stadler, G., Ghattas, O.: A fast and scalable method for A-optimal design of experiments for infinite-dimensional Bayesian nonlinear inverse problems. SIAM J. Sci. Comput. **38**(1), A243–A272 (2016)
3. Alipour, A., Zareian, F.: Study rayleigh damping in structures; uncertainties and treatments. In: Proceedings of the 14th World Conference on Earthquake Engineering, Beijing, China, pp. 1–8 (2008)
4. Bauer, I., Bock, H.G., Körkel, S., Schlöder, J.P.: Numerical methods for optimum experimental design in DAE systems. J. Comput. Appl. Math. **120**(1–2), 1–25 (2000)
5. Bergmeir, C., Benítez, J.M.: On the use of cross-validation for time series predictor evaluation. Inf. Sci. **191**, 192–213 (2012)
6. Chianeh, H.A., Stigter, J., Keesman, K.J.: Optimal input design for parameter estimation in a single and double tank system through direct control of parametric output sensitivities. J. Process Control **21**(1), 111–118 (2011)
7. Ciarlet, P.G.: Mathematical Elasticity. Studies in Mathematics and its Applications, vol. 20. North-Holland Publishing Co., Amsterdam (1988)
8. Donaldson, J.R., Schnabel, R.B.: Computational experience with confidence regions and confidence intervals for nonlinear least squares. Technometrics **29**(1), 67–82 (1987)
9. Evans, L.C.: Partial Differential Equations. Graduate Studies in Mathematics, vol. 19, 2nd edn. American Mathematical Society, Providence (2010)
10. Fletcher, R.: Practical Methods of Optimization, 2nd edn. A Wiley-Interscience Publication. John Wiley & Sons Ltd, Chichester (1987)

11. Franceschini, G., Macchietto, S.: Model-based design of experiments for parameter precision: state of the art. Chem. Eng. Sci. **63**(19), 4846–4872 (2008)
12. Gally, T., Groche, P., Hoppe, F., Kuttich, A., Matei, A., Pfetsch, M.E., Rakowitsch, M., Ulbrich, S.: Identification of model uncertainty via optimal design of experiments applied to a mechanical press. Optim. Eng. (2021). https://doi.org/10.1007/s11081-021-09600-8
13. Hiriart-Urruty, J.B., Lewis, A.S.: The Clarke and Michel-Penot subdifferentials of the eigenvalues of a symmetric matrix. Comput. Optim. Appl. **13**(1), 13–23 (1999)
14. Hoffmann, K.: An Introduction to Stress Analysis Using Strain Gauges (1987)
15. Hughes, T.J.R.: The Finite Element Method. Prentice Hall Inc., Hoboken (1987)
16. James, G., Witten, D., Hastie, T., Tibshirani, R.: An Introduction to Statistical Learning, vol. 112. Springer, Heidelberg (2013)
17. Jauberthie, C., Bournonville, F., Coton, P., Rendell, F.: Optimal input design for aircraft parameter estimation. Aerosp. Sci. Technol. **10**(4), 331–337 (2006)
18. Kolvenbach, P.: Robust optimization of PDE-constrained problems using second-order models and nonsmooth approaches. Ph.D. thesis, TU Darmstadt (2018)
19. Körkel, S., Kostina, E., Bock, H.G., Schlöder, J.P.: Numerical methods for optimal control problems in design of robust optimal experiments for nonlinear dynamic processes. Optim. Methods Softw. **19**(3–4), 327–338 (2004)
20. Mallapur, S., Platz, R.: Uncertainty quantification in the mathematical modelling of a suspension strut using Bayesian inference. Mech. Syst. Signal Process. **118**, 158–170 (2019). https://doi.org/10.1016/j.ymssp.2018.08.046
21. Mehra, R.: Optimal input signals for parameter estimation in dynamic systems-survey and new results. IEEE Trans. Autom. Control **19**(6), 753–768 (1974)
22. Morelli, E.A., Klein, V.: Optimal input design for aircraft parameter estimation using dynamic programming principles. In: Proceedings of the 17th Atmospheric Flight Mechanics Conference (1990). https://doi.org/10.2514/6.1990-2801
23. Neitzel, I., Pieper, K., Vexler, B., Walter, D.: A sparse control approach to optimal sensor placement in PDE-constrained parameter estimation problems. Numer. Math. **143**(4), 943–984 (2019)
24. Otani, S.: Nonlinear dynamic analysis of reinforced concrete building structures. Can. J. Civ. Eng. **7**(2), 333–344 (1980)
25. Puig, V., Quevedo, J., Escobet, T., Nejjari, F., de las Heras, S.: Passive robust fault detection of dynamic processes using interval models. IEEE Trans. Control Syst. Technol. **16**(5), 1083–1089 (2008)
26. Rohrbach, C.: Handbuch für elektrisches Messen mechanischer Größen. VDI-Verlag, Düsseldorf (1967)
27. Simani, S., Fantuzzi, C., Patton, R.J.: Model-based fault diagnosis techniques. In: Model-Based Fault Diagnosis in Dynamic Systems Using Identification Techniques, pp. 19–60. Springer (2003)
28. Stigter, J.D., Keesman, K.J.: Optimal parametric sensitivity control of a fed-batch reactor. Automatica **40**(8), 1459–1464 (2004)
29. Willsky, A.S.: Detection of abrupt changes in dynamic systems. In: Detection of Abrupt Changes in Signals and Dynamical Systems, pp. 27–49. Springer (1985)
30. Wriggers, P.: Nonlinear Finite Element Methods. Springer Science & Business Media, Hoboken (2008)
31. Yuen, K.V., Kuok, S.C.: Bayesian methods for updating dynamic models. Appl. Mech. Rev. **64**(1) (2011). https://doi.org/10.1115/1.4004479

Robust Topology Optimization of Truss-Like Space Structures

Michael Hartisch[✉], Christian Reintjes, Tobias Marx, and Ulf Lorenz

University of Siegen, Unteres Schloss 3, 57072 Siegen, Germany
{michael.hartisch,christian.reintjes,tobias.marx,
ulf.lorenz}@uni-siegen.de

Abstract. Due to the additional design freedom and manufacturing possibilities of additive manufacturing compared to traditional manufacturing, topology optimization via mathematical optimization gained importance in the initial design of complex high-strength lattice structures. We consider robust topology optimization of truss-like space structures with multiple loading scenarios. A typical dimensioning method is to identify and examine a suspected worst-case scenario using experience and component-specific information and to incorporate a factor of safety to hedge against uncertainty. We present a quantified programming model that allows us to specify expected scenarios without having explicit knowledge about worst-case scenarios, as the resulting optimal structure must withstand all specified scenarios individually. This leads to less human misconduct, higher efficiency and, thus, to cost and time savings in the design process. We present three-dimensional space trusses with minimal volume that are stable for up to 100 loading scenarios. Additionally, the effect of demanding a symmetric structure and explicitly limiting the diameter of truss members in the model is discussed.

Keywords: Robust truss topology optimization · Truss-like space structures · Quantified programming · Symmetric structures

1 Introduction

Robust Truss Topology Optimization (RTTO) deals with the structural design optimization problem to find a truss that is stable subjected to loading, material or geometric uncertainty. We examine loading uncertainty [1,2]. Most research in this area considers perturbations of loads and focuses on robust compliance topology optimization often resulting in semidefinite programs [4,7]. We aim at minimizing the volume of truss-like structures under multiple-load uncertainty and assume a static system in order to provide an initial design proposal. A robust mixed integer program arises that can be tackled with high performance standard solvers. We utilize the expressiveness of Quantified Mixed-Integer Linear Programming (QMIP) [5] to model the robust optimization problem.

Our approach simultaneously considers sizing and topology optimization and utilizes the so-called ground structure [9], see Fig. 1, which is given by a set of

© The Author(s) 2021
P. F. Pelz and P. Groche (Eds.): ICUME 2021, LNME, pp. 296–306, 2021.
https://doi.org/10.1007/978-3-030-77256-7_23

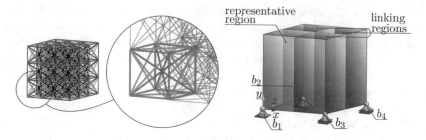

Fig. 1. (Left) The ground structure; (Right) Symmetry and bearing positions

vertices (fixed nodal points) $V \in \mathbb{R}^3$ and a set of edges (possible structural members) $E \subseteq V \times V$. The resulting structure must be adequately dimensioned for any of the anticipated loading scenarios. Therefore, the challenging task of identifying, analyzing, and quantifying the worst-case scenario [7], which thus far highly depends upon practical engineering experience, can be bypassed, leading to less human misconduct, higher efficiency and, thus, to cost and time savings.

In structural mechanics symmetry is often exploited to effectively optimize and analyze structural systems [8]. From the viewpoint of mathematical optimization, enforcing symmetry is often computationally beneficial, as it results in a reduction of free variables. In general, however, demanding symmetry results in suboptimal solutions, as optimal solutions can be asymmetric even if the design domain, the external loads, and the boundary conditions are symmetric [12]. We optionally consider two vertical planes of symmetry, as shown in Fig. 1. Hence, specifying the structure of the representative region suffices.

We introduce the robust optimization model in Sect. 2. In Sect. 3 we discuss two three-dimensional examples: a ground structure with 296 members and 128 loading scenarios as well as a 1720-member ground structure with 8 loading scenarios, before we summarize and conclude in Sect. 4.

2   Robust Truss Topology Optimization

We utilize QMIP, which is a formal extension of Mixed Integer Programming (MIP) where variables are either existentially or universally quantified resulting in a robust multistage optimization problem. For more details, we refer to [5]. A solution is a strategy for assigning existentially quantified variables such that a linear constraint system is fulfilled. In particular, in a solution it is ensured that even for the worst-case assignment of universally quantified variables the system holds. The following model features two quantifier changes and therefore also can be interpreted as adjustable robust optimization problem with right-hand side uncertainty [15]. The model aims at *finding a truss-like space structure with minimal volume such that for any anticipated loading scenario the structure remains in a static equilibrium position.* Restated in the quantification context:

$$\exists \text{ structure} \quad \forall \text{ loading scenarios} \quad \exists \text{ static equilibrium,} \tag{1}$$

i.e., does a truss-like structure *exist* such that *for all* anticipated loading scenarios a static equilibrium *exists*. Within the ground structure (see Fig. 1), given by the undirected graph $G = (V, E)$, edges must be selected at which straight and prismatic members should be placed and for each member its cross-sectional area must be determined. The locations of vertices, i.e., nodal points, are fixed in space, to allow a preprocessing of the spatial and angular relationships between edges and vertices. Additionally, a set of bearings $B \subset V$ must be specified, see Fig. 1. We also want to be able to demand symmetrical structures and therefore introduce the function $R : E \rightarrow E$, for which $R(e)$ maps to the edge representing e. In contrast to the established design variable linking technique [10], only variables that characterize representative members need to be deployed in order to enforce that members at edges e and $R(e)$ are equally dimensioned. We use $R(e) = e$ if no symmetry is demanded.

As a minimum cross-sectional area is essential due to manufacturing restrictions, we use the combination of a binary variable x_e and a continuous variable a_e in order to indicate the existence of a member at edge $e \in E$ (with specified minimum area) and its potential additional cross-sectional area. The minimum area $A_{\min}$ can either be a fixed value or optionally adhere to design rules VDI 3405-3-4, VDI 3405-3-3 [13,14]. In the latter case, the minimum area of each member is computed separately depending on its length and its spatial orientation, cf. [11]. Note that these design rules are only enforced locally for each edge: if multiple members with identical beam axes form one long structural member, post-processing is necessary. Both $\mathbf{x}$ and $\mathbf{a}$ are first stage existential variables, as they represent the selected structure. Binary universal variables $\mathbf{y}$ are used to specify the loading scenario, consisting of $C \in \mathbb{N}$ loading cases: the binary universal variable y_i indicates whether *loading case i* is active, while the variable vector $\mathbf{y}$ indicates the selected *loading scenario*. For each anticipated loading scenario the structure given by $\mathbf{x}$ and $\mathbf{a}$ must have the following properties: each nodal point as well as the entire structure must be in a static equilibrium position and the longitudinal stress within each structural member—induced by the normal force per cross-sectional area—must not exceed the member's yield strength. The existential variables n_e and r_b represent the normal force in a structural member at e and the bearing reaction force at bearing b, respectively. The variables used in our model are given in Table 1 and the parameters are listed in Table 2. We use bold letters when referring to a vector or matrix.

Table 1. Variables

symbol	stage	description
$\mathbf{x} \in \{0,1\}^E$	$1(\exists)$	x_e: indicator, whether a structural member is present at edge e
$\mathbf{a} \in \mathbb{Q}_+^E$	$1(\exists)$	a_e: additional cross-sectional area of a structural member e
$\mathbf{y} \in \{0,1\}^C$	$2(\forall)$	y_i: indicator whether load case s is active
$\mathbf{r} \in \mathbb{Q}^{B \times 3}$	$3(\exists)$	r_b^d: bearing reaction force at b in spatial direction $d \in \{x, y, z\}$
$\mathbf{n} \in \mathbb{Q}^E$	$3(\exists)$	n_e: normal force in structural beam present at edge e

Table 2. Sets, Parameters and Functions

symbol	description
V	set of vertices
$E \subseteq V \times V$	set of edges
$I : V \to 2^E$	$I(v) = \{e \in E \mid v \in e\}$: set of edges incident to vertex v
$B \subseteq V$	set of bearings
$L_e \in \mathbb{Q}_+$	length of edge e
$A_{\min} \in \mathbb{Q}_+$	minimum cross-sectional area of a member
$A_{\max} \in \mathbb{Q}_+$	maximum cross-sectional area of a member
$\sigma_y \in \mathbb{Q}_+$	yield strength of the cured material
$S \geq 1$	factor of safety
$C \in \mathbb{N}_+$	number of considered load cases
$\mathbf{F}_i \in \mathbb{Q}^{V \times 3}$	$F_{i,v}^d$: external force at vertex v in direction $d \in \{x, y, z\}$ in load case i
$\mathbf{V}(v, v') \in \mathbb{Q}^3$	vector from $v \in V$ to $v' \in V$ (corresponding to lever arm)
$R : E \to E$	$R(e)$: edge representing edge e due to symmetry

$$\min \sum_{e \in E} L_e \left(A_{\min} \cdot x_{R(e)} + a_{R(e)} \right) \tag{2}$$

$$\text{s.t. } \exists \mathbf{x} \in \{0,1\}^E \ \mathbf{a} \in \mathbb{Q}_+^E \ \ \forall \mathbf{y} \in \{0,1\}^C \ \ \exists \mathbf{r} \in \mathbb{Q}^{B \times 3} \ \mathbf{n} \in \mathbb{Q}^E: \tag{3}$$

$$S|n_e| \leq \sigma_y \left(A_{\min} x_{R(e)} + a_{R(e)} \right) \qquad \forall e \in E \tag{4}$$

$$\sum_{e \in I(b)} n_e^d + \sum_{i=1}^C y_i F_{i,b}^d + r_b^d = 0 \qquad \forall b \in B, \, d \in \{x, y, z\} \tag{5}$$

$$\sum_{e \in I(v)} n_e^d + \sum_{i=1}^C y_i F_{i,v}^d = 0 \qquad \forall v \in V \setminus B, \, d \in \{x, y, z\} \tag{6}$$

$$a_{R(e)} \leq (A_{\max} - A_{\min}) x_{R(e)} \qquad \forall e \in E \tag{7}$$

$$\sum_{v \in V} \sum_{i=1}^C \mathbf{V}(b, v) \times y_i \mathbf{F}_{i,v} + \sum_{b' \in B} \mathbf{V}(b, b') \times \mathbf{r}_{b'} = 0 \qquad \forall b \in B \tag{8}$$

$$\sum_{v \in V} \sum_{i=1}^C y_i \mathbf{F}_{i,v} + \sum_{b \in B} \mathbf{r}_b = 0 \tag{9}$$

The Objective Function (2) aims at minimizing the volume of the structure. Note that the objective value does not reflect the *exact* volume of the corresponding structure: overlapping parts of members at vertices are not merged but counted multiple times. The Quantification Sequence (3) defines the variables' domain and order, as outlined in Expression (1). If $R(e) \neq e$, $\mathbf{x}$ and $\mathbf{a}$ variables are deployed only for edges in the image of E under function R, i.e., only for the representative edges. Constraint (4) ensures that the local longitudinal stress must not exceed the member's yield strength considering a factor of safety. In particular, modifying the cross-sectional area given by $A_{min} + a_{R(e)}$, alters the stress in a member. Constraint (4) is linearized by writing the constraint once with the left-hand side $+n_e$ and once with $-n_e$. Constraints (5) and (6) ensure the static equilibrium at each vertex. The decomposition n_e^d of the normal force n_e into each spatial direction $d \in \{x, y, z\}$ is obtained by multiplying n_e with $\sin(\theta)$, where θ is the angle between the member and the corresponding d-axis. As the ground structure is fixed in space those coefficients can be preprocessed. With Constraint (7), a_e must be zero if no member is present at edge e. Constraints (8) and (9) define the equilibrium of moments by resolution of the external forces and ensure, in combination with Constraints (5) and (6), that the resulting structure is always a static system of purely axially loaded members. In particular, the cross product $\mathbf{V}(b, v) \times \mathbf{F}_{i,v}$ characterizes the moment induced by external force $\mathbf{F}_{i,v}$ on bearing b with lever arm $\mathbf{V}(b, v)$. Analogously, $\mathbf{V}(b, b') \times \mathbf{r}_{b'}$ is the moment about bearing b caused by the bearing reaction force at b'.

The resulting structure is ensured to be adequately dimensioned for any of the 2^C loading scenarios resulting from the combination of C loading cases. However, if the resulting structure only needs to be adequately dimensioned for each individual loading case, the model can be altered by enforcing $\sum_{i=1}^{C} y_i = 1$ on the universal variables. In this case only C loading scenarios, which correspond to the loading cases, are of interest. As such a constraint cannot simply be added to the constraint system, see [6], this case is implemented using a single integer universal variable specifying the load case, which is then transformed into existential indicator variables.

In order to illustrate our approach we use a 40-member planar rectangular grid with fixed bearing at the bottom left, floating bearing at the bottom right, and four loading cases color-coded in Fig. 2. Note that the dimensions and color-coded loading cases of all figures in this work are not to scale.

In Fig. 2a the optimal solution for $A_{min} = 0\,$mm is displayed, which is stable for any combination of the loading case. Figure 2b contains the optimal solution if each single member must be dimensioned according to VDI 3405-3-3, 3405-3-4. For the case that the loading cases can only occur individually the optimal solution is shown in Fig. 2c. Figure 2d displays the optimal solution if we addi-

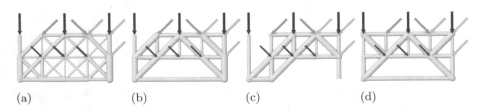

(a)   (b)   (c)   (d)

Fig. 2. Optimal solutions for (a) $A_{min} = 0\,$mm, without symmetry, combined cases, (b) $A_{min} \equiv A_{VDI}$, without symmetry, combined cases, (c) $A_{min} \equiv A_{VDI}$, without symmetry, single cases, (d) $A_{min} \equiv A_{VDI}$, with symmetry, single cases

tionally demand symmetry around the vertical mid-axis. Note that in neither case one has to explicitly deal with any kind of worst-case scenario, but can be assured that a solution is in a static equilibrium in *every* scenario. Most importantly, in general one cannot assume the worst-case scenario to be the one where all loading cases are active, as compensation of forces might occur.

3   Computational Experiments

We conduct experiments on three-dimensional ground structures with artificial loading cases. The first example showcases that a large number of scenarios can be considered, while the second example demonstrates that large three-dimensional ground structures can be used. We assume a basic vertex distance of 10 mm, two fixed bearings on the two lower left corners, and two floating bearings on the lower right corners. The considered material is fine polyamide PA 2200, with layer thickness of 0.12 mm for Selective Laser Sintering (SLS) and a yield strength of $\sigma_y = 45 \pm 0\,\mathrm{N\,mm}^{-2}$. For each QMIP instance we built the corresponding Deterministic Equivalent Problem (DEP) [3] and used CPLEX to solve the arising MIP instance[1]. For each instance we limit the runtime to 240 h.

296-member Space Truss with 128 Loading Scenarios
We assume seven loading cases and are interested in structures that are able to withstand each of the 2^7 loading scenarios resulting from combining the single loading cases. We examine the optimization results for several values of $A_{\min}$, but refer to the minimum diameter $D_{\min}$ for presentation reasons. The individual loading cases are given in Fig. 3a–3g. Additionally, optimal structures for each individual case are displayed for better comprehensibility. In Fig. 3h the best found robust solution for $D_{\min} = 1\,\mathrm{mm}$ is displayed, which is stable in any of the 2^7 loading scenarios.
Table 3 contains the objective values of the best found solutions, the best lower bounds, and the corresponding optimality gap for different settings.

Table 3. Computational results for the 296-member space truss

$D_{\min}$ in mm	without symmetry			with symmetry		
	best found in mm^3	bound in mm^3	gap in %	best found in mm^3	bound in mm^3	gap in %
0	3693	3693	0	4367	4367	0
1	3783	3697	2.28	4400	4399	0.01
2	4987	3771	24.37	5246	4674	10.90
3	7793	4286	45.00	8679	5860	32.48
4	13008	5330	59.02	12931	8045	37.79
D_{VDI}	6317	3954	37.40	5906	5218	11.65

---

[1] All experiments were run on an Intel(R) Core(TM) i7-4790 with 3.60 GHz and 32 GB RAM with CPLEX 12.9.0 running on default but restricted to a single thread.

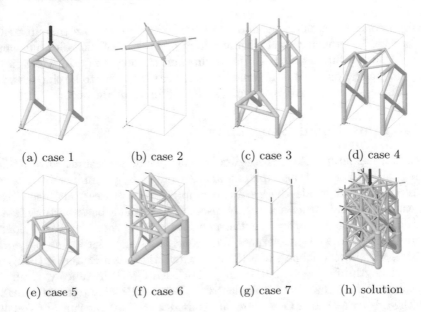

(a) case 1   (b) case 2   (c) case 3   (d) case 4

(e) case 5   (f) case 6   (g) case 7   (h) solution

Fig. 3. Loading cases (a-g) and robust solution (h) for $R(e) = e$ and $D_{\min} = 1\,\mathrm{mm}$

For $D_{\min} = 0\,\mathrm{mm}$ the DEP of the QMIP instance can be preprocessed to be a Linear Programming (LP) problem, as the binary $\mathbf{x}$ variables can be fixed to 1. The corresponding optimal solutions where found within 2 and 0.5 h for the non-symmetrical and symmetrical case, respectively. For all other instances, except for the symmetric instance with $D_{\min} = 1\,\mathrm{mm}$, the optimality gap was not closed sufficiently within 240 h. Table 3 indicates, that for increasing $D_{\min}$ it becomes computationally more expensive to obtain small optimality gaps, which is partially due to the worsening of the corresponding LP-relaxation. Four best found solutions for various values of $D_{\min}$ are shown in Fig. 4.

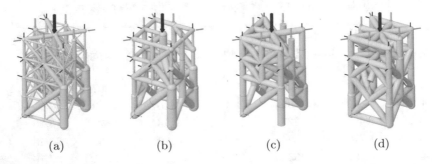

(a)   (b)   (c)   (d)

Fig. 4. Non-symmetric solutions for (a) $D_{\min} = 0\,\mathrm{mm}$, (b) $D_{\min} \equiv D_{\mathrm{VDI}}$, (c) $D_{\min} = 3\,\mathrm{mm}$, and symmetric solution for (d) $D_{\min} = 3\,\mathrm{mm}$

The solutions differ considerably: with increasing D_{min} the number of structural members tends to decrease and when additionally enforcing the VDI design rules, long members—in particular diagonal members—are avoided. Note that only the structure given in Fig. 4a is optimal. In Fig. 4d the drawback of demanding symmetry becomes apparent: the sufficient triangular structure at the bottom is no longer feasible.

Symmetry. Demanding a symmetrical structure reduces the number of continuous **a** and binary **x** variables from 296 to 94. This results in a computational benefit which is reflected in Table 3; in particular when comparing the optimality gap. Obviously, the volume of the *optimal* symmetrical structure cannot be lower than for the one without symmetry. Nevertheless, in some cases the incumbent symmetrical solution has lower volume when the time limit was reached. The solutions for $D_{min} = 0$ mm and $D_{min} = 1$ mm indicate the price of demanding a symmetric structure: the volume of the symmetric solution increased by about $\frac{1}{6}$ for this instance.

Minimal Diameter. The manufacturable minimal diameter depends on the manufacturing process and is obviously always larger than zero. The computational results for $D_{min} = 0$ mm, however, invite to use these quickly obtained solutions, but the resulting structures exhibit several structural members with extremely small diameter of < 0.1 mm. Hence, solving this LP formulation is computational efficient but unsuitable for direct application in engineering. However, in order to utilize this stable solution one can inflate small members until they have the desired diameter. Table 4 shows the resulting volumes of the structures when— starting from the optimal solution for $D_{min} = 0$ mm—the diameter of affected members is increased to the actual value of D_{min}. In all cases the volume dramatically exceeds the corresponding best structure found during the optimization process, cf. Table 3. Therefore, when disregarding the runtime, solving the model with explicitly stated minimum diameter is preferable to postprocessing the quickly obtained optimal solutions for $D_{min} = 0$ mm.

Table 4. Truss volumes when inflating members to D_{min} based on LP solution

D_{min}	0 mm (LP)	1 mm	2 mm	3 mm	4 mm	VDI
Without Symmetry	3693		4443	9208	18883	33106	16262
With Symmetry	4367		4963	9522	19599	34319	17503

1720-member Space Truss with 8 Loading Scenarios

We assume a 90 mm × 30 mm × 50 mm ground structure with 10 mm basic vertex distance and 8 loading cases. We are interested in *symmetrical* structures that are stable in each *individual* loading case.

In Table 5 computational results for various values of D_{min} are presented. Only for $D_{min} = 0$ mm the optimal solution was found (in 104 min).

Table 5. Computational results for the 1720-member space truss

D_{min}	0 mm	1 mm	2 mm	3 mm	4 mm	VDI
best found in mm^3	6436	7096	16964	35124	61133	15318
lower bound in mm^3	6436	6445	7435	11027	17606	7630
gap in %	0.00	9.18	56.17	68.60	71.20	50.19

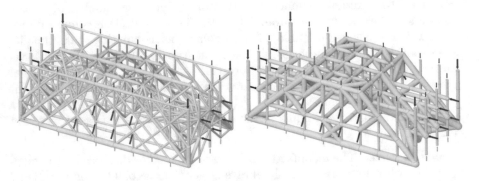

Fig. 5. $D_{min} = 1$ mm (left) and $D_{min} \equiv D_{VDI}$ (right) with eight loading scenarios

In Fig. 5 the eight (colored) loading cases as well as the best found solutions for $D_{min} = 1$ mm and $D_{min} \equiv D_{VDI}$ with 479 and 320 members, respectively, are displayed. For comparison: the optimal solution for $D_{min} = 0$ mm exhibits 1004 members. Hence, changing the minimal diameter significantly alters the structure's topology.

4   Conclusion

We utilized quantified programming to introduce a robust formulation for the shape and topology optimization of truss-like structures. Instead of determining and optimizing a worst-case scenario based on engineering experience and prone to human error, our approach allows to state loading cases while the resulting structure is ensured to be stable, even in the (unknown) worst-case. A distinction can be made as to whether the loading cases can only occur individually or whether they can occur in any combination. We presented results on a 296-member space truss, considering the combination of seven loading cases resulting in 128 scenarios. A 1720-member space truss with 8 individually occurring loading cases demonstrates the applicability of our approach for large-scale structures. Additionally, we highlighted advantages and disadvantages of explicitly enforcing a minimal cross-sectional area of structural members and a symmetric structure. Future work has to deal with the distortion of the objective value due to overlapping members and the development of problem specific heuristics.

References

1. Achtziger, W., Stolpe, M.: Truss topology optimization with discrete design variables - guaranteed global optimality and benchmark examples. Struct. Multidisciplinary Optim. **34**(1), 1–20 (2007)
2. Ben-Tal, A., Nemirovski, A.: Robust truss topology design via semidefinite programming. SIAM J. Optim. **7**(4), 991–1016 (1997)
3. Bertsimas, D., Brown, D., Caramanis, C.: Theory and applications of robust optimization. SIAM Rev. **53**(3), 464–501 (2011)
4. Gally, T., Gehb, C.M., Kolvenbach, P., Kuttich, A., Pfetsch, M.E., Ulbrich, S.: Robust truss topology design with beam elements via mixed integer nonlinear semidefinite programming. Appl. Mech. Mater. **807**, 229–238 (2015)
5. Hartisch, M.: Quantified Integer Programming with Polyhedral and Decision-Dependent Uncertainty. Ph.D. thesis, University of Siegen, Germany (2020)
6. Hartisch, M., Ederer, T., Lorenz, U., Wolf, J.: Quantified integer programs with polyhedral uncertainty set. In: Computers and Games - 9th International Conference, CG 2016, pp. 156–166. Springer (2016)
7. Kanno, Y., Guo, X.: A mixed integer programming for robust truss topology optimization with stress constraints. Int. J. Numerical Methods Eng. **83**(13), 1675–1699 (2010)
8. Marsden, J.E., Ratiu, T.S.: Introduction to mechanics and symmetry: a basic exposition of classical mechanical systems, vol. 17. Springer Science & Business Media (2013)
9. Ohsaki, M.: Optimization of Finite Dimensional Structures. CRC Press, Cambridge(2016)
10. Rao, S.S.: Engineering Optimization: Theory and Practice. John Wiley & Sons, Hoboken (2019)
11. Reintjes, C., Lorenz, U.: Bridging mixed integer linear programming for truss topology optimization and additive manufacturing. In: Optimization and Engineering: Special Issue on Technical Operations Research (2020)
12. Stolpe, M.: On some fundamental properties of structural topology optimization problems. Struct. Multidisciplinary Optim. **41**(5), 661–670 (2010)
13. Richtlinien, V.D.I.: VDI 3405-3-3: Additive manufacturing processes, rapid manufacturing - design rules for part production using laser sintering and laser beam melting. Standard, VDI-Gesellschaft Produktion und Logistik, Düsseldorf, GER (2015)
14. Richtlinien, V.D.I.: VDI 3405-3-4: Additive manufacturing processes - design rules for part production using material extrusion processes. Standard, VDI - Gesellschaft Produktion und Logistik, Düsseldorf, GER (2019)
15. Yanıkoğlu, İ, Gorissen, B.L., den Hertog, D.: A survey of adjustable robust optimization. Eur. J. Oper. Res. **277**(3), 799–813 (2019)

Author Index